Exact Analysis of Structures with Periodicity using U-Transformation

Exact Analysis of Structures with Periodicity using U-Transformation

H C Chan
University of Hong Kong

C W Cai
Zhongshan University, China

Y K Cheung
University of Hong Kong

World Scientific
Singapore • New Jersey • London • Hong Kong

Published by

World Scientific Publishing Co. Pte. Ltd.

P O Box 128, Farrer Road, Singapore 912805

USA office: Suite 1B, 1060 Main Street, River Edge, NJ 07661

UK office: 57 Shelton Street, Covent Garden, London WC2H 9HE

British Library Cataloguing-in-Publication Data
A catalogue record for this book is available from the British Library.

EXACT ANALYSIS OF STRUCTURES WITH PERIODICITY USING U-TRANSFORMATION

ISBN 981-02-3642-5

This book is printed on acid-free paper.

Printed in Singapore by Uto-Print

Exact Analysis of Structures with Periodicity Using U-Transformation

H. C. Chan
C. W. Cai
Y. K. Cheung

Authors

H.C. Chan (CHAN Hon Chuen) is Professor and Head, Department of Civil Engineering at the University of Hong Kong. He received his Ph.D. from the Imperial College of Science and Technology, University of London in 1965.

C.W. Cai (CAI Chengwu) is Professor in the Department of Mechanics at Zhongshan University, China. He graduated from the Department of Mathematics and Mechanics of Peking University in 1960.

Y.K. Cheung (CHEUNG Yau Kai) is Professor and Pro-Vice-Chancellor of the University of Hong Kong. He received his Ph.D. and D. Sc. from the University of Wales Swansea in 1964 and 1973 respectively. He was also awarded a D.E. by the University of Adelaide in 1982.

PREFACE

This book introduces a new analytical method, U-transformation method, for the exact analysis of structures with periodic property.

In a broad sense, it is fair to say that the U-transformation method is originated from the mode method and yet it is different from the latter. In fact it may be called a mode subspace method. Performing U-transformation on a cyclic periodic structure is equivalent to projecting the displacement vector on the orthogonal mode subspaces in the unitary space. Hence the governing equations for a cyclic periodic structure and the continuity conditions between the substructures could be uncoupled. Consequently the governing equation in the U-space will be turned into N independent equations in N subspaces, where N is the total number of substructures. Besides, the order of the subspace is the same as the degrees of freedom of a single substructure. The difference in the N problems in the N subspaces may be reflected by taking a different value in one of the parameters in each subspace. In analytical method, that means the N governing equations need to be solved only once.

The U-transformation method has many advantages over the mode transformation method. When applying the mode method, it is necessary to work out the mode shapes of the structure. That is a task involving a lot of calculations and computing costs. In the end only numerical solutions would be obtained. In applying the U-transformation method, one can leave out calculating for the mode shapes in a periodic structure. This is because it is possible to give in advance some normalized orthogonal basis for each mode subspace. They form the columns of the U-matrix in the U-transformation. This is the physical background of the uncoupling of the governing equations and the continuity conditions of a periodic structure. For cyclic periodic structures, the U-transformation method has many obvious advantages over the numerical method. When the number of substructures in a periodic structure is altered, it will require a complete re-calculation to obtain a new solution by any numerical method. But with the U-transformation method, it is not necessary to re-analyze the problem again because the number of substructures is merely one of the parameters in the analytical solution. When there is a change in the number of substructures in a periodic structure, the expression of the analytical solution will not change.

By applying the U-transformation method for the analysis of cyclic periodic structures, either analytical approach or numerical approach may be adopted. However, the aim of this book is to illustrate how to obtain analytical solutions.

This book also provides many practical as well as theoretical exact explicit solutions for a large range of problems. Many of these solutions are new results that have just appeared in international journals for the first time together with some which have not been published before. The reason why the analytical method is used is not only because it will save computation costs, but also because it will enable the relationships between various parameters of the geometrical and physical properties in the problem to be reflected clearly. These relationships will probably often become the bases for optimum design or investigation. It is not a gross overstatement to say that analytical methods still have their great value to justify their existence even in this super-computer age of to-day. The analytical solutions given in this book will serve the following four distinctive purposes if not anything else:

1. It has been demonstrated that the U-transformation method can be applied to obtain exact analytical solutions for periodic structures. These exact analytical solutions could not be easily obtained or have not been obtained by other methods before. This book has illustrated how the U-transformation method can be used to derive the exact solutions for a number of problems of finite and infinite periodic structures under arbitrary static or dynamic loading.

2. With an exact analytical solution for a problem as a reference standard, it is possible to investigate the errors in approximate models for the problem. The analytical solution also gives the relations between the various parameters that contribute towards the errors. Hence it will enable a decision to be made as to under what conditions the approximate model may be acceptable. For instance, a continuous beam on periodic elastic supports may sometimes be simplified as a beam on elastic foundation. As another example, a stiffened plate with equal distance ribs may be treated as an orthotropic plate. This book has given detail analysis and discussions on the errors incurred by employing these two models. An explicit solution has also been given.

3. When a regular uniform continuous structure is sub-divided into a regular equal finite element mesh, for instance, a rectangular plate is sub-divided into a rectangular mesh, the sub-divided mesh may be treated as a periodic structure. That will allow it to be analyzed by the U-transformation method to yield an analytical solution for the finite element equations. Hence, that would facilitate the study of the rate of convergence as well as the convergence of the finite element solution when the number of finite elements approaches infinity. Furthermore, an explicit expression for the main body of the errors may be derived. This could not be achieved with the other methods.

4. It has been demonstrated how to apply the U-transformation method to the analysis of nearly periodic structures. Here only problems containing a few substructures with different properties are discussed and only for the case when

there is disturbance in the parameters of stiffness and masses. The aim is to study the mode localization phenomena. This phenomenon is very sensitive to the variation of a particular parameter in the sub-structure. It is therefore necessary to obtain a rigorous solution. If an approximate method is employed, it may lead to significant disagreement. It is best to employ analytical methods for this kind of problem.

The first five chapters in this book deal with the fundamental concept and theory of the U-transformation method. In addition to attempting to explain how the U-transformation method can be employed for the analysis of periodic structures, the authors of this book also try to explain how this method may be used in the study of two very important aspects of theoretical mechanics as mentioned in 3 and 4 above. Chapter 6 of this book deals with the exact solution of finite element equation. The chapter also gives an example for the case when the element stiffness matrix contains a stiffness regulative factor. It is possible to adjust the value of the stiffness regulative factor such that the error is minimum so that the rate of convergence may be enhanced. The mode localization is discussed in Chapter 7. Some solutions and results which have not been solved by using other methods are given in this chapter. These include the conditions that localized mode may occur in one dimension nearly periodic structures, and the theoretical results of the threshold value that causes the localized mode of 3-D nearly periodic structures.

As a mathematical tool, the U-transformation method has now been well systematized. The authors deeply believe that the problems that can be analyzed with the U-transformation method could be far more than those mentioned in this book.

The authors would like to express their sincere gratitude to the University of Hong Kong for providing research grants to support this work and to Professor J.K. Liu and Dr. Y. Li for their assistance in the preparation of this book and typing of the manuscript. While it is not possible to make individual acknowledgment to all the sources of references and materials adopted in this book, it is hoped that full credit is noted through this acknowledgment. Though every effort has been made to ensure exactness and to eliminate errors as far as possible, should errors be found, the authors would greatly appreciate comments and notifications from the learned readers.

H.C. Chan
C.W. Cai
Y.K. Cheung

CONTENTS

Chapter 1

UNCOUPLING OF THE DYNAMIC EQUATION FOR CYCLIC PERIODIC STRUCTURES AND U-TRANSFORMATION

The uncoupling of the dynamic equation for structural analysis is usually effected by adopting the mode method which will resolve the dynamic equation into a number of independent equations of single degree of freedom. Theoretically the method is a general as well as exact method. Surely it can be employed to analyze cyclic periodic structures. However, when the number of degrees of freedom is very large, it is impractical, impossible and unnecessary to find all the modes of the structure exactly.

Usually only a few low frequency modes are determined and the rest of the higher order frequency modes are neglected. That is to say, the displacement vector of the structural system will be resolved into a series of the mode shapes. Only the first few lower order frequency terms are considered and the effect of the higher order frequency terms are neglected. Therefore the mode method in practice is in fact an approximate method.

If we apply the general mode method to a cyclic periodic structure and treat it like any ordinary structure, the efficiency will be very low. To determine the modes of the whole structure will require very great effort in calculation and will be costly. Thomas[1] has proved that for cyclic periodic structures, all modes could be represented by rotating modes and each rotating mode has the characteristic that every mode of a substructure has the same amplitude when compared with the mode of the previous and adjacent substructure, but the phase is increased by a constant. According to the periodicity of cyclic periodic structures, the mode phase difference between two adjacent substructures should be $2m\pi/N$ (where $m = 1,2,3...,N$ and N is the total number of substructures).

The modes with the same phase difference in adjacent substructures will form a mode subspace and there will be N mode subspace for a cyclic periodic structure with N substructures. As a result of this fact, we can regard the mode space of a cyclic periodic structure as a direct sum of N mode subspaces. The dimensions of the mode subspace and the degrees of freedom in a single substructure is the same and the displacement vector in each mode subspace has rotating mode characteristics.

Hence we can obtain a set of orthogonal basis (basic vector is not the same as mode) corresponding to the mode subspace without carrying out any calculation. They constitute the column elements of a U-matrix.

Due to the property of orthogonality between the mode subspaces, the U-transformation method can be employed to uncouple the governing equation together with the boundary and continuity conditions for cyclic periodic structures into N independent equations. The number of unknowns in each individual equation (mode subspace coordinate) is the same as the degrees of freedom in a substructure.

This analytical technique is an exact method which does not require any preliminary or auxiliary calculations. The following section will explain how to apply this principle to derive rigorously the mathematical formulation of the U-transformation method for uncoupling the governing equation of cyclic periodic structures.

1.1 Dynamic Equation

A cyclic periodic structure may be defined as one which is made up of N substructures each having the same geometric and physical properties. The N substructures are rotationally symmetric about some axis, *i.e.,* when the cyclic periodic structure with N substructures rotates an angle $2m\pi/N$ (m=1,2,...,N) about the axis of symmetry, the position of the structure is not changed where $2\pi/N$ is referred to as the period.

In order to express the governing equation for cyclic periodic structures, the discrete equation will be considered in this chapter. The difficulty caused by the mathematics models corresponding to different structural forms can be avoided. In a general way, the finite element equation for cyclic periodic structures without damping may be expressed as [2]

$$\mathbf{M}\ddot{\mathbf{X}} + \mathbf{K}\mathbf{X} = \mathbf{F} \tag{1.1.1}$$

where a superior dot denotes differentiation with respect to the time variable t, **K** and **M** are stiffness and mass matrices and **F** and **X** are loading and displacement vectors respectively. Generally they can be written as

$$\mathbf{F} = \left[\mathbf{F}_1\ \mathbf{F}_2 \cdots \mathbf{F}_N\right]^T, \qquad \mathbf{X} = \left[\mathbf{X}_1\ \mathbf{X}_2 \cdots \mathbf{X}_N\right]^T \tag{1.1.2}$$

and

$$\mathbf{K} = \begin{bmatrix} \mathbf{K}_{11}, \mathbf{K}_{12}, \cdots, \mathbf{K}_{1N} \\ \mathbf{K}_{21}, \mathbf{K}_{22}, \cdots, \mathbf{K}_{2N} \\ \cdots\cdots\cdots\cdots \\ \mathbf{K}_{N1}, \mathbf{K}_{N2}, \cdots, \mathbf{K}_{NN} \end{bmatrix}, \quad \mathbf{M} = \begin{bmatrix} \mathbf{M}_{11}, \mathbf{M}_{12}, \cdots, \mathbf{M}_{1N} \\ \mathbf{M}_{21}, \mathbf{M}_{22}, \cdots, \mathbf{M}_{2N} \\ \cdots\cdots\cdots\cdots \\ \mathbf{M}_{N1}, \mathbf{M}_{N2}, \cdots, \mathbf{M}_{NN} \end{bmatrix} \tag{1.1.3}$$

where N denotes the total number of substructures. The order of submatrices $\mathbf{K}_{rs}, \mathbf{M}_{rs}$ (r,s=1,2,...,N) and vector components $\mathbf{F}_r, \mathbf{X}_r$ (r=1,2,...,N) is in agreement with the degrees of freedom of a substructure. With no loss of generality, let J denote the number of degrees of freedom for a single substructure. The stiffness and mass matrices for the cyclic periodic structures are symmetric , cyclic matrices, *i. e.*,

$$\mathbf{K}_{rs} = \mathbf{K}_{sr}^{T}, \quad r,s=1,2,...,N$$

$$\mathbf{K}_{11} = \mathbf{K}_{22} = \cdots = \mathbf{K}_{NN},$$

$$\mathbf{K}_{1,s} = \mathbf{K}_{2,s+1} = \cdots = \mathbf{K}_{N-s+1,N} = \mathbf{K}_{N-s+2,1} = \cdots = \mathbf{K}_{N,s-1}, \quad s=2,3,...,N \tag{1.1.4}$$

$$\mathbf{M}_{rs} = \mathbf{M}_{sr}^{T}, \quad r,s=1,2,...,N$$

$$\mathbf{M}_{11} = \mathbf{M}_{22} = \cdots = \mathbf{M}_{NN},$$

$$\mathbf{M}_{1,s} = \mathbf{M}_{2,s+1} = \cdots = \mathbf{M}_{N-s+1,N} = \mathbf{M}_{N-s+2,1} = \cdots = \mathbf{M}_{N,s-1}, \quad s=2,3,...,N \tag{1.1.5}$$

where $[\ ]^T$ denotes the transposed matrix of $[\ ]$.

1.2 Mode Subspaces of Cyclic Periodic Structures

The equation for natural vibration can be obtained by substituting equations $\mathbf{F} = \mathbf{0}$ and

$$\mathbf{X} = \mathbf{X}e^{i\omega t} \tag{1.2.1}$$

into equation (1.1.1), as

$$\mathbf{KX} = \omega^2 \mathbf{MX}. \tag{1.2.2}$$

This is an eigenvalue problem where the eigenvalue ω^2 represents the square of natural frequency and eigenvector $\mathbf{X}$ the natural mode. Because **K**, **M** in equation (1.2.2) are nonnegative matrices, the eigenvalues are nonnegative real numbers as a general rule for any structure.

The natural vibration of rotationally periodic structures has been studied by Thomas[1]. It is shown that, for each rotating mode, the deflection of one substructure has the same amplitude as, and a constant phase difference from, the deflection of the preceding substructure. Because of the cyclic periodicity, the phase difference between two adjacent substructures must be equal to an integer times the period of the cyclic periodic structures, *i.e.*,

$$\Delta\psi = \frac{2s\pi}{N}, \quad s=1,2,...,N \tag{1.2.3}$$

where $\Delta\psi$ denotes the phase difference between two adjacent substructures. Then the modes for the whole structure can be expressed in terms of the modes for first substructure , *i.e.,*

$$\mathbf{X} = \mathbf{U}_s\mathbf{Z}_s, \quad s=1,2,...,N \tag{1.2.4}$$

where

$$\mathbf{U_s} = \frac{\mathbf{1}}{\sqrt{\mathbf{N}}}\left[\mathbf{I_J},\ e^{is\psi}\,\mathbf{I_J},\ e^{i2s\psi}\,\mathbf{I_J},\cdots,\ e^{i(N-1)s\psi}\,\mathbf{I_J}\right]^{T} \qquad s=1,2,...,N \tag{1.2.5}$$

in which $\mathbf{I_J}$ denotes the unit matrix of order J , and ψ denotes the period , *i.e.*,

$$\psi = 2\pi/N\,. \tag{1.2.6}$$

$\mathbf{Z}_s$ represents the mode of the first substructure for phase difference $s\psi$. The constant factor $1/\sqrt{N}$ is added in order to normalize the column vectors of matrix $\mathbf{U}_s$. The integers N and J denote the total number of substructures and the number of degrees of freedom for a single substructure respectively. It may be verified that the N matrices defined by equation (1.2.5) satisfy the normalized orthogonal condition (see Appendix), *i.e.*,

$$\overline{\mathbf{U}}_s^T \mathbf{U}_r = \begin{cases} \mathbf{I}_J, & r = s \\ \mathbf{0}, & r \neq s \end{cases} \tag{1.2.7}$$

where a superior bar denotes complex conjugation. Let

$$\mathbf{U} = \left[\mathbf{U}_1, \mathbf{U}_2, \ldots, \mathbf{U}_N\right]. \tag{1.2.8}$$

Applying orthogonality of matrices $\mathbf{U}_s$ (s=1,2,...,N) shown in equation (1.2.7) to the matrix **U**, yields

$$\overline{\mathbf{U}}^T \mathbf{U} = \mathbf{I} \tag{1.2.9}$$

where **I** is a unit matrix of order N times J. **U** satisfying equation (1.2.9) is referred to as unitary matrix or U-matrix . Equation (1.2.9) indicates N times J column vectors of matrix **U** are a set of normalized orthogonal basis in the unitary space with N times J dimensions.

The matrix **U** defined by equations (1.2.8) and (1.2.5) has the following important property (see Appendix)

$$\overline{\mathbf{U}}^T \mathbf{K} \mathbf{U} = \begin{bmatrix} \mathbf{k}_1 & & & \mathbf{0} \\ & \mathbf{k}_2 & & \\ & & \ddots & \\ \mathbf{0} & & & \mathbf{k}_N \end{bmatrix} \tag{1.2.10}$$

$$\overline{\mathbf{U}}^T \mathbf{M} \mathbf{U} = \begin{bmatrix} \mathbf{m}_1 & & & \mathbf{0} \\ & \mathbf{m}_2 & & \\ & & \ddots & \\ \mathbf{0} & & & \mathbf{m}_N \end{bmatrix} \tag{1.2.11}$$

where the matrices on the right hand sides of equations (1.2.10) and (1.2.11) are quasi-diagonal matrices and the submatrices of order J can be expressed as

$$\mathbf{k}_s = \sum_{r=1}^{N} \mathbf{K}_{1r}\, e^{i(r-1)s\psi} \qquad s=1,2,\ldots,N$$

$$\mathbf{m}_s = \sum_{r=1}^{N} \mathbf{M}_{1r}\, e^{i(r-1)s\psi} \qquad s=1,2,\ldots,N \tag{1.2.12}$$

in which $\mathbf{K}_{1r}, \mathbf{M}_{1r}$ denote the submatrices of $\mathbf{K}$, $\mathbf{M}$ as shown in equation (1.1.3).

The equations (1.2.10) and (1.2.11) indicate that a cyclic matrix and the quasi-diagonal matrix are U-similar matrices with each other.

Every complex vector $\mathbf{X}$ in unitary space can be expressed as

$$\mathbf{X=UZ} \tag{1.2.13}$$

where

$$\mathbf{Z} = \{\mathbf{Z}_1, \mathbf{Z}_2, \cdots, \mathbf{Z}_N\}^T \tag{1.2.14}$$

and $\mathbf{Z}_s$ (s=1,2,...,N) is a complex vector of order J.

Substituting equation (1.2.13) into equation (1.2.2) and then premultiplying by $\overline{\mathbf{U}}^T$ yields

$$\mathbf{k}_s \mathbf{Z}_s = \omega^2 \mathbf{m}_s \mathbf{Z}_s \qquad s=1,2,\ldots,N. \tag{1.2.15}$$

Equation (1.2.15) represents N independent eigenvalue problems. If ω^2 and $\mathbf{Z}_s$ are eigenpairs then ω^2 is the square of natural frequency and $\mathbf{U}_s \mathbf{Z}_s$ is the natural mode for whole cyclic periodic structure. Noting the definitions of $\mathbf{k}_s$ and $\mathbf{m}_s$ shown in equation (1.2.12), it is obvious that

$$\mathbf{k}_{N-s} = \overline{\mathbf{k}}_s, \quad \mathbf{m}_{N-s} = \overline{\mathbf{m}}_s, \qquad s=1,2,\ldots,(N-1)/2 \text{ (when N is odd)}$$
$$\text{or } N/2 \text{ (when N is even)} \tag{1.2.16}$$

So there are some relations to the eigenpairs of s-th and (N-s)-th eigenproblems in equation (1.2.15), *i.e.*, the eigenvalues are the same and eigenvectors are conjugate to each other. The eigenproblems for s=1, 2, 3, ..., (n-1)/2 (when N is odd) or N/2 (when N is even) and N in equation (1.2.15) need be considered only. The matrices $\mathbf{k}_s$ and $\mathbf{m}_s$ defined as equation (1.2.12) are Hermitian matrices, *i.e.*,

$$\overline{\mathbf{k}}_s^T = \mathbf{k}_s, \qquad \overline{\mathbf{m}}_s^T = \mathbf{m}_s. \tag{1.2.17}$$

There are J real eigenvalues and corresponding J orthogonal eigenvectors for equation (1.2.15) in which s takes a fixed integer. For the structure considered, there are N times J normalized orthogonal eigenvectors which can be expressed as

$$\mathbf{X} = \mathbf{U}_s \mathbf{Z}_{sj} \qquad s=1,2,...,N, \quad j=1,2,...,J \tag{1.2.18}$$

where $\mathbf{Z}_{sj}$ denotes j-th normalized eigenvector of equation (1.2.15) and $\mathbf{U}_s \mathbf{Z}_{sj}$, j=1,2,...,J represent the rotating modes with phase difference $s\psi$ for two adjacent substructures. They denote the orthogonal basis of the mode subspace. It is obvious that J column vectors of matrix $\mathbf{U}_s$ are also a set of orthogonal basis in the same mode subspace, but each column vector of $\mathbf{U}_s$ doesn't represent any mode in a general way. Both orthogonal bases are equivalent. It must be pointed out that $\mathbf{U}_s$ (s=1,2,...,N) have been known and $\mathbf{Z}_{sj}$ need be found from equation (1.2.15). It is important to note that a set of orthogonal basis of each mode subspace has been obtained without any computation and these mode subspaces satisfy the orthogonality condition as shown in equation (1.2.7). This fact indicates that the dynamic equation for cyclic periodic structures can be uncoupled in each mode subspace by the unitary transformation shown in equation (1.2.13), because the column vectors of the unitary matrix **U** are made up of the orthogonal bases of the orthogonal mode subspaces.

1.3 U-Transformation

According to the expansion theorem, every possible displacement vector may be expressed as a series of natural modes, *i.e.*,

$$\mathbf{X} = \sum_{s=1}^{N} \sum_{j=1}^{J} a_{sj} \mathbf{U}_s \mathbf{Z}_{sj} \tag{1.3.1}$$

where a_{sj} denotes the mode co-ordinate, and **X** represents the possible displacement vector for the cyclic periodic structure with N substructures and each substructure having J degrees of freedom. Equation (1.3.1) can be rewritten as

$$\mathbf{X} = \sum_{s=1}^{N} \mathbf{U}_s \mathbf{q}_s \tag{1.3.2}$$

where

$$\mathbf{q}_s = \sum_{j=1}^{J} a_{sj} \mathbf{Z}_{sj} \qquad s=1,2,...,N \tag{1.3.3}$$

are referred to as mode subspace co-ordinates. Its number of degrees of freedom is in agreement with that in a single substructure, *i.e.,* J. $\mathbf{q}_s$ s=1,2,...,N denote a set of generalized displacements. The right hand side of equation (1.3.2) may be regarded as a series of mode subspaces and rewritten in a linear transformation form as

$$\mathbf{X} = \mathbf{U}\mathbf{q} \tag{1.3.4}$$

where **U** has been defined by equations (1.2.8) and (1.2.5) and **q** is a vector with vector components $\mathbf{q}_1, \mathbf{q}_2, \cdots, \mathbf{q}_N$, *i.e.,*

$$\mathbf{q} = \left\{\mathbf{q}_1, \mathbf{q}_2, \cdots, \mathbf{q}_N\right\}^T \tag{1.3.5}$$

Because the coefficient matrix **U** is an unitary matrix therefore the linear transformation (1.3.4) is referred to as the U-transformation. Substituting equation (1.2.5) into equation (1.3.2) yields the U-transformation in terms of the components as

$$\mathbf{X}_k = \frac{1}{\sqrt{N}} \sum_{s=1}^{N} e^{i(k-1)s\psi} \mathbf{q}_s \qquad k=1,2,...,N \tag{1.3.6}$$

with $\psi = 2\pi/N$.

Noting the equation (1.2.9) and premultiplying both sides of equation (1.3.4) by $\overline{\mathbf{U}}^T$ yields the inverse transformation as

$$\mathbf{q} = \overline{\mathbf{U}}^T \mathbf{X} \tag{1.3.7}$$

with its component form

$$\mathbf{q}_s = \frac{1}{\sqrt{N}} \sum_{k=1}^{N} e^{-i(k-1)s\psi} \mathbf{X}_k \qquad s=1,2,...,N. \tag{1.3.8}$$

Substituting equation (1.3.4) into equation (1.1.1) and then premultiplying by $\overline{\mathbf{U}}^T$, yields the uncoupled equation in each mode subspace as [3]

$$\mathbf{m}_s \ddot{\mathbf{q}}_s + \mathbf{k}_s \mathbf{q}_s = \mathbf{f}_s \qquad s=1,2,...,N \tag{1.3.9}$$

where $\mathbf{k}_s, \mathbf{m}_s$ have been defined in equation (1.2.12) and

$$\mathbf{f}_s = \overline{\mathbf{U}}_s^T \mathbf{F} = \frac{1}{\sqrt{N}} \sum_{k=1}^{N} e^{-i(k-1)s\psi} \mathbf{F}_k \qquad s=1,2,...,N. \tag{1.3.10}$$

$\mathbf{f}_s$ denotes the projected vector of **F** on the s-th mode subspace.

Consider now the initial condition as

$$\mathbf{X}_{t=0} = \mathbf{X}_0, \qquad \dot{\mathbf{X}}_{t=0} = \dot{\mathbf{X}}_0, \tag{1.3.11}$$

where $\mathbf{X}_0, \dot{\mathbf{X}}_0$ denote the initial displacement and velocity vectors respectively. Substituting equation (1.3.4) into initial condition (1.3.11) and premultiplying by $\overline{\mathbf{U}}^T$ yields

$$\mathbf{q}_{t=0} = \mathbf{q}_0, \qquad \dot{\mathbf{q}}_{t=0} = \dot{\mathbf{q}}_0, \tag{1.3.12}$$

with

$$\mathbf{q}_0 = \overline{\mathbf{U}}^T \mathbf{X}_0, \qquad \dot{\mathbf{q}}_0 = \overline{\mathbf{U}}^T \dot{\mathbf{X}}_0. \tag{1.3.13}$$

The initial condition (1.3.12) with equation (1.3.13) may be rewritten in the component form as

$$\mathbf{q}_{s,t=0} = \overline{\mathbf{U}}_s^T \mathbf{X}_0, \qquad \dot{\mathbf{q}}_{s,t=0} = \overline{\mathbf{U}}_s^T \dot{\mathbf{X}}_0, \quad s=1,2,...,N. \tag{1.3.14}$$

Equation (1.3.9) must be solved subject to the initial condition (1.3.14) for each mode subspace co-ordinate.

Upon examining the mathematical characteristics of equations (1.3.9) and (1.3.14), it is apparent that $\mathbf{q}_{N-s} = \overline{\mathbf{q}}_s$ and $\mathbf{q}_N$, $\overline{\mathbf{q}}_{N/2}$ (if N is even) are real vectors. Therefore $\mathbf{X}_k$ shown in equation (1.3.6) must be a real vector and the number of independent equations shown in equation (1.3.9) is equal to (N+1)/2 if N is odd or (N/2)+1 if N is even . Equation (1.3.6) may be rewritten as

$$\mathbf{X}_k = \frac{1}{\sqrt{N}} \sum_{s=1}^{N} \mathrm{Re}\left[e^{i(k-1)s\psi}\,\mathbf{q}_s\right] \qquad k=1,2,\ldots,N \tag{1.3.15}$$

where the symbol Re denotes the real-part operator.

Chapter 2

ANALYTICAL SOLUTIONS FOR CYCLIC PERIODIC STRUCTURES

The physical meaning and mathematical formulation of the U-transformation have been described in the previous chapter. The dynamic equation for cyclic periodic structures can be uncoupled in each mode subspace by U-transformation. It may be employed to reduce computation effort in numerical analysis of cyclic periodic structures. However, it is more important that it makes it possible to derive the analytical solutions of periodic structures. The following examples [4] will illustrate how this can be done.

2.1 Regular Polygon Frames

Consider the plane frame shown in Figure 2.1.1 which is made up of N similar beams with all nodes simply supported. The bending moment can be transmitted from a beam to the adjacent beams. The structure considered can be regarded as a cyclic periodic structure with N substructures. Each substructure is a simply supported beam. In each beam a local co-ordinate system oxy is established with the origin o at its mid-span and x axis along the beam as shown in Figure 2.1.1.

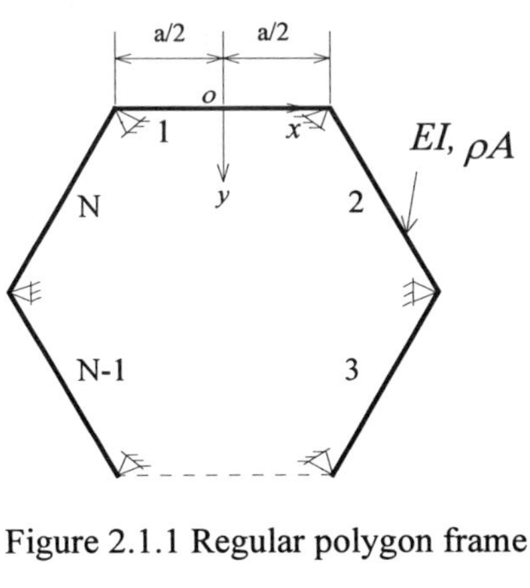

Figure 2.1.1 Regular polygon frame

The dynamic equations for all substructures are of the same form, *i.e.*,

$$EI\frac{\partial^4 W_k(x,t)}{\partial x^4}+\rho A\frac{\partial^2 W_k(x,t)}{\partial t^2}=F_k(x,t), \qquad -\frac{a}{2}<x<\frac{a}{2},$$

$$k=1,2,\ldots,N \tag{2.1.1}$$

where EI denotes flexural rigidity, ρ A denotes mass per unit length, "a" denotes the span length and W_k, F_k denote the transverse displacement and loading functions for the k-th substructure. The variables x, t of functions W_k and F_k are co-ordinate and time variables. The restrained condition at nodes can be expressed as

$$W_k\big|_{x=a/2}=0, \qquad W_k\big|_{x=-a/2}=0, \qquad k=1,2,\ldots,N. \tag{2.1.2}$$

Continuity across the supports requires the following conditions to be satisfied:

$$\frac{\partial W_k}{\partial x}\Big|_{x=a/2}=\frac{\partial W_{k+1}}{\partial x}\Big|_{x=-a/2}, \qquad \frac{\partial^2 W_k}{\partial x^2}\Big|_{x=a/2}=\frac{\partial^2 W_{k+1}}{\partial x^2}\Big|_{x=-a/2},$$

$$k=1,2,\ldots,N. \tag{2.1.3}$$

with $W_{N+1}\equiv W_1$.

Equations (2.1.1) to (2.1.3) may be arranged in matrix form. Let column vectors **W** and **F** denote the ordered sets of transverse displacements and loading functions, *i.e.*,

$$\mathbf{W}=[W_1,W_2,\cdots,W_N]^T, \qquad \mathbf{F}=[F_1,F_2,\cdots,F_N]^T. \tag{2.1.4}$$

With this notation, equation (2.1.1) takes the form

$$EI\frac{\partial^4 \mathbf{W}}{\partial x^4}+\rho A\frac{\partial^2 \mathbf{W}}{\partial t^2}=\mathbf{F} \tag{2.1.5}$$

and the conditions (2.1.2) and (2.1.3) become

$$\mathbf{W}\Big|_{x=a/2}=0, \quad \mathbf{W}\Big|_{x=-a/2}=0,$$

$$\frac{\partial \mathbf{W}}{\partial x}\Big|_{x=a/2}=\boldsymbol{\varepsilon}\frac{\partial \mathbf{W}}{\partial x}\Big|_{x=-a/2}, \quad \frac{\partial^2 \mathbf{W}}{\partial x^2}\Big|_{x=a/2}=\boldsymbol{\varepsilon}\frac{\partial^2 \mathbf{W}}{\partial x^2}\Big|_{x=-a/2} \tag{2.1.6}$$

where $\boldsymbol{\varepsilon}$ is a cyclic matrix of order N defined as

$$\boldsymbol{\varepsilon}=\begin{bmatrix} 0 & 1 & & & \\ & 0 & 1 & & \\ & & \ddots & \ddots & \\ & & & 0 & 1 \\ 1 & 0 & \cdots & & 0 \end{bmatrix} \tag{2.1.7}$$

in which the empty elements are zeros.

In order to uncouple the boundary conditions in equation (2.1.6), the U-transformation is needed. Let

$$\mathbf{W}=\mathbf{U}\mathbf{q} \tag{2.1.8a}$$

or

$$W_k=\frac{1}{\sqrt{N}}\sum_{m=1}^{N} e^{i(k-1)m\psi}q_m\,, \qquad k=1,2,\ldots,N \tag{2.1.8b}$$

where q_m is m-th component of vector $\mathbf{q}$ and referred to as m-th mode subspace co-ordinate. Matrix $\mathbf{U}$ has been defined as equations (1.2.8) and (1.2.5) in which the unit matrix I_J should be replaced by one. ψ denotes the period of cyclic periodic structure, *i.e.,* $\psi=2\pi/N$.

Substituting equation (2.1.8a) into equations (2.1.5) and (2.1.6), and then premultiplying by $\overline{\mathbf{U}}^T$ result in

$$EI\frac{\partial^4 \mathbf{q}}{\partial x^4}+\rho A\frac{\partial^2 \mathbf{q}}{\partial t^2}=\mathbf{f} \tag{2.1.9}$$

and

$$\mathbf{q}\Big|_{x=a/2}=0, \qquad \mathbf{q}\Big|_{x=-a/2}=0,$$

$$\frac{\partial \mathbf{q}}{\partial x}\Big|_{x=a/2} = \Phi \frac{\partial \mathbf{q}}{\partial x}\Big|_{x=-a/2}, \quad \frac{\partial^2 \mathbf{q}}{\partial x^2}\Big|_{x=a/2} = \Phi \frac{\partial^2 \mathbf{q}}{\partial x^2}\Big|_{x=-a/2} \tag{2.1.10}$$

where

$$\mathbf{f} = \overline{\mathbf{U}}^{T}\mathbf{F} \tag{2.1.11}$$

and

$$\Phi = \overline{\mathbf{U}}^{T}\boldsymbol{\varepsilon}\,\mathbf{U} = \begin{bmatrix} e^{i\psi} & & & 0 \\ & e^{i2\psi} & & \\ & & \ddots & \\ 0 & & & e^{iN\psi} \end{bmatrix} \tag{2.1.12}$$

where Φ is a diagonal matrix of order N. As a result the equations (2.1.9) and (2.1.10) are made up of N independent equations as follows:

$$EI\frac{\partial^4 q_m}{\partial x^4} + \rho A \frac{\partial^2 q_m}{\partial t^2} = f_m, \qquad m=1,2,...,N, \tag{2.1.13}$$

$$q_m\Big|_{x=a/2}=0, \qquad q_m\Big|_{x=-a/2}=0,$$

$$\frac{\partial q_m}{\partial x}\Big|_{x=a/2} = e^{im\psi}\frac{\partial q_m}{\partial x}\Big|_{x=-a/2}, \quad \frac{\partial^2 q_m}{\partial x^2}\Big|_{x=a/2} = e^{im\psi}\frac{\partial^2 q_m}{\partial x^2}\Big|_{x=-a/2},$$

$$m=1,2,...,N \tag{2.1.14}$$

where f_m denotes m-th component of vector **f** and it is defined as

$$f_m = \overline{\mathbf{U}}_m^{T}\mathbf{F} = \frac{1}{\sqrt{N}}\sum_{k=1}^{N} e^{-i(k-1)m\psi}F_k, \qquad m=1,2,...,N. \tag{2.1.15}$$

The dynamic equations (2.1.1)-(2.1.3) have been uncoupled in each mode subspace as shown in equations (2.1.13) and (2.1.14) by U-transformation. Equation (2.1.13) must be solved subject to the boundary condition (2.1.14) for each mode subspace co-ordinate. This can be done by considering a specific loading condition.

2.1.1 Static Problem

Let a concentrated load of magnitude P, which is independent of time, act at the mid-point of the first beam , with no loads imposed on the remaining (N-1) beams, *i.e.*,

$$F_1 = P\delta(x), \quad F_2 = F_3 = \cdots = F_N = 0 , \tag{2.1.16}$$

where $\delta(x)$ denotes the Dirac delta function. Substituting equation (2.1.16) into equation (2.1.15) yields

$$f_m = \frac{P}{\sqrt{N}}\delta(x) , \qquad m=1,2,...,N. \tag{2.1.17}$$

Noting the generalized displacements q_m (m=1,2,...,N) are independent of time, equation (2.1.13) takes the form

$$EI\frac{d^4 q_m}{dx^4} = \frac{P}{\sqrt{N}}\delta(x) , \qquad -\frac{a}{2} < x < \frac{a}{2}, \qquad m=1,2,...,N. \tag{2.1.18}$$

The general solution for q_m of equation (2.1.18) may be expressed as

$$q_m = (P / EI\sqrt{N})(C_{m0} + C_{m1}x + C_{m2}x^2 + C_{m3}x^3 - x^3/12), \quad -a/2 \le x \le 0$$

$$q_m = (P / EI\sqrt{N})(C_{m0} + C_{m1}x + C_{m2}x^2 + C_{m3}x^3 + x^3/12), \quad 0 \le x \le a/2 \tag{2.1.19}$$

where C_{m0} to C_{m3} are four constants of integration. They can be determined by using boundary condition (2.1.14) as

$$C_{m0} = \frac{a^3}{384}(\frac{7-\cos m\psi}{2+\cos m\psi}), \quad C_{m2} = -\frac{a}{32}(\frac{5+\cos m\psi}{2+\cos m\psi}),$$

$$C_{m1} = -i\frac{a^2}{64}(\frac{\sin m\psi}{2+\cos m\psi}), \quad C_{m3} = \frac{i}{16}(\frac{\sin m\psi}{2+\cos m\psi}),$$

$$m=1,2,...,N. \qquad (2.1.20)$$

The deflection in each substructure may now be found by substituting equations (2.1.19), (2.1.20) into equation (2.1.8b). Since both continuous system and discrete system for cyclic periodic structures have the same mathematical characteristics, *i.e.,* $q_m(x) = \bar{q}_{N-m}(x)$, therefore equation (2.1.8b) can be rewritten as

$$W_k = \frac{1}{\sqrt{N}}\sum_{m=1}^{N} \mathrm{Re}[e^{i(k-1)m\psi} q_m], \quad k=1,2,...,N. \qquad (2.1.21)$$

The deflection in the first substructure may be expressed as

$$W_1 = \frac{1}{\sqrt{N}}\sum_{m=1}^{N} \mathrm{Re}[q_m]. \qquad (2.1.22)$$

Substituting equation (2.1.19) with equation (2.1.20) into equation (2.1.22) yields

$$W_1(x) = \begin{cases} \frac{Pa^3}{384EI}\frac{1}{N}\sum_{m=1}^{N}[\frac{7-\cos m\psi}{2+\cos m\psi} - \frac{12(5+\cos m\psi)}{2+\cos m\psi}(\frac{x}{a})^2 - 32(\frac{x}{a})^3], & -\frac{a}{2} \le x \le 0 \\ \frac{Pa^3}{384EI}\frac{1}{N}\sum_{m=1}^{N}[\frac{7-\cos m\psi}{2+\cos m\psi} - \frac{12(5+\cos m\psi)}{2+\cos m\psi}(\frac{x}{a})^2 + 32(\frac{x}{a})^3], & 0 \le x \le \frac{a}{2} \end{cases}$$

$$(2.1.23)$$

This deflection function is applicable to the loaded beam substructure only. For other substructures, the deflection functions can be obtained similarly without any particular difficulty.

The maximum deflection occurs at mid-point of the loaded beam and its magnitude is

$$W_{max} = W_1(0) = \frac{Pa^3}{384EI}\frac{1}{N}\sum_{m=1}^{N}\frac{7-\cos m\psi}{2+\cos m\psi}. \quad (2.1.24)$$

The bending moment for the loaded beam may be obtained from equation (2.1.23) as

$$M = -EI\frac{d^2W_1}{dx^2} = \frac{Pa}{16}\frac{1}{N}\sum_{m=1}^{N}[\frac{5+\cos m\psi}{2+\cos m\psi} \pm 8\frac{x}{a}], \quad \begin{matrix} -a/2 \le x \le 0 \\ 0 \le x \le a/2 \end{matrix}. \quad (2.1.25)$$

The maximum moment also occurs at mid-span with magnitude

$$M_{max} = \frac{Pa}{16}\frac{1}{N}\sum_{m=1}^{N}\frac{5+\cos m\psi}{2+\cos m\psi}. \quad (2.1.26)$$

Some numerical results for maximum deflection and moment are given in Table 2.1.1 .

Table 2.1.1 Deflections and moments of regular polygon frames subjected to concentrated load P at mid-point of beam substructure

Number of substructures N	3	5	7	9
$W_{max}/(Pa^3/384EI)$	4.00000	4.18182	4.19512	4.19608
$M_{max}/(Pa/16)$	2.66667	2.72727	2.73171	2.73203

2.1.2 Natural Vibration

The natural vibration equation with boundary condition may be obtained by inserting

$$f_m = 0, \quad q_m = Q_m(x)e^{i\omega t} \quad m=1,2,...,N \quad (2.1.27)$$

in equations (2.1.13) and (2.1.14) , as

$$\frac{d^4 Q_m}{dx^4} = \beta^4 Q_m, \qquad m=1,2,...,N \tag{2.1.28}$$

with

$$Q_m(\frac{a}{2}) = 0, \quad Q_m(-\frac{a}{2}) = 0,$$

$$Q'_m(\frac{a}{2}) = e^{im\psi} Q'_m(-\frac{a}{2}), \quad Q''_m(\frac{a}{2}) = e^{im\psi} Q''_m(-\frac{a}{2}), \quad m=1,2,...,N \tag{2.1.29}$$

where a prime denotes differentiation with respect to *x*, and

$$\beta^4 = \frac{\rho A}{EI} \omega^2 . \tag{2.1.30}$$

Homogeneous equation (2.1.28) and boundary condition (2.1.29) represent an eigenvalue problem of the self-adjoint differential equation with positive real eigenvalue. The eigenvalue β^4 is a frequency parameter which is related to the natural frequency ω by equation (2.1.30). The eigenfunction $Q_m(x)$ represents the amplitude function of m-th mode subspace co-ordinate q_m .The corresponding natural mode may be obtained from equation (2.1.21) in which $q_m = \overline{q}_{N-m} = Q_m$ with other mode subspace co-ordinates vanishing. The eigenproblem shown in equations (2.1.28) and (2.1.29) is similar to equation (1.2.15), *i.e.,* for m-th and (N-m)th eigenproblems in equation (2.1.28) the eigenvalues are the same and eigenfunctions are conjugate each other. Noting that both cases of m=N and m=0 in equations (2.1.28) and (2.1.29) are equivalent, the cases of $0 \le m\psi \le \pi$ need be considered only for equations (2.1.28) and (2.1.29).

The general solution for Q_m of homogeneous equation (2.1.28) may be expressed as

$$Q_m(x) = C_1 \sin\beta x + iC_2 \cos\beta x + C_3 \text{sh}\beta x + iC_4 \text{ch}\beta x \tag{2.1.31}$$

where $C_1 - C_4$ are four constants of integration. Substituting equation (2.1.31) into boundary condition (2.1.29) yields

$$-C_1 \sin\alpha + iC_2 \cos\alpha - C_3 \mathrm{sh}\alpha + iC_4 \mathrm{ch}\alpha = 0$$

$$C_1 \sin\alpha + iC_2 \cos\alpha + C_3 \mathrm{sh}\alpha + iC_4 \mathrm{ch}\alpha = 0$$

$$C_1 \cos\alpha(1 - e^{im\psi}) - iC_2 \sin\alpha(1 + e^{im\psi}) + C_3 \mathrm{ch}\alpha(1 - e^{im\psi})$$

$$+ iC_4 \mathrm{sh}\alpha(1 + e^{im\psi}) = 0$$

$$-C_1 \sin\alpha(1 + e^{im\psi}) - iC_2 \cos\alpha(1 - e^{im\psi}) + C_3 \mathrm{sh}\alpha(1 + e^{im\psi}) + iC_4 \mathrm{ch}\alpha(1 - e^{im\psi}) = 0, \tag{2.1.32}$$

where

$$\alpha = \beta a / 2 \tag{2.1.33}$$

Let the coefficient determinant of the simultaneous equations (2.1.32) be equal to zero and then expanding this determinant yields the frequency equation:

$$\cos m\psi(\mathrm{sh}2\alpha - \sin 2\alpha) + \mathrm{ch}2\alpha \sin 2\alpha - \mathrm{sh}2\alpha \cos 2\alpha = 0,$$

$$m=1,2,\ldots,N. \tag{2.1.34}$$

The natural frequency corresponding to the frequency parameter α can be expressed as

$$\omega = \frac{4}{a^2}\sqrt{\frac{EI}{\rho A}}\alpha^2 \tag{2.1.35}$$

where α denotes a positive root of equation (2.1.34) for given m. For fixed value of $m\psi$, there are infinite numbers of positive root of equation (2.1.34), denoted by α_{mj} (j=1,2,..., ∞) with $\alpha_{mj} \le \alpha_{m,j+1}$ (j=1,2,..., ∞). For given positive integer j, α_{mj} (m=1,2,...,N) fall inside the limits of j-th frequency band. The upper and lower limits of j-th frequency band can be determined by $\alpha_{N/2,j}$ and $\alpha_{N,j}$, when N approaches infinity. Upon substituting $\alpha = \alpha_{mj}$ into equation (2.1.32) the non-trivial solution for C_1 to C_4 can be found as

$$C_1 = \mathrm{sgn}[B(\alpha)]\mathrm{sh}\alpha\sqrt{\mathrm{ch}\alpha|B(\alpha)|}\ ,$$

$$C_2 = -\mathrm{ch}\alpha\sqrt{\mathrm{sh}\alpha|\sin\alpha(\mathrm{sh}\alpha\cos\alpha - \mathrm{ch}\alpha\sin\alpha)|}\ ,$$

$$C_3 = -\mathrm{sgn}[B(\alpha)]\sin\alpha\sqrt{\mathrm{ch}\alpha|B(\alpha)|}\ ,$$

$$C_4 = \cos\alpha\sqrt{\mathrm{sh}\alpha|\sin\alpha(\mathrm{sh}\alpha\cos\alpha - \mathrm{ch}\alpha\sin\alpha)|}\ ; \quad 0 \le m\psi \le \pi \qquad (2.1.36)$$

where sgn denotes signum function, $\alpha = \alpha_{mj}$ and

$$B(\alpha) = \cos\alpha(\mathrm{ch}\alpha\sin\alpha + \mathrm{sh}\alpha\cos\alpha)\,. \qquad (2.1.37)$$

The eigenfunction corresponding to frequency parameter α_{mj} can be expressed as

$$Q_{mj}(x) = C_1\sin\frac{2\alpha}{a}x + iC_2\cos\frac{2\alpha}{a}x + C_3\mathrm{sh}\frac{2\alpha}{a}x + iC_4\mathrm{ch}\frac{2\alpha}{a}x\ ;$$
$$0 \le m\psi \le \pi \qquad (2.1.38)$$

with $\alpha = \alpha_{mj}$ and C_1 to C_4 as defined in equation (2.1.36) . It is well known that the eigenfunction can be multiplied by an arbitrary complex multiplier, which is omitted in equation (2.1.38).

By substituting $q_m = \overline{q}_{N-m} = CQ_{mj}$ and $q_s = 0$ $(s \ne m, N\text{-}m)$ into equation (2.1.8b) , the natural mode can be obtain

$$W_k = \frac{2}{\sqrt{N}}\mathrm{Re}[e^{i(k-1)m\psi}CQ_{mj}] \qquad k=1,2,\ldots,N \qquad (2.1.39)$$

where C denotes an arbitrary complex factor. Letting C=1 and C=i , two independent modes can be found as

$$W_k^1 = \mathrm{Re}[e^{i(k-1)m\psi} Q_{mj}] \quad \text{and} \quad W_k^2 = \mathrm{Im}[e^{i(k-1)m\psi} Q_{mj}],$$

$$k=1,2,\ldots,N \tag{2.1.40}$$

where the arbitrary real multiplier has been omitted .The symbol Im denotes the imaginary-part of the function in the brackets. It can be proved that for m=N and m=N/2 (if N is even), the one of both shown in equation (2.1.40) must be equal to zero, *i.e.,* there is one mode only corresponding to the eigenvalues α_{Nj} and $\alpha_{N/2,j}$ (if N is even).

2.1.3 Forced Vibration

Consider the harmonic vibration of the regular polygon frames subjected to harmonic forces. In general, the loading and deflection functions can be expressed as

$$F_k(x,t) = F_k(x)e^{i\omega t}, \qquad W_k(x,t) = W_k(x)e^{i\omega t}, \qquad k=1,2,\ldots,N. \tag{2.1.41}$$

Substituting equation (2.1.41) into equation (2.1.1) yields

$$\frac{d^4 W_k(x)}{dx^4} - \beta^4 W_k(x) = \frac{F_k(x)}{EI}, \qquad k=1,2,\ldots,N \tag{2.1.42}$$

where

$$\beta^4 = \frac{\rho A}{EI}\omega^2. \tag{2.1.43}$$

The restrained and continuous conditions are of the same form as equations (2.1.2) and (2.1.3). The uncoupled equation in terms of the mode subspace coordinates can be expressed as

$$\frac{d^4 q_m(x)}{dx^4} - \beta^4 q_m(x) = \frac{f_m(x)}{EI}, \qquad m=1,2,\ldots,N \tag{2.1.44}$$

and the restrained and continuous conditions become

$$q_m(-\frac{a}{2})=0, \quad q_m(\frac{a}{2})=0,$$

$$q'_m(\frac{a}{2})=e^{im\psi}q'_m(-\frac{a}{2}), \quad q''_m(\frac{a}{2})=e^{im\psi}q''_m(-\frac{a}{2}), \qquad m=1,2,...,N \tag{2.1.45}$$

where f_m has been defined as equation (2.1.15). The solution for $q_m(x)$ of equation (2.1.44) must be solved subject to equation (2.1.45).

Consider now a specific loading condition. Let a concentrated harmonic force with amplitude P act at the mid-point of first beam substructure, with no loads imposed on the remaining (N-1) substructures, *i.e.*,

$$F_1 = P\delta(x), \qquad F_2 = F_3 = \cdots = F_N = 0 \tag{2.1.46}$$

By substituting equation (2.1.46) into equation (2.1.15) the generalized loads can be obtained as

$$f_m(x) = \frac{P}{\sqrt{N}}\delta(x), \qquad m=1,2,...,N. \tag{2.1.47}$$

The solution for q_m of equations (2.1.44) (2.1.47) and (2.1.45) can be expressed as

$$q_m = \begin{cases} \dfrac{P}{2EI\beta^3\sqrt{N}}[(C_{m1}+\dfrac{1}{2})\sin\beta x + C_{m2}\cos\beta x + (C_{m3}-\dfrac{1}{2})\mathrm{sh}\beta x + C_{m4}\mathrm{ch}\beta x], \\ \qquad\qquad -\dfrac{a}{2} \le x \le 0 \\ \dfrac{P}{2EI\beta^3\sqrt{N}}[(C_{m1}-\dfrac{1}{2})\sin\beta x + C_{m2}\cos\beta x + (C_{m3}+\dfrac{1}{2})\mathrm{sh}\beta x + C_{m4}\mathrm{ch}\beta x], \\ \qquad\qquad 0 \le x \le \dfrac{a}{2} \end{cases}$$

$$m=1,2,...,N \tag{2.1.48}$$

where

$$C_{m1} = i\sin m\psi\ \mathrm{sh}\alpha\ (\cos\alpha - \mathrm{ch}\alpha)/D_m(\alpha),$$

$$C_{m2} = \sin\alpha\,[\mathrm{sh}\alpha + \mathrm{ch}2\alpha\sin\alpha - \mathrm{sh}2\alpha\cos\alpha + (\mathrm{sh}\alpha - \sin\alpha)\cos m\psi]/D_m(\alpha)\,,$$

$$C_{m3} = i\sin m\psi\,\sin\alpha(\mathrm{ch}\alpha - \cos\alpha)\,/\,D_m(\alpha)\,,$$

$$C_{m4} = \mathrm{sh}\alpha\,[\sin\alpha + \mathrm{sh}\alpha\cos 2\alpha - \mathrm{ch}\alpha\sin 2\alpha + (\sin\alpha - \mathrm{sh}\alpha)\cos m\psi]/D_m(\alpha)\,,$$

$$D_m(\alpha) = \mathrm{ch}2\alpha\sin 2\alpha - \mathrm{sh}2\alpha\cos 2\alpha + (\mathrm{sh}2\alpha - \sin 2\alpha)\cos m\psi\,,$$

$$\alpha = \beta a/2\,. \tag{2.1.49}$$

The amplitude of the deflection for each substructure may be found by substituting equations (2.1.48) and (2.1.49) into equation (2.1.21) , and then the amplitude function for the loaded substructure is

$$W_1(x) = \begin{cases} \dfrac{P}{2EI\beta^3}\dfrac{1}{N}\sum\limits_{m=1}^{N}(\dfrac{1}{2}\sin\beta x + C_{m2}\cos\beta x - \dfrac{1}{2}\mathrm{sh}\beta x + C_{m4}\mathrm{ch}\beta x), & -\dfrac{a}{2} \le x \le 0 \\ \dfrac{P}{2EI\beta^3}\dfrac{1}{N}\sum\limits_{m=1}^{N}(-\dfrac{1}{2}\sin\beta x + C_{m2}\cos\beta x + \dfrac{1}{2}\mathrm{sh}\beta x + C_{m4}\mathrm{ch}\beta x), & 0 \le x \le \dfrac{a}{2} \end{cases}$$

$$\tag{2.1.50}$$

with C_{m2} and C_{m4} defined as in equation (2.1.49).

When the frequency of the excitation approaches the natural frequency, *i.e.,* α approaches the root of the frequency equation (2.1.34), some of $D_1(\alpha)$,..., $D_N(\alpha)$ will be approaching zero and then the values of deflection functions become infinity except at the nodes of natural vibration.

2.2 Prism Shells with Regular Polygon Cross Section

Consider a prism shell with regular polygon cross section and two ends simply supported, as shown in Figure 2.2.1(a). On the assumption that the ridges of the prism shell are stiffened by stiffening ribs which are not flexible but can be twisted freely, the substructure is a rectangular plate with four edges simply supported as shown in Figure 2.2.1 (b) . In each substructure, establish a set of local Cartesian coordinates oxy with the origin at one of the corner points.

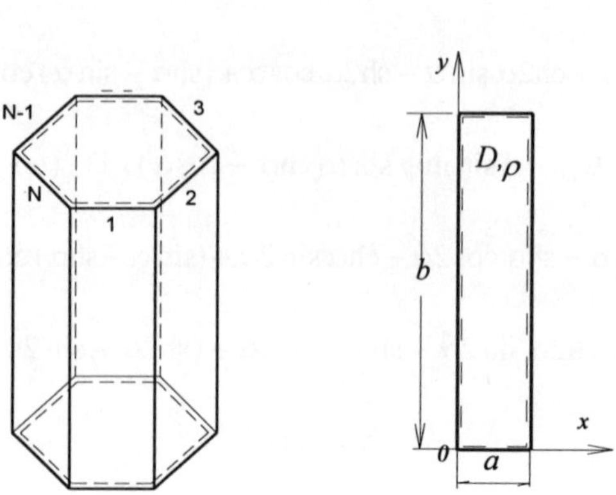

Figure 2.2.1. Prism shell with regular polygon cross section
(a) Prism shell; (b) Substructure with local co-ordinates

The governing equation for the transverse displacement W_k of the k-th plate substructure is

$$D\left(\frac{\partial^4 W_k}{\partial x^4}+2\frac{\partial^4 W_k}{\partial x^2 \partial y^2}+\frac{\partial^4 W_k}{\partial y^4}\right)+\rho\frac{\partial^2 W_k}{\partial t^2}=F_k\,, \quad k=1,2,\ldots,N \qquad (2.2.1)$$

where D, ρ denote the flexural rigidity, mass per unit area of the plate and F_k denotes the loading function of the k-th substructure, N denotes the total number of substructures. The boundary conditions of two end sections can be expressed as

$$W_{k|y=0}=0, \qquad W_{k,yy|y=0}=0,$$

$$W_{k|y=b}=0, \qquad W_{k,yy|y=b}=0, \qquad k=1,2,\ldots,N \qquad (2.2.2)$$

and the restrained, continuous conditions at intersections of two adjacent substructures are

$$W_{k|x=0}=0, \qquad W_{k|x=a}=0,$$

$$W_{k,x|x=a} = W_{k+1,x|x=0}, \qquad W_{k,xx|x=a} = W_{k+1,xx|x=0}, \quad k=1,2,...,N \qquad (2.2.3)$$

where a and b denote the length and width of the plate substructure, $W_{N+1} \equiv W_1$ and the subscripts x and y denote the partial differentiation with respect to x and y, respectively.

Applying the U-transformation

$$q_m = \frac{1}{\sqrt{N}} \sum_{k=1}^{N} e^{-i(k-1)m\psi} W_k, \qquad m=1,2,...,N \qquad (2.2.4)$$

to equations (2.2.1)-(2.2.3), *i.e.,* multiplying both sides of equations (2.2.1) to (2.2.3) by $e^{-i(k-1)m\psi}/\sqrt{N}$ and then summation for k from 1 to N, yield

$$D\left(\frac{\partial^4 q_m}{\partial x^4} + 2\frac{\partial^4 q_m}{\partial x^2 \partial y^2} + \frac{\partial^4 q_m}{\partial y^4}\right) + \rho\frac{\partial^2 q_m}{\partial t^2} = f_m, \qquad m=1,2,...,N \qquad (2.2.5)$$

$$q_{m|y=0} = 0, \qquad q_{m,yy|y=0} = 0,$$

$$q_{m|y=b} = 0, \qquad q_{m,yy|y=b} = 0, \qquad (2.2.6)$$

and

$$q_{m|x=0} = 0, \qquad q_{m|x=a} = 0,$$

$$q_{m,x|x=a} = e^{im\psi} q_{m,x|x=0}, \qquad q_{m,xx|x=a} = e^{im\psi} q_{m,xx|x=0}, \qquad (2.2.7)$$

with $\psi = 2\pi/N$ and

$$f_m = \frac{1}{\sqrt{N}} \sum_{k=1}^{N} e^{-i(k-1)m\psi} F_k. \qquad (2.2.8)$$

It is obvious that the mode subspace coordinates q_1, q_2, ..., q_N are independent of each other.

2.2.1 Natural Vibration

The natural vibration equation may be obtained by substituting

$$q_m(x,y,t) = Q_m(x,y)\,e^{i\omega t} \quad \text{and} \quad f_m = 0, \qquad m=1,2,...,N \tag{2.2.9}$$

into equation (2.2.5), giving

$$\frac{\partial^4 Q_m}{\partial x^4} + 2\frac{\partial^4 Q_m}{\partial x^2 \partial y^2} + \frac{\partial^4 Q_m}{\partial y^4} = \beta^4 Q_m\,, \qquad m=1,2,...,N \tag{2.2.10}$$

where Q_m denotes the amplitude function of the m-th mode subspace coordinate and

$$\beta^4 = \frac{\rho\omega^2}{D}, \tag{2.2.11}$$

Corresponding boundary conditions take the same form as equations (2.2.6) and (2.2.7), *i.e.,*

$$Q_m(x,0) = 0, \qquad Q_{m,yy}(x,0) = 0,$$

$$Q_m(x,b) = 0, \qquad Q_{m,yy}(x,b) = 0, \qquad 0 \le x \le a \tag{2.2.12}$$

and

$$Q_m(0,y) = 0, \qquad Q_m(a,y) = 0,$$

$$Q_{m,x}(a,y) = e^{im\psi} Q_{m,x}(0,y), \qquad Q_{m,xx}(a,y) = e^{im\psi} Q_{m,xx}(0,y),$$

$$0 \le y \le b \tag{2.2.13}$$

In view of the simply supported conditions at y=0 and b, Levy's method of solution may be applied with the solution form:

$$Q_m(x,y) = X_{mn}(x)\sin\frac{n\pi y}{b} \tag{2.2.14}$$

which has satisfied the simply supported condition (2.2.12). Substituting equation (2.2.14) into equation (2.2.10) yields

$$\frac{d^4 X_{mn}}{d x^4} - 2\frac{n^2\pi^2}{b^2}\frac{d^2 X_{mn}}{d x^2} + (\frac{n^4\pi^4}{b^4} - \beta^4) X_{mn} = 0\,. \tag{2.2.15}$$

By inserting equation (2.2.14) in equation (2.2.13), the boundary condition corresponding to equation (2.2.15) may be obtained as

$$X_{mn}(0) = 0, \qquad X_{mn}(a) = 0\,,$$

$$X'_{mn}(a) = e^{im\psi} X'_{mn}(0), \quad X''_{mn}(a) = e^{im\psi} X''_{mn}(0) \tag{2.2.16}$$

The homogeneous equation (2.2.15) with homogeneous boundary condition (2.2.16) represent an eigenvalue problem of self-adjoint differential equation. The general solution for X_{mn} of homogeneous equation (2.2.15) may be expressed as

$$X_{mn}(x) = A_{mn}\,\text{sh}\,\lambda_n x + B_{mn}\,\text{ch}\,\lambda_n x + C_{mn}\sin\mu_n x + D_{mn}\cos\mu_n x \tag{2.2.17}$$

where $A_{mn}, B_{mn}, C_{mn}, D_{mn}$ are four constants of integration and λ_n, μ_n are defined as

$$\lambda_n = \sqrt{\frac{n^2\pi^2}{b^2} + \beta^2}, \qquad \mu_n = \sqrt{\beta^2 - \frac{n^2\pi^2}{b^2}}\,. \tag{2.2.18}$$

Substituting equation (2.2.17) into boundary condition (2.2.16) yields

$$B_{mn} + D_{mn} = 0\,,$$

$$A_{mn}\,\text{sh}\,\lambda_n a + B_{mn}\,\text{ch}\,\lambda_n a + C_{mn}\sin\mu_n a + D_{mn}\cos\mu_n a = 0\,,$$

$$A_{mn}\lambda_n(\text{ch}\,\lambda_n a - e^{im\psi}) + B_{mn}\lambda_n\,\text{sh}\,\lambda_n a + C_{mn}\mu_n(\cos\mu_n a - e^{im\psi}) - D_{mn}\mu_n\sin\mu_n a = 0\,,$$

$$A_{mn}\lambda_n^2\sin\lambda_n a + B_{mn}\lambda_n^2(\mathrm{ch}\lambda_n a - e^{im\psi}) - C_{mn}\mu_n^2\sin\mu_n a - D_{mn}\mu_n^2(\cos\mu_n a - e^{im\psi}) = 0 . \tag{2.2.19}$$

Let the coefficient determinant of the simultaneous equations (2.2.19) be equal to zero. Then expanding this determinant will yield the frequency equation

$$\mu_n\,\mathrm{sh}\,\lambda_n a\cos\mu_n a - \lambda_n\,\mathrm{ch}\,\lambda_n a\sin\mu_n a + \cos m\psi(\lambda_n\sin\mu_n a - \mu_n\,\mathrm{sh}\,\lambda_n a) = 0$$

$$m=1,2,3,\ldots,N, \quad n=1,2,3,\ldots,\infty . \tag{2.2.20}$$

Since λ_n, μ_n are dependent on the frequency parameter β therefore equation (2.2.20) is a transcendental equation of β for given m and n.

Consider now two extreme cases of m=N and m=N/2 (if N is even). Firstly substituting m=N ($m\psi = 2\pi$) into equation (2.2.20) yields

$$\mu_n\,\mathrm{sh}\,\lambda_n a(\cos\mu_n a - 1) - \lambda_n\sin\mu_n a(\mathrm{ch}\,\lambda_n a - 1) = 0 . \tag{2.2.21}$$

Because of the phase difference $m\psi = 2\pi$, the corresponding modes of all substructures are the same for both amplitude and phase . The equation (2.2.21) can be rewritten as

$$-4\sin\frac{\mu_n a}{2}\mathrm{sh}\frac{\lambda_n a}{2}(\mu_n\mathrm{ch}\frac{\lambda_n a}{2}\sin\frac{\mu_n a}{2} + \lambda_n\mathrm{sh}\frac{\lambda_n a}{2}\cos\frac{\mu_n a}{2}) = 0 \tag{2.2.22}$$

which is equivalent to

$$\sin\frac{\mu_n a}{2} = 0 \tag{2.2.23a}$$

or

$$\mu_n\mathrm{ch}\frac{\lambda_n a}{2}\sin\frac{\mu_n a}{2} + \lambda_n\mathrm{sh}\frac{\lambda_n a}{2}\cos\frac{\mu_n a}{2} = 0 . \tag{2.2.23b}$$

The solution for β of equation (2.2.23a) may be expressed as

$$\beta^2 = \pi^2(\frac{s^2}{a^2} + \frac{n^2}{b^2}), \qquad n=1,2,3,.... \quad s=2,4,6,... \tag{2.2.24}$$

Noting equation (2.2.11), the corresponding natural frequencies can be found as

$$\omega = \pi^2(\frac{s^2}{a^2} + \frac{n^2}{b^2})\sqrt{\frac{D}{\rho}}, \qquad n=1,2,3,.... \quad s=2,4,6... \tag{2.2.25}$$

The corresponding natural modes for each substructure are in agreement with those having even number (*i.e.*,s) of half waves in x-direction for plate substructure with all edges simply supported. It can be proved that the roots for ω of the frequency equation (2.2.23b) are corresponding to the modes which are symmetrical about the center line (x=a/2) for plate substructure with a pair of opposite sides simply supported and the other pair of opposite sides clamped.

Secondly consider the case that the mode of one substructure is of opposite phase to the mode of the preceding substructure, *i.e.*, m=N/2 ($m\psi = \pi$). Substituting $m\psi = \pi$ into equation (2.2.20) yields

$$\mu_n \, \text{sh}\, \lambda_n a(\cos\mu_n a + 1) - \lambda_n \sin\mu_n a(\text{ch}\, \lambda_n a + 1) = 0 \tag{2.2.26}$$

which can be rewritten as

$$4\cos\frac{\mu_n a}{2}\text{ch}\frac{\lambda_n a}{2}(\mu_n \text{sh}\frac{\lambda_n a}{2}\cos\frac{\mu_n a}{2} - \lambda_n \text{ch}\frac{\lambda_n a}{2}\sin\frac{\mu_n a}{2}) = 0 \,. \tag{2.2.27}$$

This equation is equivalent to

$$\cos\frac{\mu_n a}{2} = 0 \tag{2.2.28a}$$

or

$$\mu_n \text{sh}\frac{\lambda_n a}{2}\cos\frac{\mu_n a}{2} - \lambda_n \text{ch}\frac{\lambda_n a}{2}\sin\frac{\mu_n a}{2} = 0 \,. \tag{2.2.28b}$$

The solution for β of equation (2.2.28a) is

$$\beta^2 = \pi^2\left(\frac{s^2}{a^2}+\frac{n^2}{b^2}\right) \qquad n=1,2,3,... \qquad s=1,3,5,... \qquad (2.2.29)$$

The corresponding natural frequencies can be obtained by inserting equation (2.2.29) in left hand side of equation (2.2.11), as

$$\omega = \pi^2\left(\frac{s^2}{a^2}+\frac{n^2}{b^2}\right)\sqrt{\frac{D}{\rho}} \qquad n=1,2,3,.... \qquad s=1,3,5,.. \qquad (2.2.30)$$

The corresponding natural modes for substructures are in agreement with those having odd number of half wave in x-direction for plate substructure with four edges simply supported. It may be proved that the roots for ω of equation (2.2.28b) are corresponding to the modes which are anti-symmetric about center line (x=a/2) for plate substructure with a pair of opposite sides simply supported and another pair of opposite sides clamped.

2.2.2 Buckling Load

The prism shell shown in Figure 2.2.1 subjected to an uniformly distributed compression in y-direction is considered. The buckling differential equation for each substructure takes the same form, *i.e.,*

$$D\left(\frac{\partial^4 W_k}{\partial x^4}+2\frac{\partial^4 W_k}{\partial x^2 \partial y^2}+\frac{\partial^4 W_k}{\partial y^4}\right)+\sigma\frac{\partial^2 W_k}{\partial y^2}=0 \qquad k=1,2,...,N \qquad (2.2.31)$$

where σ denotes the pressure per unit length.

The boundary condition and restrained, continuous conditions are the same as given in equations (2.2.2) and (2.2.3) respectively. Applying the U transformation (2.2.4) to equations (2.2.31), (2.2.2) and (2.2.3) result in

$$D\left(\frac{\partial^4 q_m}{\partial x^4}+2\frac{\partial^4 q_m}{\partial x^2 \partial y^2}+\frac{\partial^4 q_m}{\partial y^4}\right)+\sigma\frac{\partial^2 q_m}{\partial y^2}=0, \qquad m=1,2,...,N \qquad (2.2.32)$$

together with equations (2.2.6) and (2.2.7) as given previously.

The homogeneous equations (2.2.32) (2.2.6) and (2.2.7) represent an eigenvalue problem of partial differential equation and its minimum eigenvalue denotes the critical load. The corresponding eigenfunction represents the buckling shape.

Let the buckling shape takes the form (if $a \cong b$)

$$q_m(x,y) = X_m(x)\sin\frac{\pi y}{b}. \tag{2.2.33}$$

Substituting equation (2.2.33) into equations (2.2.32) and (2.2.7) yields

$$\frac{d^4 X_m}{dx^4} - 2\frac{\pi^2}{b^2}\frac{d^2 X_m}{dx^2} + \left(\frac{\pi^4}{b^4} - \frac{\sigma}{D}\frac{\pi^2}{b^2}\right)X_m = 0 \tag{2.2.34}$$

and

$$X_m(0) = 0, \qquad X_m(a) = 0,$$

$$X_m'(a) = e^{im\psi}X_m'(0), \quad X_m''(a) = e^{im\psi}X_m''(0). \tag{2.2.35}$$

The boundary condition (2.2.6) has been satisfied previously. As compared with the eigenproblem of natural vibration , it is obvious that by inserting n=1 and $\beta^2 = (\pi/b)\sqrt{\sigma/D}$ in equation (2.2.15), both equations (2.2.34) and (2.2.15) are the same. As a result, there exists the following relation

$$\beta_{min}^2 = \frac{\pi}{b}\sqrt{\frac{\sigma_{cr}}{D}} \tag{2.2.36}$$

where β_{min} denotes the minimum root of the frequency equation (2.2.20) and σ_{cr} denotes the critical load.

Consider the simple case of a=b. Let

$$K_{cr} = \frac{\sigma_{cr}a^2}{D\pi^2} \tag{2.2.37}$$

where K_{cr} is a non-dimensional buckling parameter. When N is even, β_{min} can be obtained from equation (2.2.29) as

$$\beta_{min}^2 = \frac{2\pi^2}{a^2}. \tag{2.2.38}$$

The critical load can be found by substituting equation (2.2.38) into equation (2.2.36), giving

$$\sigma_{cr} = \frac{4\pi^2}{a^2} D. \tag{2.2.39}$$

Then by inserting this result in equation (2.2.37), it will give

$$K_{cr} = 4. \tag{2.2.40}$$

When the total number of substructures is odd, the minimum positive root for β^2 of equation (2.2.20) can be found by some numerical methods. A few numerical results are given in Table 2.2.1.

Table 2.2.1. Critical load of prism shells subjected to axial compression

N	even number	3	5	7	9	11	∞
K_{cr}	4.0000	4.7905	4.2867	4.1466	4.0887	4.0594	4.000

It can be concluded that when N is an even number, the critical load is minimum and when N is equal to three, the critical load is maximum.

Chapter 3

LINEAR PERIODIC STRUCTURES - DISCRETE SYSTEMS

Essentially the U-transformation method is a mode subspace method for cyclic periodic structures. This method can not be immediately applied to analysis of linear periodic structures. However if the extreme boundary conditions of the linear periodic structure meet certain requirements, it may be treated as a cyclic periodic system which is equivalent to the linear periodic system. Then we can analyze the equivalent system with cyclic periodicity instead of the original system.

3.1 Periodic Mass-Spring System with Single Degree of Freedom Substructure

In order to apply the U-transformation method, an equivalent system with a cyclic periodicity property must be formed. The equivalent system is dependent on the extreme boundary conditions of the linear periodic system. The following examples will illustrate how the equivalent system can be created according to the specific extreme boundary conditions.

3.1.1 Periodic Mass-Spring System with Fixed Extreme Ends

Consider a periodic mass spring system with fixed extreme ends and n mass points as shown in Figure 3.1.1(a) where M denotes the mass of each mass point, K denotes the stiffness constant of the coupling spring. Assuming that the mass points can move only along the longitudinal direction, x_k and F_k denote the longitudinal displacement and load for mass point k. The fixed extreme end conditions can be expressed as $x_0 = 0$ and $x_{n+1} = 0$. The equivalent system with cyclic periodicity must satisfy these restrained conditions. Such an equivalent system can be obtained by producing a symmetrical image of the original system and loading about one fixed end as shown in Figure 3.1.1(b). In order to form a cyclic periodic system, it is necessary that two fixed ends of original system may be replaced by the concentrated mass without any restriction and two ends for the equivalent system will be imaginarily put together and treated as one point in mathematics. Because of the symmetry of the equivalent system, the displacements of symmetric centers must be equal to zero, *i.e.*, $x_0 = 0$ and $x_{n+1} = 0$. As a result the fixed end conditions for original system may be satisfied automatically. Therefore both systems shown in Figure 3.1.1(a) and (b) are equivalent.

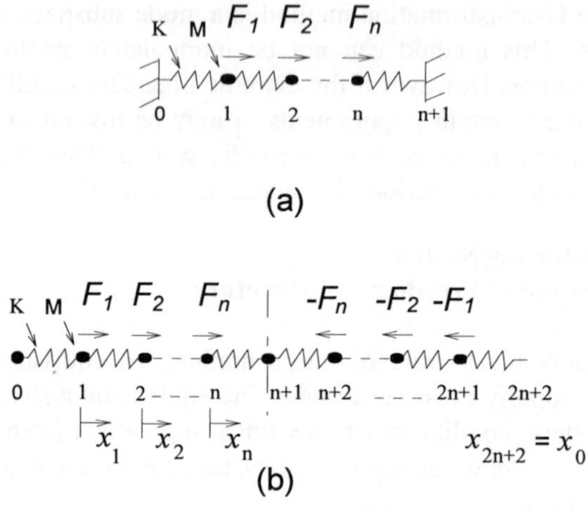

Figure 3.1.1 Periodic mass spring system with fixed extreme ends;
(a) Original system; (b) Equivalent system

For the cyclic periodic system, the dynamic differential equations for all mass points are of the same form, *i.e.*,

$$T_{k+1} - T_k + F_k = M\ddot{x}_k \tag{3.1.1}$$

where T_{k+1} and T_k denote the spring tensions on right and left sides respectively of mass point k (see Fig.3.1.2) which can be expressed as

$$T_{k+1} = K(x_{k+1} - x_k), \quad T_k = K(x_k - x_{k-1}) \; . \tag{3.1.2}$$

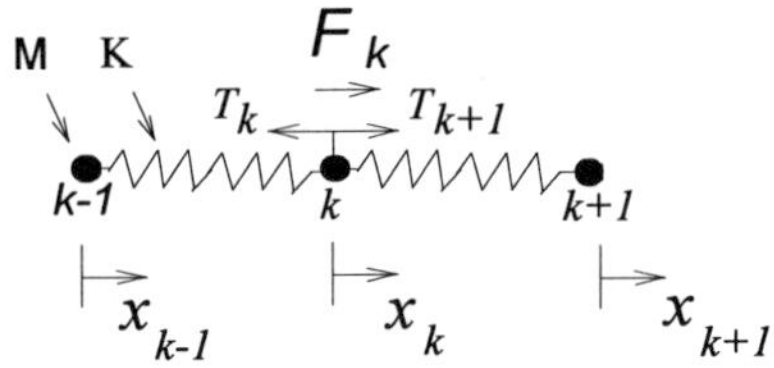

Figure 3.1.2 Mass point k

Substituting equation (3.1.2) into equation (3.1.1) yields

$$M\ddot{x}_k - K(x_{k+1} - 2x_k + x_{k-1}) = F_k, \qquad k=1,2,...,N \tag{3.1.3}$$

where N denotes the total number of substructures for cyclic periodic system, *i.e.*, N=2n+2 and $x_{N+1} \equiv x_1$.

In equation (3.1.3), the loads must satisfy the symmetric condition, *i.e.*,

$$F_{N-k} = -F_k \ , \qquad k=1,2,...,n$$

$$F_N = F_{n+1} = 0 \ . \tag{3.1.4}$$

The initial condition may be expressed as

$$x_k(0) = x_{k,0} \quad \text{and} \quad \dot{x}_k(0) = \dot{x}_{k,0} \ , \qquad k=1,2,...,N \tag{3.1.5}$$

where x_{k0} and $\dot{x}_{ko}$ denote initial displacement and velocity of particle k, which should also satisfy the symmetric condition, *i.e.*,

$$x_{(N-k),0} = -x_{k,0}, \qquad \dot{x}_{(N-k),0} = -\dot{x}_{k,0} \qquad k=1,2,...,n$$

$$x_{(n+1),0} = 0, \qquad \dot{x}_{(n+1),0} = 0,$$

$$x_{N,0} = 0, \qquad \dot{x}_{N,0} = 0 \ . \tag{3.1.6}$$

Applying the U-transformation

$$x_k = \frac{1}{\sqrt{N}} \sum_{m=1}^{N} e^{i(k-1)m\psi} q_m \tag{3.1.7a}$$

or

$$q_m = \frac{1}{\sqrt{N}} \sum_{k=1}^{N} e^{-i(k-1)m\psi} x_k \tag{3.1.7b}$$

to equation(3.1.3), *i.e.*, multiplying by $e^{-i(k-1)m\psi}/\sqrt{N}$ and then summarizing for k from 1 to N, yields

$$M\ddot{q}_m + 2K(1-\cos m\psi)q_m = f_m \qquad m=1,2,...,N \tag{3.1.8}$$

where $\psi = 2\pi/N$ and

$$f_m = \frac{1}{\sqrt{N}} \sum_{k=1}^{N} e^{-i(k-1)m\psi} F_k \ . \tag{3.1.9}$$

Substituting equation (3.1.4) into equation (3.1.9), results in

$$f_m = \frac{-2i\,e^{im\psi}}{\sqrt{N}} \sum_{k=1}^{n} \sin(km\psi) F_k \ , \qquad m=1,2,...,N. \tag{3.1.10}$$

It can be verified that

$$f_{n+1} = 0, \qquad f_N = 0,$$

$$f_{N-m} = \overline{f_m} \ , \qquad m=1,2,...,n. \tag{3.1.11}$$

Corresponding initial condition can be obtained by inserting t=0 and equation (3.1.5) in equation (3.1.7b) yielding

$$q_m(0) = \frac{1}{\sqrt{N}} \sum_{k=1}^{N} e^{-i(k-1)m\psi} x_{k0} \ ,$$

$$\dot{q}_m(0) = \frac{1}{\sqrt{N}} \sum_{k=1}^{N} e^{-i(k-1)m\psi} \dot{x}_{k0} \tag{3.1.12}$$

Then substituting equation (3.1.6) into equation (3.1.12) yields

$$q_m(0) = \frac{-2ie^{im\psi}}{\sqrt{N}} \sum_{k=1}^{n} \sin(km\psi) x_{k0} \ ,$$

$$\dot{q}_m(0) = \frac{-2ie^{im\psi}}{\sqrt{N}} \sum_{k=1}^{n} \sin(km\psi) \dot{x}_{k0} \ , \qquad m=1,2,...,N. \tag{3.1.13}$$

As a result

$$q_{n+1}(0) = \dot{q}_{n+1}(0) = 0, \qquad q_N(0) = \dot{q}_N(0) = 0 \, ,$$

$$q_{N-m}(0) = \bar{q}_m(0), \qquad \dot{q}_{N-m}(0) = \bar{\dot{q}}_m(0), \qquad m=1,2,...,n \tag{3.1.14}$$

which are similar to equation (3.1.11).

Upon examining the mathematical characteristics of equations (3.1.8), (3.1.11) and (3.1.14), it is apparent that

$$q_{n+1}(t) = 0, \qquad q_N(t) = 0 \, ,$$

$$q_{N-m}(t) = \bar{q}_m(t), \qquad m=1,2,...,n. \tag{3.1.15}$$

In fact, only the first n mode subspace coordinates need be found.

At first, consider the natural vibration. Substituting $f_m = 0$ and $q_m = Q_m e^{i\omega t}$ ($x_k = X_k e^{i\omega t}$) into equation (3.1.8) yields

$$\left[2K(1-\cos m\psi) - M\omega^2\right] Q_m = 0 \, , \qquad m=1,2,...,n \tag{3.1.16}$$

where ω denotes the natural frequency, Q_m denotes amplitude of the m-th mode subspace coordinate, and $\psi = \pi/(n+1)$. The natural frequencies may be found, from equation (3.1.16), as

$$\omega_m = \sqrt{2(1-\cos\frac{m\pi}{n+1})}\sqrt{\frac{K}{M}} \qquad m=1,2,...,n. \tag{3.1.17}$$

Some exact results are given in Table 3.1.1.

Table 3.1.1. Natural frequencies for mass-spring systems with fixed extreme ends

Degrees of freedom n	ω_1	ω_2	ω_3	ω_4	ω_5
1	$\sqrt{2}$				
2	1	$\sqrt{3}$			
3	$\sqrt{2-\sqrt{2}}$	$\sqrt{2}$	$\sqrt{2+\sqrt{2}}$		
5	$\sqrt{2-\sqrt{3}}$	1	$\sqrt{2}$	$\sqrt{3}$	$\sqrt{2+\sqrt{3}}$
Multiplier			$\sqrt{K/M}$		

It must be pointed out that because of $\omega_{N-m} = \omega_m$, the frequencies ω_m (m=1,2,...,n) are double eigenvalues therefore there are two independent modes corresponding to one natural frequency ω_m for the equivalent system. If and only if the mode satisfies the symmetric condition, *i.e.*, $x_{N-k} = -x_k$ (k=1,2,...,n) and $x_N = -x_{n+1} = 0$, it is the real mode for the original system. By using the symmetric condition, equation (3.1.7b) may be expressed as

$$q_m = \frac{-2i\,e^{im\psi}}{\sqrt{N}} \sum_{k=1}^{n} \sin(km\psi)\,x_k \qquad m=1,2,...,N. \tag{3.1.18}$$

Obviously the m-th mode subspace coordinate must include the complex factor $ie^{im\psi}$. Therefore the mode in terms of mode subspace coordinates may be expressed as

$$Q_m = i\,e^{im\psi}\cdot A, \quad Q_{N-m} = \overline{Q_m}, \quad Q_s = 0, \qquad s \neq m,\ N-m \tag{3.1.19}$$

where "A" is an arbitrary real number. By replacing q_m with Q_m in equation (3.1.7a) the natural mode may be found as

$$X_k = A\sin(km\psi)\,, \qquad k=1,2,...,N \tag{3.1.20}$$

with $\psi = \pi/(n+1)$. It is obvious that the mode shown in equation (3.1.20) satisfies the restricted condition of the original system, *i.e.*, $X_0 \equiv X_N = 0, X_{n+1} = 0$, and then equation (3.1.20) represents the natural mode of the original system. Some exact natural modes are given in Table 3.1.2, in which the modes have been normalized according to the first displacement being equal to one, *i.e.*, taking $A=1/\sin m\psi$.

Table 3.1.2 Natural modes for mass-spring systems with fixed extreme ends

Degrees of freedom n	Natural modes $[X_1, X_2, \ldots, X_n]$				
	1st	2nd	3rd	4th	5th
1	[1]				
2	[1,1]	[1,-1]			
3	$[1,\sqrt{2},1]$	[1,0,-1]	$[1,-\sqrt{2},1]$		
5	$[1,\sqrt{3},2,\sqrt{3},1]$	[1,1,0,-1,-1]	[1,0,-1,0,1]	[1,-1,0,1,-1]	$[1,-\sqrt{3},2,-\sqrt{3},1]$

Moreover , consider the dynamic response of mass-spring systems subjected to arbitrary excitation. The dynamic equation in terms of the mode subspace coordinates has been obtained as equations (3.1.8) and (3.1.9) . Noting the equation (3.1.17), the dynamic equation may be rewritten as

$$\ddot{q}_m + \omega_m^2 q_m = \frac{f_m}{M}\,, \qquad m=1,2,...,N \tag{3.1.21}$$

which is the vibration equation with a single degree of freedom. The solution for q_m of equation (3.1.21) with initial condition (3.1.13) can be expressed as a Duhamel integral, *i.e.*,

$$q_m(t) = \frac{1}{M\omega_m}\int_0^t f_m(\tau)\sin\omega_m(t-\tau)d\tau + q_m(0)\cos\omega_m t + \frac{\dot{q}_m(0)}{\omega_m}\sin\omega_m t\,, \tag{3.1.22}$$

where f_m, $q_m(0)$, $\dot{q}_m(0)$ are dependent on the load , initial displacement and velocity of the original system as shown in equations (3.1.10) and (3.1.13) respectively. The dynamic response for displacement of original system can be found by substituting equation (3.1.22) into the U-transformation (3.1.7a). This can be illustrated by considering a specific loading and initial conditions.

As an example, a mass-spring system subjected to harmonic excitation with zero initial condition is considered now. Let

$$F_k(t) = F_{k0} \sin \omega t\,, \qquad k=1,2,..,n, \tag{3.1.23}$$

$$x_{k0} = 0, \quad \dot{x}_{k0} = 0\,, \qquad k=1,2,..,n. \tag{3.1.24}$$

Inserting equations (3.1.23) and (3.1.24) in equations (3.1.10) and (3.1.13) respectively, yield

$$f_m = \frac{-2i\,e^{im\psi}}{\sqrt{N}}\left[\sum_{k=1}^{n} F_{k0}\sin(km\psi)\right]\sin \omega t\ , \qquad m=1,2,...,N \tag{3.1.25}$$

and

$$q_m(0) = 0, \qquad \dot{q}_m(0) = 0\,, \quad m=1,2,...,N. \tag{3.1.26}$$

Substituting equations (3.1.25) and (3.1.26) into equation (3.1.22) and then integrating with respect to τ yields

$$q_m(t) = \frac{-2i\,e^{im\psi}}{M\omega_m\sqrt{N}}\left[\sum_{k=1}^{n} F_{k0}\sin(km\psi)\right]\frac{\omega \sin \omega_m t - \omega_m \sin \omega t}{\omega^2 - \omega_m^2}\,,$$

$$m=1,2,...,N \tag{3.1.27}$$

with N=2n+2 and $\psi = \pi/(n+1)$.

Noting equation (3.1.15), the U-transformation (3.1.7a) becomes

$$x_k = \frac{2}{\sqrt{2n+2}}\sum_{m=1}^{n} \mathrm{Re}\left[e^{i(k-1)m\psi}\, q_m(t)\right]. \tag{3.1.28}$$

Substituting the response functions for mode subspace coordinates shown in equation (3.1.27) into equation (3.1.28) yields

$$x_k(t) = \frac{2}{M(n+1)} \sum_{m=1}^{n} \left\{ \frac{\sin km\psi}{\omega_m} \left[\sum_{j=1}^{n} F_{j0} \sin(jm\psi) \right] \frac{\omega \sin \omega_m t - \omega_m \sin \omega t}{\omega^2 - \omega_m^2} \right\},$$

$$k=1,2,..,N \qquad (3.1.29)$$

with ω_m defined as in equation (3.1.17). Obviously, the response functions of displacements satisfy the fixed end conditions of the original system, *i.e.*, $x_0 \equiv x_N = 0$ and $x_{n+1} = 0$.

3.1.2 Periodic Mass-Spring System with Free Extreme Ends

Consider a periodic mass-spring system with free extreme ends and n particles as shown in Figure 3.1.3(a), where the symbols denote the same meaning as those in Figure 3.1.1. In order to form an equivalent system with cyclic periodicity, it is necessary to extend the original system by its symmetrical image and apply anti-symmetrical loading on the corresponding extended particles. In such a system the vibration displacements will be anti-symmetric about the center line if the initial displacements and velocities are also anti-symmetric about the center line. For any pair of symmetrical particles, both longitudinal displacements are the same. As a result $x_n = x_{n+1}$ and $x_1 = x_{2n}$, which indicate that the connecting spring between both particles n and n+1 does not transmit any force and the particle 2n can be imaginarily connected to particle 1 by the same spring as the other one, *i.e.*, particle 2n can be regarded as the preceding particle of particle 1. Therefore the extended system shown in Figure 3.1.3 (b) may be regarded as a cyclic periodic system.

The dynamic equation for each particle takes the same form as equation (3.1.3) where the total number of particles for equivalent system are 2n instead of 2n+2; *i.e.*, N=2n. The uncoupled equation in terms of mode subspace coordinates also is the same as equation (3.1.8) with equation (3.1.9). For the system considered, the loads must satisfy the anti-symmetrical condition, *i.e.*,

$$F_{N-k+1} = F_k , \quad k=1,2,...,n \qquad (3.1.30)$$

with N=2n.

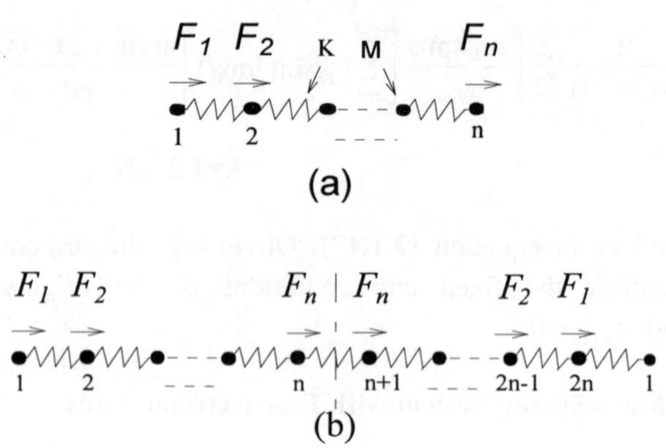

Figure 3.1.3 Periodic mass-spring system with free extreme ends
(a) Original system ; (b) Equivalent system

Substituting this condition into equation (3.1.9), results in

$$f_m = \frac{2e^{i\frac{1}{2}m\psi}}{\sqrt{2n}} \sum_{k=1}^{n} \cos\left[(k-\frac{1}{2})m\psi\right] F_k , \tag{3.1.31}$$

with $\psi = \frac{\pi}{n}$.

It can be verified that

$$f_{2n-m} = \bar{f}_m \quad \text{and} \quad f_n = 0 \ . \tag{3.1.32}$$

Since the longitudinal displacements of particles in the equivalent system should also be anti-symmetrical, *i.e.*,

$$x_{2n-k+1} = x_k \ , \qquad k=1,2,...,n. \tag{3.1.33}$$

So that the mode subspace coordinates must have similar property to that shown in equation (3.1.32), *i.e.*,

$$q_{2n-m} = \bar{q}_m \quad \text{and} \quad q_n = 0 \, . \tag{3.1.34}$$

The mode subspace coordinates q_m for m=1,2,...,n-1,2n have to found in equation (3.1.8) and equation (3.1.31).

Consider now the natural vibration. The natural frequencies can be obtained from equation (3.1.16) with $\psi = \pi/n$, which yields

$$\omega_m = \sqrt{2(1-\cos\frac{m\pi}{n})}\sqrt{\frac{K}{M}}, \qquad m=0,1,...,n-1. \tag{3.1.35}$$

When m=0 (or m=2n), $\omega_m = 0$. The corresponding mode is the rigid body mode, *i.e.*, $x_1 = x_2 = \cdots = x_{2n}$ .Some exact results are given in Table 3.1.3.

When the modes satisfy the anti-symmetrical condition (3.1.33), equation (3.1.7b) becomes

$$q_m = \frac{2\,e^{i\frac{1}{2}m\psi}}{\sqrt{2n}}\sum_{k=1}^{n}\cos\left[(k-\frac{1}{2})m\psi\right]x_k, \qquad m=1,2,...2n \tag{3.1.36}$$

with $\psi = \pi/n$.

Table 3.1.3 Natural frequencies for mass-spring systems with free extreme ends

Degrees of freedom n	ω_0	ω_1	ω_2	ω_3	ω_4	ω_5
1	0					
2	0	$\sqrt{2}$				
3	0	1	$\sqrt{3}$			
4	0	$\sqrt{2-\sqrt{2}}$	$\sqrt{2}$	$\sqrt{2+\sqrt{2}}$		
6	0	$\sqrt{2-\sqrt{3}}$	1	$\sqrt{2}$	$\sqrt{3}$	$\sqrt{2+\sqrt{3}}$
Multiplier	$\sqrt{K/M}$					

The mth mode subspace coordinate must include a complex factor $e^{i\frac{1}{2}m\psi}$. Therefore the natural mode in terms of mode subspace coordinates may be expressed as

$$Q_m = A e^{i\frac{1}{2}m\psi}, \quad Q_{2n-m} = \overline{Q}_m, \quad Q_s = 0, \qquad s \neq m,\ 2n-m \qquad (3.1.37)$$

where Q_m denotes the amplitude of the m-th mode subspace coordinate and "A" is an arbitrary real number. By substituting $q_m = Q_m$ (m=1,2,...,2n) into the U-transformation (3.1.7a), the natural mode may be obtained as

$$X_k = A\cos\left[(k-\frac{1}{2})\frac{m\pi}{n}\right], \qquad k=1,2,...,2n. \qquad (3.1.38)$$

In order to make the first displacement equal to one, let $A = 1/\cos(m\pi/2n)$. The exact natural modes for some systems are given in Table 3.1.4. Compare Tables 3.1.3 and 3.1.4 with Tables 3.1.1 and 3.1.2, it is apparent that the non zero frequencies are the same for both systems but the corresponding natural modes are different.

Table 3.1.4 Natural modes for mass-spring systems with free extreme ends

Degrees of freedom n	Natural modes $[X_1, X_2, \cdots, X_n]$ 0 order	1st	2nd	3rd	4th	5th
1	[1]					
2	[1,1]	[1,-1]				
3	[1,1,1]	[1,0,-1]	[1,-2,1]			
4	[1,1,1,1]	$[1,\sqrt{2}-1,1-\sqrt{2},-1]$	[1,-1,-1,1]	[1,-1,1,-1]		
6	[1,1,1,1,1,1]	λ_1	[1,0,-1,-1,0,1]	[1,-1,-1,1,1,-1]	[1,-2,1,1,-2,1]	λ_5

$$\lambda_1 = [1, \sqrt{2(2-\sqrt{3})}, 2-\sqrt{3}, \sqrt{3}-2, -\sqrt{2(2-\sqrt{3})}, -1]$$

$$\lambda_5 = [1, -\sqrt{2(2+\sqrt{3})}, 2+\sqrt{3}, -\sqrt{3}-2, \sqrt{2(2+\sqrt{3})}, -1]$$

3.1.3 Periodic Mass-Spring System with One End Fixed and the Other Free

Consider a periodic mass spring system with fixed and free extreme ends and n particles as shown in Figure 3.1.4(a) , where the symbols denote the same meanings as those in Figure 3.1.1. Firstly, the system is extended at the free end by its symmetrical image and apply the anti-symmetrical loading on the corresponding extended part as shown in Figure 3.1.4(b). Then the new system may be treated as a mass-spring system with both extreme ends fixed as stated in section 3.1.1. The equivalent system with cyclic periodicity and 4n+2 degrees of freedom is shown in Figure 3.1.4(c) .

For the loads of the equivalent system, there are the following relations.

$$F_{N-k} = F_{2n+k+1} = -F_{2n-k+1} = -F_k \ , \quad k=1,2,...,n,$$

$$F_{2n+1} = F_N = 0 \tag{3.1.39}$$

where N=4n+2.

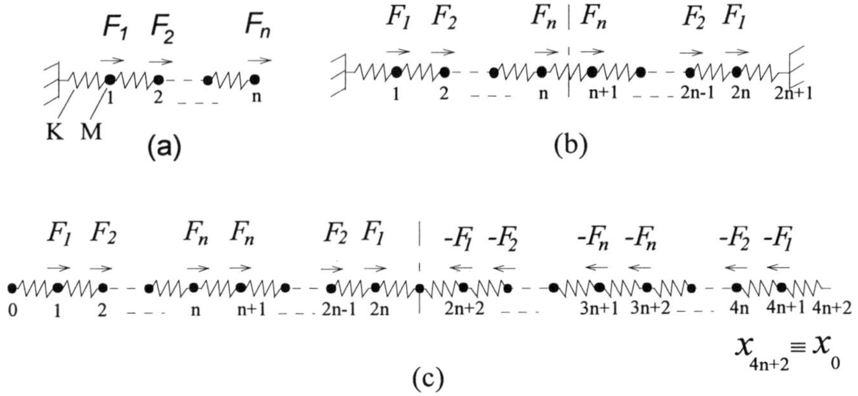

Figure 3.1.4 Periodic mass spring system with fixed and free ends
(a) Original system ; (b) Transition system; (c) Equivalent system

Substituting equation (3.1.39) into equation (3.1.9) yields

$$f_m = \frac{-4ie^{im\psi}}{\sqrt{4n+2}} \sum_{k=1}^{n} \sin(km\psi) F_k \ , \quad m=1,3,5,...,4n+1,$$

$$f_m = 0 \ , \quad m=2,4,6,..,4n+2 \text{ and } 2n+1 \tag{3.1.40}$$

with $\psi = \pi/(2n+1)$. There are only n conjugate pairs which are non zero for generalized loads, *i.e.*,

$$f_{N-m} = \bar{f}_m \neq 0 \qquad m=1,3,5,...,2n-1. \tag{3.1.41}$$

Since the displacements for the equivalent system have the same property as those of the loads shown in equation (3.1.39) , the relation similar to equation (3.1.40) can be obtained from equation (3.1.7b), *i.e.*,

$$q_m = \frac{-4ie^{im\psi}}{\sqrt{4n+2}} \sum_{k=1}^{n} \sin(km\psi) x_k \,, \qquad m=1,3,5,...,4n+1,$$

$$q_m = 0, \qquad m=2,4,6,..,4n+2 \text{ and } 2n+1. \tag{3.1.42}$$

In equation (3.1.8), only if m=1,3,...,2n-1, n equations with single degree of freedom need be considered .

Consider now the natural vibration. By inserting $\psi = \pi/(2n+1)$ in equation (3.1.16), the natural frequency can be obtained as follows:

$$\omega^2 = 2(1 - \cos\frac{m\pi}{2n+1})\frac{K}{M}\,, \qquad m=1,3,...,2n-1. \tag{3.1.43}$$

Altogether n different natural frequencies may be expressed as

$$\omega_j = \sqrt{2(1 - \cos\frac{(2j-1)\pi}{2n+1})}\sqrt{\frac{K}{M}}\,, \qquad j=1,2,..,n. \tag{3.1.44}$$

Some numerical results are given in Table 3.1.5.

Equation (3.1.42) indicates that the *m*-th mode subspace coordinate must include the complex factor $ie^{im\psi}$. The mode corresponding to natural frequency ω_j may be expressed as

$$Q_{2j-1} = ie^{i(2j-1)\psi}A, \qquad Q_{N-2j+1} = \bar{Q}_{2j-1}, \qquad Q_m = 0,$$
$$m \neq 2j\text{-}1, N\text{-}2j\text{+}1 \tag{3.1.45}$$

where N=4n+2 and “A” represents an arbitrary real number.

Table 3.1.5 Natural frequencies for mass-spring systems with fixed and free extreme ends

Degrees of freedom n	ω_1	ω_2	ω_3	ω_4	ω_5
1	1				
2	0.618034	1.618034			
3	0.445042	1.246980	1.801938		
5	0.284630	0.830830	1.309721	1.682507	1.918986
Multiplier	$\sqrt{K/M}$				

Substituting $q_m = Q_m$ (m=1.2,...N) into equation (3.1.7a), yields the j-th natural mode

$$X_k = A \sin \frac{k(2j-1)\pi}{2n+1}, \qquad k=1,2,...,4n+2. \tag{3.1.46}$$

It is obvious that the modes satisfy the boundary condition of the original system, *i.e.*,

$$X_0 \equiv X_{4n+2} = 0, \qquad X_n = X_{n+1}. \tag{3.1.47}$$

In order to make the displacement of the free end equal to one, let $A = \{\sin[n(2j-1)\pi/(2n+1)]\}^{-1}$. Some numerical results for natural modes are given in Table 3.1.6.

3.2 Periodic Mass-Spring System with Two Degrees of Freedom Substructures

In this section, we will consider the mass-spring system with two degrees of freedom subsystems. The system considered includes two kinds of springs with stiffness constants K_1 and K_2 as shown in Figure 3.2.1(a) where M denotes the mass of each particle and two extreme ends are free. The serial numbers of the particles are made up of two integer numbers, in which the first number denotes the serial number of the particle in the subsystem and the second number denotes the ordinal number of the subsystem. In a similar manner as employed previously to form the equivalent system in section 3.1.2, the equivalent system for the present case can be obtained as shown in Figure 3.2.1(b) where the original system is extended by its symmetrical image and the loads for the extended system is anti-symmetric about the center line. Besides the particles (1,1) and (2,2n) must imaginarily be connected by using the spring K_2. Such a system may be regarded as a cyclic periodic system with 2n subsystems. Each subsystem includes two particles with mass M and two springs with stiffness constants K_1 and K_2 as shown in Figure 3.2.2 in which t_j and T_j denote the tensions of spring K_1 and K_2 in the j-th subsystem.

Table. 3.1.6 Natural modes for mass-spring systems with fixed and free extreme ends

Degrees of freedom n	Natural modes $[X_1, X_2, \ldots, X_n]^T$ 1st	2nd	3rd	4th	5th
1	$\{1\}$				
2	$\begin{Bmatrix}0.618034\\1.000000\end{Bmatrix}$	$\begin{Bmatrix}-1.618034\\1.000000\end{Bmatrix}$			
3	$\begin{Bmatrix}0.445042\\0.801938\\1.000000\end{Bmatrix}$	$\begin{Bmatrix}-1.246980\\-0.554958\\1.000000\end{Bmatrix}$	$\begin{Bmatrix}1.801938\\-2.246980\\1.000000\end{Bmatrix}$		
5	$\begin{Bmatrix}0.284630\\0.546200\\0.763521\\0.918986\\1.000000\end{Bmatrix}$	$\begin{Bmatrix}-0.830830\\-1.088156\\-0.594351\\0.309721\\1.000000\end{Bmatrix}$	$\begin{Bmatrix}1.309721\\0.372786\\-1.203616\\-0.715370\\1.000000\end{Bmatrix}$	$\begin{Bmatrix}-1.682507\\1.397877\\0.521109\\-1.830830\\1.000000\end{Bmatrix}$	$\begin{Bmatrix}1.918986\\-3.228707\\3.513337\\-2.682507\\1.000000\end{Bmatrix}$

For the j-th subsystem, the dynamic equation takes the form

$$M\ddot{x}_{1,j} = t_j - T_{j-1} + F_{1,j} \,,$$

$$M\ddot{x}_{2,j} = T_j - t_j + F_{2,j} \,, \quad j=1,2,\ldots,2n \tag{3.2.1}$$

where x_{1j}, x_{2j} denote the longitudinal displacements and F_{1j}, F_{2j} denote the longitudinal loads for particles (1,j) and (2,j).

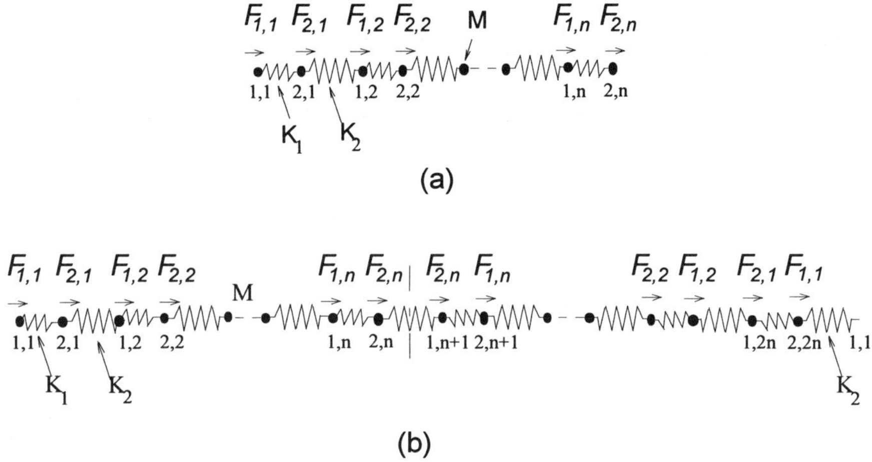

Figure 3.2.1 Periodic mass spring system with two kinds of spring and free extreme ends (a) Original system; (b) Equivalent system.

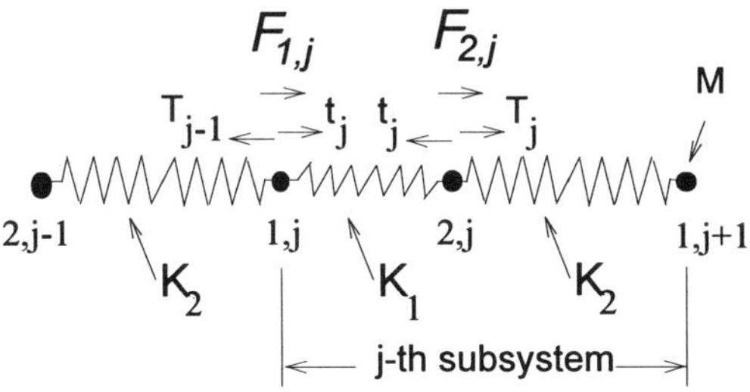

Figure 3.2.2 Subsystem

According to Hooke's law, the restoring forces of springs K_1 and K_2 may be expressed as

$$t_j = K_1(x_{2,j} - x_{1,j}), \quad T_j = K_2(x_{1,j+1} - x_{2,j}),$$

$$j=1,2,\ldots,2n \tag{3.2.2}$$

where $x_{1,2n+1} = x_{1,1}$.

Substituting equation (3.2.2) into equation (3.2.1) yields

$$M\ddot{x}_{1,j} + (K_1 + K_2)x_{1,j} - K_1 x_{2,j} - K_2 x_{2,j-1} = F_{1,j},$$

$$M\ddot{x}_{2,j} + (K_1 + K_2)x_{2,j} - K_1 x_{1,j} - K_2 x_{1,j+1} = F_{2,j}, \quad j=1,2,\ldots,2n. \tag{3.2.3}$$

Applying the U-transformation

$$x_{k,j} = \frac{1}{\sqrt{2n}} \sum_{m=1}^{2n} e^{i(j-1)m\psi} q_{k,m}, \quad k=1,2 \tag{3.2.4a}$$

and its inverse transformation

$$q_{k,m} = \frac{1}{\sqrt{2n}} \sum_{j=1}^{2n} e^{-i(j-1)m\psi} x_{k,j}, \quad k=1,2 \tag{3.2.4b}$$

to equation (3.2.3) results in

$$M\ddot{q}_{1,m} + (K_1 + K_2)q_{1,m} - K_1 q_{2,m} - K_2 e^{-im\psi} q_{2,m} = f_{1,m},$$

$$M\ddot{q}_{2,m} + (K_1 + K_2)q_{2,m} - K_1 q_{1,m} - K_2 e^{im\psi} q_{1,m} = f_{2,m},$$

$$m=1,2,\ldots,2n \tag{3.2.5}$$

with $\psi = \pi / n$ and

$$f_{k,m} = \frac{1}{\sqrt{2n}} \sum_{j=1}^{2n} e^{-i(j-1)m\psi} F_{k,j}, \quad k=1,2, \quad m=1,2,\ldots,2n. \tag{3.2.6}$$

The loads for the equivalent system with 2n subsystems must satisfy the following condition

$$F_{2,2n-j+1} = F_{1,j}, \qquad F_{1,2n-j+1} = F_{2,j}, \qquad j=1,2,...,n. \tag{3.2.7}$$

By substituting equation (3.2.7) into equation (3.2.6), it can be verified that

$$f_{1,m} = e^{im\psi}\bar{f}_{2,m}, \qquad m=1,2,...,2n. \tag{3.2.8}$$

In a similar way as that employed to derive equation (3.2.8), the similar relation for mode subspace coordinates can be obtained as

$$q_{1,m} = e^{im\psi}\bar{q}_{2,m}, \qquad m=1,2,...,2n. \tag{3.2.9}$$

This equation is equivalent to the anti-symmetry condition of displacements for the equivalent system.

Consider now the natural vibration [5]. By substituting $q_{k,m}(t) = Q_{k,m}\sin\omega t$ (k=1,2) and $f_{k,m}(t) = 0$ (k=1,2) into equation(3.2.5), the natural vibration equation may be obtained as

$$\begin{gathered}(K_1 + K_2 - M\omega^2)Q_{1,m} - (K_1 + K_2 e^{-im\psi})Q_{2,m} = 0 \quad , \\ -(K_1 + K_2 e^{im\psi})Q_{1,m} + (K_1 + K_2 - M\omega^2)Q_{2,m} = 0 \, , \\ m=1,2,...,2n.\end{gathered} \tag{3.2.10}$$

where ω denotes the natural frequency and $Q_{1,m}Q_{2,m}$ denote the amplitudes of $q_{1m,}\,q_{2m}$. If there is a nontrivial solution for $Q_{1,m}Q_{2,m}$ of equation (3.2.10), the coefficient determinant must be equal to zero, *i.e.*,

$$\begin{vmatrix} K_1 + K_2 - M\omega^2 & -(K_1 + K_2 e^{-im\psi}) \\ -(K_1 + K_2 e^{im\psi}) & K_1 + K_2 - M\omega^2 \end{vmatrix} = 0\,. \tag{3.2.11}$$

Expanding this determinant yields the frequency equation:

$$M^2\omega^4 - 2(K_1 + K_2)M\omega^2 + 2K_1K_2(1 - \cos m\psi) = 0\,. \tag{3.2.12}$$

The solution for ω^2 of the quadratic equation (3.2.12) may be expressed as

$$\omega^2_{\substack{m,1\\m,2}} = \frac{K_1 + K_2 \mp \sqrt{K_1^2 + K_2^2 + 2K_1K_2\cos m\psi}}{M}, \quad m=1,2,...,2n \tag{3.2.13}$$

with $\psi = \pi / n$.

When $\omega = \omega_{m,r}$ (r=1,2), the ratio between $Q_{1,m}$ and $Q_{2,m}$ may be obtained from equation (3.2.10) as

$$\frac{Q_{1,m}}{Q_{2,m}} = \frac{K_1 + K_2 e^{-im\psi}}{K_1 + K_2 - M\omega^2_{m,r}} = \frac{\rho e^{i\alpha m\psi}}{K_1 + K_2 - M\omega^2_{m,r}} \tag{3.2.14}$$

where

$$\rho = \sqrt{K_1^2 + K_2^2 + 2K_1K_2\cos m\psi}, \tag{3.2.15}$$

$$\alpha = -\frac{1}{m\psi}\text{arctg}\frac{K_2\sin m\psi}{K_1 + K_2\cos m\psi} \quad (\text{if } K_1 > K_2),$$

$$m=1,2,...,n,...,2n. \tag{3.2.16}$$

The equation (3.2.13) may be rewritten as

$$K_1 + K_2 - M\omega^2_{m,1} = \rho,$$

$$K_1 + K_2 - M\omega^2_{m,2} = -\rho. \tag{3.2.17}$$

Substituting equation (3.2.17) into equation (3.2.14) results in

$$\frac{Q_{1,m}}{Q_{2,m}} = \begin{cases} e^{i\alpha m\psi} \ (\text{when } \omega = \omega_{m,1}), \\ -e^{i\alpha m\psi} (\text{when } \omega = \omega_{m,2}). \end{cases} \tag{3.2.18}$$

As a result, the mode in terms of mode subspace coordinates can be expressed as

$$Q_{1,m} = A e^{i\beta m\psi} e^{i\alpha m\psi}, \quad Q_{2,m} = A e^{i\beta m\psi},$$

$$Q_{1,2n-m} = \overline{Q}_{1,m}, \qquad Q_{2,2n-m} = \overline{Q}_{2,m},$$

$$Q_{1,s} = Q_{2,s} = 0 \ , \qquad s \neq m, 2n-m, \ \ \text{when } \omega = \omega_{m,1} \tag{3.2.19}$$

where "A" denotes an arbitrary real number and β is the real number to be determined. If the corresponding mode satisfies anti-symmetric condition, its mode subspace coordinates must satisfy equation (3.2.9), *i.e.*,

$$Q_{1,m} = e^{im\psi}\,\overline{Q}_{2,m} \ . \tag{3.2.20}$$

Substituting equation(3.2.19) into equation(3.2.20) results in

$$\beta = \frac{1-\alpha}{2} \tag{3.2.21}$$

and then

$$Q_{1,m} = A\, e^{i\frac{1+\alpha}{2}m\psi}, \qquad Q_{2,m} = A\, e^{i\frac{1-\alpha}{2}m\psi},$$

$$Q_{1,2n-m} = \overline{Q}_{1,m} \ , \qquad Q_{2,2n-m} = \overline{Q}_{2,m} \ , \tag{3.2.22}$$

with other mode subspace coordinates vanishing. The natural mode corresponding to natural frequency $\omega_{m,1}$ may be obtained by substituting equation (3.2.22) into equation (3.2.4a) as

$$X_{1,j} = A\cos[\left(j - \tfrac{1-\alpha}{2}\right)m\psi],$$

$$X_{2,j} = A\cos[\left(j - \tfrac{1+\alpha}{2}\right)m\psi], \qquad j=1,2,\ldots,2n. \tag{3.2.23}$$

Consider the natural mode which is corresponding to the natural frequency $\omega_{m,2}$. Noting the result given in equation (3.2.18), the amplitudes of mode subspace coordinates may be expressed as

$$Q_{1,m} = A\, e^{i\beta m\psi}\, e^{i\alpha m\psi}, \quad Q_{2,m} = -A\, e^{i\beta m\psi},$$

$$Q_{1,2n-m} = \overline{Q}_{1,m}, \quad Q_{2,2n-m} = \overline{Q}_{2,m},$$

$$Q_{1,s} = Q_{2,s} = 0, \qquad s \neq m, \ 2n-m \tag{3.2.24}$$

instead of equation (3.2.19).

Substituting equation (3.2.24) into (3.2.20) results in

$$\beta = \frac{1-\alpha}{2} + \frac{\pi}{2m\psi} \tag{3.2.25}$$

with $\psi = \pi / n$.

Inserting equation (3.2.25) in equation (3.2.24) yields

$$Q_{1,m} = iA\, e^{i\frac{1+\alpha}{2}m\psi}, \qquad Q_{2,m} = -iA\, e^{i\frac{1-\alpha}{2}m\psi},$$

$$Q_{1,2n-m} = \overline{Q}_{1,m}, \quad Q_{2,2n-m} = \overline{Q}_{2,m} \tag{3.2.26}$$

and other mode subspace coordinates are equal to zero. Substituting equation (3.2.26) into the U-transformation (3.2.4a), yields the natural mode:

$$X_{1,j} = -A \sin[(j - \tfrac{1-\alpha}{2})m\psi],$$

$$X_{2,j} = A \sin[(j - \tfrac{1+\alpha}{2})m\psi], \qquad j=1,2,\ldots,2n \tag{3.2.27}$$

which is corresponding to the natural frequency $\omega_{m,2}$.

It must be pointed out that some of the natural frequencies for the equivalent system are not natural frequency of the original system, *i.e.*, some of the solutions shown in equations (3.2.23) and (3.2.27) are trivial solutions. There are 2n natural frequencies and modes altogether for original system with 2n particles. The validity of the above formulas is demonstrated with a simple example.

Consider the simple case of n=1 and $K_1 > K_2$. The natural frequencies for equivalent system can be obtained from equation (3.2.13) as

$$\omega_{1,1} = \sqrt{\frac{2K_2}{M}}, \quad \omega_{1,2} = \sqrt{\frac{2K_1}{M}},$$

$$\omega_{2,1} = 0, \quad \omega_{2,2} = \sqrt{\frac{2(K_1 + K_2)}{M}} \tag{3.2.28}$$

The parameter α defined in equation (3.2.16) may be found to be

$$\alpha = 0, \qquad m=1,2. \tag{3.2.29}$$

The modes corresponding to $\omega_{m,1}$ (m=1,2) and $\omega_{m,2}$ (m=1,2) can be found from equations (3.2.23) and (3.2.27) respectively, as shown in Table 3.2.1.

It is apparent that $\omega_{1,1}$ and $\omega_{2,2}$ are not natural frequencies for the original system. The natural frequencies and modes for the original system may be expressed as

$$\omega = 0 \quad \text{and} \quad \begin{Bmatrix} X_{1,1} \\ X_{2,1} \end{Bmatrix} = \begin{Bmatrix} 1 \\ 1 \end{Bmatrix};$$

$$\omega = \sqrt{2K_1 / M} \quad \text{and} \quad \begin{Bmatrix} X_{1,1} \\ X_{2,1} \end{Bmatrix} = \begin{Bmatrix} 1 \\ -1 \end{Bmatrix}. \tag{3.2.30}$$

Table 3.2.1 Natural frequencies and modes for example (n=1)

Natural frequencies ω	Natural modes $X_{1,1}$	$X_{2,1}$	$X_{1,2}$	$X_{2,2}$
$\omega_{1,1} = \sqrt{2K_2 / M}$	0	0	0	0
$\omega_{2,1} = 0$	1	1	1	1
$\omega_{1,2} = \sqrt{2K_1 / M}$	1	-1	-1	1
$\omega_{2,2} = \sqrt{2(K_1 + K_2) / M}$	0	0	0	0

3.3 Transverse Vibration of Plane Trusses

In this section the slightly more complex periodic structures - plane trusses will be discussed. The type of plane truss considered is the Warren truss [6], which is built up with basic units consisting of four sets of bars pin-jointed at the nodes and subjected to transverse nodal loads. The natural vibration and dynamic response are studied for the simply supported truss with lumped masses at the nodes.

3.3.1 Governing Equations

Consider the plane truss shown in Figure 3.3.1 which is made up of four sets of bars pin-jointed at the nodes so that only axial forces but no bending moments and shear forces can exist on the cross-sections of the bars. The masses of the bars are taken as lumped at the nodes. Two lumped masses denoted by M_1 and M_2 are attached to every lower and upper nodes, respectively. The bars in the longitudinal direction have modules of elasticity E_1, cross-sectional area A_1 and length L_1. The inclined bars have modules of elasticity E_2, cross-sectional area A_2 and length L_2.

A typical substructure is shown in Figure 3.3.2(a). Each substructure includes two nodes and four bars. The serial numbers of the nodes and the bars are made up of two integer numbers, in which the first number denotes the ordinal number of the substructure and the second number denotes the serial number of the node or bar in the substructure. In order to avoid ambiguity the serial numbers of the nodes are put in round brackets.

As mentioned in the above sections, it is necessary to create a cyclic periodic system which is equivalent to the actual system. Consider now a truss with both ends simply supported and n substructures. The transverse concentrated load acting at the nodes (j,1) and (j,2) are denoted by $F_{(j,1)}(t)$ and $F_{(j,2)}(t)$. The actual system can be replaced by an equivalent system with both ends simply supported and 2n substructures upon which equal but opposite concentrated loads $F_{(j,1)}(t)$, $F_{(j,2)}(t)$ and $-F_{(j,1)}(t)$, $-F_{(j,2)}(t)$ must be applied at the nodes (j,1), (j,2) and (2n-j+2,1), (2n-j+1,2), as shown in Figure 3.3.3 (for j=2). Such an equivalent system can be regarded as a cyclic periodic system, *i.e.*, two pairs of extreme nodes A,A' and B,B' are imaginarily jointed by hinges respectively, and the supports from the ends A,A' are removed. The simply supported condition at both ends for the actual system are satisfied automatically in its equivalent system. Two bars denoted by dotted lines shown in Figure 3.3.3 are not subjected to any loads for anti-symmetric vibration modes. Consequently, the anti-symmetric vibration for the equivalent system is not affected by such additional bars which are necessary in order to form the cyclic periodic system.

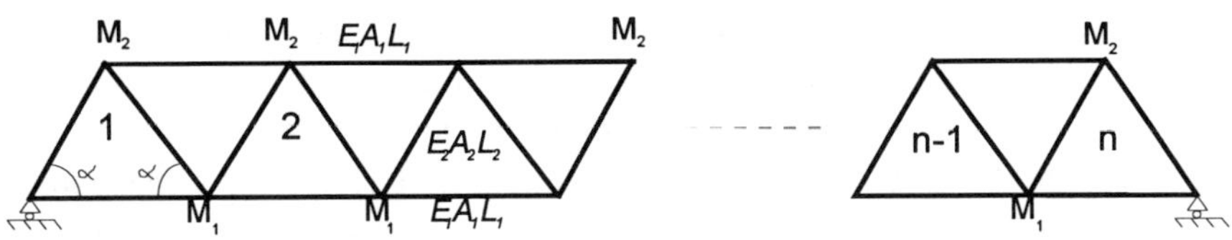

Figure 3.3.1 Plane truss with both ends simply supported

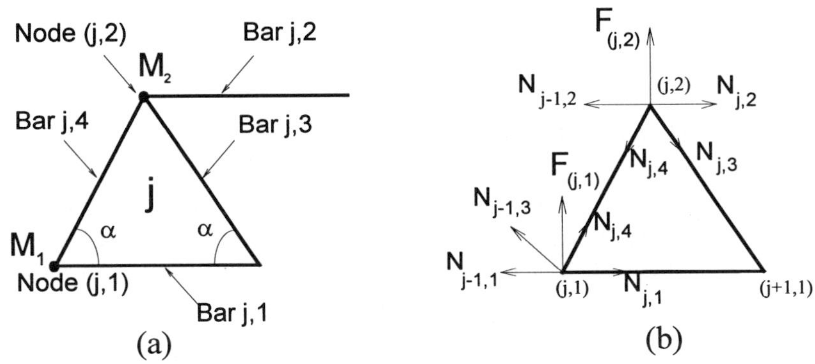

Figure 3.3.2 Substructure j

(a) Serial numbers of the nodes and the bars;

(b) The forces acting on the nodes.

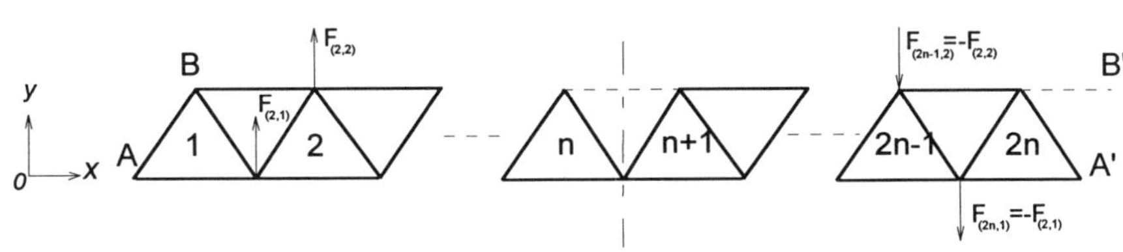

Figure 3.3.3 Equivalent system

A global co-ordinate system oxy with x and y axes along the longitudinal and transverse directions is established as shown in Figure 3.3.3. The displacement components in the x and y directions for node (j,k) are denoted by $u_{(j,k)}$ and $v_{(j,k)}$. When the longitudinal size of the truss is much greater than the transverse size, the inertia forces in the *x*-direction can be neglected. The dynamic equations for all substructures of the equivalent system are of the same form:

$$N_{j,1} - N_{j-1,1} + N_{j,4}\cos\alpha - N_{j-1,3}\cos\alpha = 0\,,$$

$$N_{j,4}\sin\alpha + N_{j-1,3}\sin\alpha + F_{(j,1)} = M_1\ddot{v}_{(j,1)}\,,$$

$$N_{j,2} - N_{j-1,2} + N_{j,3}\cos\alpha - N_{j,4}\cos\alpha = 0\,,$$

$$-N_{j,3}\sin\alpha - N_{j,4}\sin\alpha + F_{(j,2)} = M_2\ddot{v}_{(j,2)}\,,\qquad j=1,2,\ldots,N. \qquad (3.3.1)$$

Here, N=2n and $N_{0,k} \equiv N_{2n,k}$ (k=1,2,3); $N_{j,k}$ denotes the axial force on the section of the k-th bar in the j-th substructure as shown in Figure 3.3.2(b); α denotes the inclination of the inclined bar and a superior dot denotes differentiation with respect to the time variable t. The inertia forces in the x-direction have been neglected in equation (3.3.1).

The stress-displacement relation may be written as

$$N_{j,1} = K_1(u_{(j+1,1)} - u_{(j,1)})\,,$$

$$N_{j,2} = K_1(u_{(j+1,2)} - u_{(j,2)})\,,$$

$$N_{j,3} = K_2(u_{(j+1,1)}\cos\alpha - v_{(j+1,1)}\sin\alpha - u_{(j,2)}\cos\alpha + v_{(j,2)}\sin\alpha)\,,$$

$$N_{j,4} = K_2(u_{(j,2)}\cos\alpha + v_{(j,2)}\sin\alpha - u_{(j,1)}\cos\alpha - v_{(j,1)}\sin\alpha)\,,$$

$$j=1,2,\ldots,N \qquad (3.3.2)$$

where $K_1 = E_1A_1/L_1$, $K_2 = E_2A_2/L_2$, $u_{(N+1,k)} \equiv u_{(1,k)}$, $v_{(N+1,k)} \equiv v_{(1,k)}$ $(k = 1,2)$.

One can now apply the U-transformation to equations (3.3.1) and (3.3.2), *i.e.*, let

$$u_{(j,1)} = \frac{1}{\sqrt{N}} \sum_{m=1}^{N} e^{i(j-1)m\psi} q_{m,1},$$

$$v_{(j,1)} = \frac{1}{\sqrt{N}} \sum_{m=1}^{N} e^{i(j-1)m\psi} q_{m,2},$$

$$u_{(j,2)} = \frac{1}{\sqrt{N}} \sum_{m=1}^{N} e^{i(j-1)m\psi} q_{m,3},$$

$$v_{(j,2)} = \frac{1}{\sqrt{N}} \sum_{m=1}^{N} e^{i(j-1)m\psi} q_{m,4},$$

$$N_{j,k} = \frac{1}{\sqrt{N}} \sum_{m=1}^{N} e^{i(j-1)m\psi} \sigma_{m,k}, \qquad k=1,2,3,4, \qquad j=1,2,\ldots,N \qquad (3.3.3)$$

then multiplying equations (3.3.1) and (3.3.2) by $e^{-i(j-1)m\psi} / \sqrt{N}$ and finding the sum of equations for j=1,2,...,N, yields

$$\sigma_{m,1} - e^{-im\psi} \sigma_{m,1} + \sigma_{m,4} \cos\alpha - e^{-im\psi} \sigma_{m,3} \cos\alpha = 0,$$

$$\sigma_{m,4} \sin\alpha + e^{-im\psi} \sigma_{m,3} \sin\alpha + f_{m,1} = M_1 \ddot{q}_{m,2},$$

$$\sigma_{m,2} - e^{-im\psi} \sigma_{m,2} + \sigma_{m,3} \cos\alpha - \sigma_{m,4} \cos\alpha = 0,$$

$$-\sigma_{m,3} \sin\alpha - \sigma_{m,4} \sin\alpha + f_{m,2} = M_2 \ddot{q}_{m,4}, \qquad m=1,2,\ldots,N \qquad (3.3.4)$$

and

$$\sigma_{m,1} = K_1 (e^{im\psi} q_{m,1} - q_{m,1}),$$

$$\sigma_{m,2} = K_1 (e^{im\psi} q_{m,3} - q_{m,3}),$$

$$\sigma_{m,3} = K_2 (e^{im\psi} q_{m,1} \cos\alpha - e^{im\psi} q_{m,2} \sin\alpha - q_{m,3} \cos\alpha + q_{m,4} \sin\alpha),$$

$$\sigma_{m,4} = K_2 (q_{m,3} \cos\alpha + q_{m,4} \sin\alpha - q_{m,1} \cos\alpha - q_{m,2} \sin\alpha),$$

$$m=1,2,\ldots,N \tag{3.3.5}$$

where

$$f_{m,k} = \frac{1}{\sqrt{N}} \sum_{j=1}^{N} e^{-i(j-1)m\psi} F_{(j,k)}, \qquad k=1,2, \qquad m=1,2,\ldots,N. \tag{3.3.6}$$

Noting that

$$F_{(N-j+2,1)} = -F_{(j,1)}, \quad (j=2,3,\ldots,n), \qquad F_{(1,1)} = F_{(n+1,1)} = 0,$$

$$F_{(N-j+1,2)} = -F_{(j,2)}, \quad (j=1,2,3,\ldots,n).$$

one can write $f_{m,1}$ and $f_{m,2}$ as

$$f_{m,1} = \frac{-2i}{\sqrt{N}} \sum_{j=2}^{n} \sin(j-1)m\psi \, F_{(j,1)},$$

$$f_{m,2} = \frac{-2i}{\sqrt{N}} e^{im\psi/2} \sum_{j=1}^{n} \sin(j-\frac{1}{2})m\psi \, F_{(j,2)}, \qquad m=1,2,\ldots,N. \tag{3.3.7}$$

Note that $f_{N,1}, f_{N,2}$ and $f_{n,1}$ all vanish. Substituting equations(3.3.5) into equations(3.3.4), yields

$$\left[2K_1(\cos m\psi - 1) - 2K_2\cos^2\alpha\right] q_{m,1} + K_2 \cos^2\alpha(1+e^{-im\psi}) q_{m,3}$$

$$+ K_2 \sin\alpha\cos\alpha(1-e^{-im\psi}) q_{m,4} = 0$$

$$-2K_2\sin^2\alpha \, q_{m,2} + K_2\sin\alpha\cos\alpha(1-e^{-im\psi}) q_{m,3}$$

$$+ K_2\sin^2\alpha(1+e^{-im\psi}) q_{m,4} + f_{m,1} = M_1 \ddot{q}_{m,2}$$

$$K_2\cos^2\alpha(1+e^{im\psi}) q_{m,1} + K_2\sin\alpha\cos\alpha(1-e^{im\psi}) q_{m,2}$$

$$+\left[2K_1(\cos m\psi - 1) - 2K_2\cos^2\alpha\right]q_{m,3} = 0$$

$$K_2\sin\alpha\cos\alpha(1 - e^{im\psi})q_{m,1} + K_2\sin^2\alpha(1 + e^{im\psi})q_{m,2}$$

$$-2K_2\sin^2\alpha\, q_{m,4} + f_{m,2} = M_2\ddot{q}_{m,4}$$

$$m=1,2,\ldots,N. \tag{3.3.8}$$

When $m = N\ (m\psi = 2\pi)$, equations(3.3.8) are homogeneous. The non-trivial solution for $q_{N,1}$ to $q_{N,4}$ represent the rigid body motion. In the actual system the rigid body motion case is impossible; therefore the solution corresponding to rigid body motion should be neglected, *i.e.*, let $q_{N,K} = 0\ (k = 1,2,3,4)$.

Eliminating $q_{m,1}$ and $q_{m,3}$ from equations (3.3.8) gives

$$[M]\begin{Bmatrix}\ddot{q}_{m,2}\\ \ddot{q}_{m,4}\end{Bmatrix} + [K]\begin{Bmatrix}q_{m,2}\\ q_{m,4}\end{Bmatrix} = \begin{Bmatrix}f_{m,1}\\ f_{m,2}\end{Bmatrix}, \qquad m=1,2,\ldots,N-1. \tag{3.3.9}$$

where

$$[M] = \begin{bmatrix}M_1 & 0\\ 0 & M_2\end{bmatrix}, \qquad [K] = \begin{bmatrix}K_{11} & K_{12}\\ K_{21} & K_{22}\end{bmatrix}. \tag{3.3.10}$$

with the elements of the stiffness matrix [K] being

$$K_{11} = K_{22} = \frac{2K_1\sin^2\alpha\left[(1-\cos m\psi)(2\beta - \cos^2\alpha) + 4\cos^2\alpha\right]}{2\beta^2(1-\cos m\psi) + 4\beta\cos^2\alpha + \cos^4\alpha},$$

$$\overline{K}_{21} = K_{12} = \frac{-2K_1\sin^2\alpha(1+e^{-im\psi})\left[2\cos^2\alpha + \beta(1-\cos m\psi)\right]}{2\beta^2(1-\cos m\psi) + 4\beta\cos^2\alpha + \cos^4\alpha} \tag{3.3.11}$$

where $\beta = K_1 / K_2$.

The N-degree-of-freedom system yields a 2-DOF equation (3.3.9) N-1 times over when the U-transformation is used. The generalized displacements $q_{m.k}$ are mode subspace co-ordinates.

The initial condition can be written as

$$q_{m,2}(0) = \frac{1}{\sqrt{N}} \sum_{k=1}^{N} e^{-i(k-1)m\psi} v_{(k,1)}(0),$$

$$q_{m,4}(0) = \frac{1}{\sqrt{N}} \sum_{k=1}^{N} e^{-i(k-1)m\psi} v_{(k,2)}(0),$$

$$\dot{q}_{m,2}(0) = \frac{1}{\sqrt{N}} \sum_{k=1}^{N} e^{-i(k-1)m\psi} \dot{v}_{(k,1)}(0),$$

$$\dot{q}_{m,4}(0) = \frac{1}{\sqrt{N}} \sum_{k=1}^{N} e^{-i(k-1)m\psi} \dot{v}_{(k,2)}(0). \tag{3.3.12}$$

This can be derived from equations (3.3.3).

In equations (3.3.12), $v_{(k,1)}(0)$, $v_{(k,2)}(0)$ and $\dot{v}_{k,1}(0)$, $\dot{v}_{(k,2)}(0)$ denote the initial nodal displacements and velocities for k-th substructure of the equivalent system and they must be in agreement with anti-symmetric vibration. In a similar way to that in equations (3.3.7) for nodal loads, the initial conditions can be rewritten as

$$q_{m,2}(0) = \frac{-2i}{\sqrt{N}} \sum_{k=2}^{n} \sin(k-1)m\psi \, v_{(k,1)}(0),$$

$$q_{m,4}(0) = \frac{-2i}{\sqrt{N}} e^{im\psi/2} \sum_{k=1}^{n} \sin(k-\tfrac{1}{2})m\psi \, v_{(k,2)}(0),$$

$$\dot{q}_{m,2}(0) = \frac{-2i}{\sqrt{N}} \sum_{k=2}^{n} \sin(k-1)m\psi \, \dot{v}_{(k,1)}(0),$$

$$\dot{q}_{m,4}(0) = \frac{-2i}{\sqrt{N}} e^{im\psi/2} \sum_{k=1}^{n} \sin(k-\tfrac{1}{2})m\psi \, \dot{v}_{(k,2)}(0). \tag{3.3.13}$$

Incidentally, when m=n $(m\psi=\pi)$, the first of equations (3.3.9) becomes $M_1\ddot{q}_{n,2}+K_{11}q_{n,2}=0$, with initial conditions $q_{n,2}(0)=\dot{q}_{n,2}(0)=0$, the generalized displacement $q_{n,2}(t)$ also identically vanishes.

Upon noting that $f_{N-m,k}=\bar{f}_{m,k}$ (k = 1,2), $q_{N-m,k}(0)=\bar{q}_{m,k}(0)$ and $\dot{q}_{N-m,k}(0)=\bar{\dot{q}}_{m,k}(0)$ (k=2,4), it can be easily proved that the sum of the terms in each series on the right hand side of equation (3.3.3) is a real function. Consequently, the terms in the series can be replaced by their real parts.

3.3.2 Static Problem

Let the terms of inertia forces in the dynamic equation (3.3.9) vanish. The equilibrium equation takes the form

$$[K]\begin{Bmatrix} q_{m,2} \\ q_{m,4} \end{Bmatrix}=\begin{Bmatrix} f_{m,1} \\ f_{m,2} \end{Bmatrix}, \qquad m=1,2,\ldots,N-1 \tag{3.3.14}$$

and the solution for $q_{m,2}$ and $q_{m,4}$ of equation (3.3.14) can be written as

$$\begin{Bmatrix} q_{m,2} \\ q_{m,4} \end{Bmatrix}=[L]\begin{Bmatrix} f_{m,1} \\ f_{m,2} \end{Bmatrix}, \qquad m=1,2,\ldots,N-1 \tag{3.3.15}$$

with the components of the flexibility matrix $[L]_{2\times 2}$ as

$$l_{11}=l_{22}=\frac{(1-\cos m\psi)(2\beta-\cos^2\alpha)+4\cos^2\alpha}{2K_1\sin^2\alpha(1-\cos m\psi)^2},$$

$$\overline{l_{21}}=l_{12}=\frac{(1+e^{-im\psi})[2\cos^2\alpha+\beta(1-\cos m\psi)]}{2K_1\sin^2\alpha(1-\cos m\psi)^2}. \tag{3.3.16}$$

It is well known that

$$[L]=[K]^{-1}. \tag{3.3.17}$$

Substituting equations (3.3.15) and (3.3.16) into the first and third of equations (3.3.8), and solving the simultaneous equations for $q_{m,1}$ and $q_{m,3}$, gives

$$q_{m,1} = \{\cos\alpha / [2K_1 \sin\alpha(1-\cos m\psi)^2]\}\{(1-e^{-im\psi})f_{m,2} + i\sin m\psi\, f_{m,1}\},$$

$$q_{m,3} = \{\cos\alpha / [2K_1 \sin\alpha(1-\cos m\psi)^2]\}\{(1-e^{im\psi})f_{m,1} - i\sin m\psi\, f_{m,2}\},$$

$$m=1,2,..,N-1. \tag{3.3.18}$$

The static displacements can be found by substituting equations (3.3.15), (3.3.16), (3.3.18) and (3.3.7) into equation (3.3.3); they are

$$u_{(k,1)} = \frac{\cos\alpha}{2K_1\sin\alpha}\frac{1}{n}\sum_{m=1}^{2n-1}\frac{\cos(k-1)m\psi}{(1-\cos m\psi)^2}\Big[2\sin\frac{1}{2}m\psi\sum_{s=1}^{n}\sin(s-\frac{1}{2})m\psi\, F_{(s,2)}$$

$$+\sin m\psi\sum_{s=2}^{n}\sin(s-1)m\psi\, F_{(s,1)}\Big]$$

$$u_{(k,2)} = \frac{-\cos\alpha}{2K_1\sin\alpha}\frac{1}{n}\sum_{m=1}^{2n-1}\frac{\cos(k-\frac{1}{2})m\psi}{(1-\cos m\psi)^2}\Big[2\sin\frac{1}{2}m\psi\sum_{s=2}^{n}\sin(s-1)m\psi\, F_{(s,1)}$$

$$+\sin m\psi\sum_{s=1}^{n}\sin(s-\frac{1}{2})m\psi\, F_{(s,2)}\Big]$$

$$v_{(k,1)} = \frac{1}{2K_1\sin^2\alpha}\ \frac{1}{n}\sum_{m=1}^{2n-1}\frac{\sin(k-1)m\psi}{(1-\cos m\psi)^2}\Big\{\big[(1-\cos m\psi)(2\beta-\cos^2\alpha)+4\cos^2\alpha\big]$$

$$\times\sum_{s=2}^{n}\sin(s-1)m\psi\, F_{(s,1)} + 2\cos\frac{m\psi}{2}\big[2\cos^2\alpha+\beta(1-\cos m\psi)\big]\sum_{s=1}^{n}\sin(s-\frac{1}{2})m\psi\, F_{(s,2)}\Big\}$$

$$v_{(k,2)} = \frac{1}{2K_1\sin^2\alpha}\ \frac{1}{n}\sum_{m=1}^{2n-1}\frac{\sin(k-\frac{1}{2})m\psi}{(1-\cos m\psi)^2}\Big\{2\cos\frac{m\psi}{2}[2\cos^2\alpha+\beta(1-\cos m\psi)]$$

$$\times\sum_{s=2}^{n}\sin(s-1)m\psi\, F_{(s,1)} + \big[4\cos^2\alpha+(2\beta-\cos^2\alpha)(1-\cos m\psi)\big]\sum_{s=1}^{n}\sin(s-\frac{1}{2})m\psi\, F_{(s,2)}\Big\}$$

$$\tag{3.3.19}$$

It is obvious that the displacement solution is in agreement with the anti-symmetry condition, *i.e.*,

$$v_{(1,1)} = v_{(n+1,1)} = 0,$$

$$u_{(N-k+2,1)} = u_{(k,1)}, \qquad v_{(N-k+2,1)} = -v_{(k,1)},$$

$$u_{(N-k+1,2)} = u_{(k,2)}, \qquad v_{(N-k+1,2)} = -v_{(k,2)}, \qquad k=1,2,...,n. \tag{3.3.20}$$

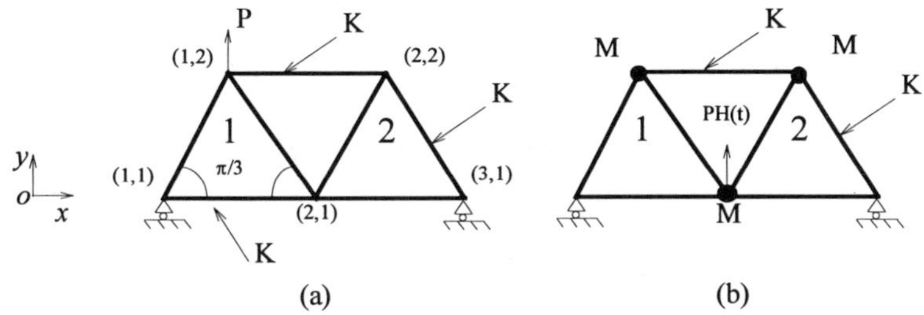

Figure 3.3.4 Truss with two substructures
(a) A static load P acting on the node (1,2);
(b) A suddenly applied load P at node (2,1).

Let us now calculate a simple example by using equations (3.3.19) and (3.3.2). Consider the example as shown in Figure 3.3.4(a) with the following data:

$$n = 2 \ (\psi = \frac{\pi}{2}), \ \ K_1 = K_2 = K \ (\beta = 1), \ \ \alpha = \frac{\pi}{3},$$

$$F_{(1,2)} = P, \qquad F_{(j,k)} = 0 \qquad j \neq 1 \ \text{ or } \ k \neq 2. \tag{3.3.21}$$

Substituting the data given in equation (3.3.21) into equation (3.3.19), yields

$$u_{(1,1)} = \frac{5\sqrt{3}}{24}\frac{P}{K}, \qquad v_{(1,1)} = 0, \qquad u_{(1,2)} = -\frac{\sqrt{3}}{12}\frac{P}{K}, \qquad v_{(1,2)} = \frac{31}{24}\frac{P}{K},$$

$$u_{(2,1)} = \frac{-\sqrt{3}}{24}\frac{P}{K}, \qquad v_{(2,1)} = \frac{P}{K}, \qquad u_{(2,2)} = \frac{\sqrt{3}}{12}\frac{P}{K}, \qquad v_{(2,2)} = \frac{13}{24}\frac{P}{K},$$

$$u_{(3,1)} = -\frac{\sqrt{3}}{8}\frac{P}{K}, \qquad v_{(3,1)} = 0. \tag{3.3.22}$$

The axial forces can be obtained by substituting equations (3.3.22) into equations (3.3.2); they are

$$N_{1,1} = -\frac{\sqrt{3}}{4}P, \qquad N_{1,2} = \frac{\sqrt{3}}{6}P, \qquad N_{1,3} = \frac{\sqrt{3}}{6}P, \qquad N_{1,4} = \frac{\sqrt{3}}{2}P,$$

$$N_{2,1} = -\frac{\sqrt{3}}{12}P, \qquad N_{2,3} = \frac{\sqrt{3}}{6}P, \qquad N_{2,4} = -\frac{\sqrt{3}}{6}P. \tag{3.3.23}$$

The solutions shown in equations (3.3.22) and (3.3.23) satisfy the equilibrium equations and boundary condition for the actual system.

3.3.3 Natural Vibration

The free vibration equation can be obtained by inserting $f_{m,1} = 0$ and $f_{m,2} = 0$ into equation(3.3.9); as

$$[M]\begin{Bmatrix} \ddot{q}_{m,2} \\ \ddot{q}_{m,4} \end{Bmatrix} + [K]\begin{Bmatrix} q_{m,2} \\ q_{m,4} \end{Bmatrix} = \begin{Bmatrix} 0 \\ 0 \end{Bmatrix}, \qquad m=1,2,..,n-1,n+1,...,N-1, \tag{3.3.24a}$$

and

$$M_2\ddot{q}_{n,4} + K_{22}q_{n,4} = 0, \qquad m=n \tag{3.3.24b}$$

with

$$q_{m,2} = Q_{m,2}e^{i\omega t}, \qquad q_{m,4} = Q_{m,4}e^{i\omega t} \tag{3.3.25}$$

in which n denotes the total number of substructures for the original system and N=2n. Substituting equation (3.3.25) into equations (3.3.24a,b) yields

$$[K]\begin{Bmatrix} Q_{m,2} \\ Q_{m,4} \end{Bmatrix} - \omega^2[M]\begin{Bmatrix} Q_{m,2} \\ Q_{m,4} \end{Bmatrix} = \begin{Bmatrix} 0 \\ 0 \end{Bmatrix}, \qquad m=1,2,..,n-1,n+1,...,N-1, \qquad (3.3.26a)$$

and

$$K_{22}Q_{n,4} - \omega^2 M_2 Q_{n,4} = 0, \qquad m=n \qquad (3.3.26b)$$

where $[Q_{m,2}, Q_{m,4}]^T$ denotes the amplitude vector and ω the angular frequency. $[K]$ and $[M]$ defined by equations (3.3.10) with (3.3.11) are Hermitian matrices. Consequently, the eigenvalues of equation (3.3.26a) are real numbers.

The characteristic equation may be found by letting the characteristic determinant vanish; *i.e.*,

$$\begin{vmatrix} K_{11} - M_1\omega^2 & K_{12} \\ K_{21} & K_{22} - M_2\omega^2 \end{vmatrix} = 0, \qquad m \neq n, 2n. \qquad (3.3.27)$$

Expanding this determinant yields the frequency equation:

$$M_1 M_2 \omega^4 - (K_{11}M_2 + K_{22}M_1)\omega^2 + K_{11}K_{22} - K_{12}K_{21} = 0. \qquad (3.3.28)$$

The characteristic roots for ω^2 of equation (3.3.28) can be expressed as

$$\omega^2_{\substack{m,1 \\ m,2}} = \frac{K_{11}M_2 + K_{22}M_1 \mp \sqrt{(K_{11}M_2 - K_{22}M_1)^2 + 4M_1M_2K_{12}K_{21}}}{2M_1M_2},$$

$$m=1,2,...,n-1,\ n+1,...,\ N-1 \qquad (3.3.29)$$

and the other eigenvalue for m=n $(K_{12} = K_{21} = 0)$ can be obtained from equation (3.3.26b) as

$$\omega_n^2 = K_{22} / M_2. \qquad (3.3.30)$$

Noting that $\omega_{N-m,1} = \omega_{m,1}$ and $\omega_{N-m,2} = \omega_{m,2}$, equations (3.3.29) and (3.3.30) give 2n-1 different frequencies in all; *i.e.*, $\omega_{m,1}$ $\omega_{m,2}$ (m=1,2,...,n-1) and ω_n. It is in agreement with the number of lumped masses for the actual system.

Consider now the specific case of

$$M_1 = M_2 = M \qquad \text{and} \qquad K_1 = K_2 = K\ (\beta = 1). \tag{3.3.31}$$

Substituting equations (3.3.11) and (3.3.31) into equation (3.3.29) yields

$$\omega^2_{\substack{m,1\\m,2}} = \frac{2\sin^2\alpha\left[A_m\left(2-\cos^2\alpha\right)+4\cos^2\alpha \mp \sqrt{2\left(2-A_m\right)\left(A_m+2\cos^2\alpha\right)}\right]}{2A_m+4\cos^2\alpha+\cos^4\alpha}\Omega^2$$

$$m=1,2,\ldots,N\text{-}1. \tag{3.3.32}$$

in which $A_m \equiv 1-\cos m\psi$ and

$$\Omega^2 = K/M. \tag{3.3.33}$$

Since $\omega^2_{n,1} = \omega^2_{n,2} = \omega^2_n$, the expression (3.3.30) is not necessary in this case.

Some numerical results for $\alpha = \pi/3$ are summarized in Table 3.3.1.

Table 3.3.1 Natural frequencies of plane Warren trusses

ω \ n	1	2	3	4	5
1st	1.154700538	0.554909659	0.322433015	0.207882984	0.143726360
2nd		1.154700538	0.774596669	0.554909659	0.416110323
3rd		1.544652978	1.154700538	0.877523964	0.688803256
4th			1.441153384	1.154700538	0.936855129
5th			1.619663446	1.379155727	1.154700538
6th				1.544652978	1.338794025
7th				1.646134405	1.485887523
8th					1.593155376
9th					1.658425219
Multiplier	$\sqrt{K/M}$				

The natural modes can be obtained from equation (3.3.26), in which ω should be replaced by $\omega_{m,1}$ and $\omega_{m,2}$ respectively, as

$$\begin{Bmatrix} Q_{m,2} \\ Q_{m,4} \end{Bmatrix}_k = \begin{Bmatrix} -K_{12} \\ K_{11} - M_1 \omega_{m,k}^2 \end{Bmatrix}, \qquad k=1,2, \quad m \neq n. \tag{3.3.34}$$

The normalized mode matrix for the mode subspace co-ordinates can be expressed as

$$[Q_m] = \begin{bmatrix} -K_{12} / D_1, & -K_{12} / D_2 \\ (K_{11} - M_1 \omega_{m,1}^2) / D_1, & (K_{11} - M_1 \omega_{m,2}^2) / D_2 \end{bmatrix}, \qquad m \neq n, \tag{3.3.35}$$

where

$$D_k = \left[M_1 K_{12} \overline{K}_{12} + M_2 \left(K_{11} - M_1 \omega_{m,k}^2 \right)^2 \right]^{\frac{1}{2}}, \qquad k=1,2. \tag{3.3.36}$$

The mode matrix $[Q_m]$ satisfies the orthonormal conditions; *i.e.*,

$$[\overline{Q}_m]^T [M][Q_m] = [I],$$

$$[\overline{Q}_m]^T [K] [Q_m] = \begin{bmatrix} \omega_{m,1}^2 & \\ & \omega_{m,2}^2 \end{bmatrix}, \qquad m \neq n. \tag{3.3.37}$$

Noting that the equations (3.3.24b) and (3.3.26b) for m=n involve only one unknown, the normalized mode is not necessary.

3.3.4 Dynamic Response

Since $q_{N,2}$, $q_{N,4}$ and $q_{n,2}$ vanish identically for the equivalent system, the dynamic equation (3.3.9) may be rewritten as

$$[M]\begin{Bmatrix} \ddot{q}_{m,2} \\ \ddot{q}_{m,4} \end{Bmatrix} + [K]\begin{Bmatrix} q_{m,2} \\ q_{m,4} \end{Bmatrix} = \begin{Bmatrix} f_{m,1} \\ f_{m,2} \end{Bmatrix}, \qquad m=1,2,..,n-1,n+1,...,N-1,$$

$$M_2 \ddot{q}_{n,4} + K_{22} q_{n,4} = f_{n,2}, \qquad m=n. \tag{3.3.38}$$

It is well known that the first of equations (3.3.38) can be solved by the mode method. Let

$$\begin{Bmatrix} q_{m,2}(t) \\ q_{m,4}(t) \end{Bmatrix} = [Q_m] \begin{Bmatrix} a_{m,1}(t) \\ a_{m,2}(t) \end{Bmatrix}, \qquad m \neq n, N. \tag{3.3.39}$$

Substituting equation (3.3.39) into equation (3.3.38) and pre-multiplying by $[\overline{Q}_m]^T$ yields the uncoupled equations as

$$\ddot{a}_{m,1} + \omega_{m,1}^2 a_{m,1} = p_{m,1}(t),$$

$$\ddot{a}_{m,2} + \omega_{m,2}^2 a_{m,2} = p_{m,2}(t), \qquad m \neq n, N \tag{3.3.40}$$

where

$$\begin{Bmatrix} p_{m,1} \\ p_{m,2} \end{Bmatrix} = [\overline{Q}_m]^T \begin{Bmatrix} f_{m,1} \\ f_{m,2} \end{Bmatrix}. \tag{3.3.41}$$

The solutions for $a_{m,1}$, $a_{m,2}$ and $q_{n,4}$ of equations (3.3.40) and the second of equations (3.3.38) can be expressed by using Duhamel's integral as

$$a_{m,k} = \frac{1}{\omega_{m,k}} \int_0^t p_{m,k}(\tau) \sin \omega_{m,k}(t-\tau)d\tau + a_{m,k}(0) \cos \omega_{m,k} t + \frac{\dot{a}_{m,k}(0)}{\omega_{m,k}} \sin \omega_{m,k} t,$$

$$m=1,2,..,n-1,n+1,..,N-1, \quad k=1,2 \tag{3.3.42}$$

and

$$q_{n,4} = \frac{1}{M_2 \omega_n} \int_0^t f_{n,2}(\tau) \sin \omega_n (t-\tau)d\tau + q_{n,4}(0) \cos \omega_n t + \frac{\dot{q}_{n,4}(0)}{\omega_n} \sin \omega_n t \tag{3.3.43}$$

in which $a_{m,1}(0), a_{m,2}(0)$ and $\dot{a}_{m,1}(0), \dot{a}_{m,2}(0)$ are dependent on the initial conditions; *i.e.*,

$$\begin{Bmatrix} a_{m,1}(0) \\ a_{m,2}(0) \end{Bmatrix} = \left[\overline{Q}_m\right]^T [M] \begin{Bmatrix} q_{m,2}(0) \\ q_{m,4}(0) \end{Bmatrix},$$

$$\begin{Bmatrix} \dot{a}_{m,1}(0) \\ \dot{a}_{m,2}(0) \end{Bmatrix} = \left[\overline{Q}_m\right]^T [M] \begin{Bmatrix} \dot{q}_{m,2}(0) \\ \dot{q}_{m,4}(0) \end{Bmatrix}, \tag{3.3.44}$$

and $q_{m,2}(0)$, $q_{m,4}(0)$, $\dot{q}_{m,2}(0)$, $\dot{q}_{m,4}(0)$ have been defined in equation (3.3.13).

The response functions $q_{m,2}(t)$ and $q_{m,4}(t)$ can be found by substituting equation (3.3.42) into equation (3.3.39). Other response functions $q_{m,1}(t)$ and $q_{m,3}(t)$ may be obtained from equation (3.3.18), in which $f_{m,1}$ and $f_{m,2}$ should be replaced by $(f_{m,1} - M_1\ddot{q}_{m,2})$ and $(f_{m,2} - M_2\ddot{q}_{m,4})$ respectively. Then the physical displacements can be found by inserting the solutions for $q_{m,k}(t)$ (k=1,2,3,4) into equations (3.3.3).

The procedure of the present method can be illustrated by considering a specific example . Consider now the truss shown in Figure 3.3.4(b), subjected to a suddenly applied constant load P at node (2,1) with zero initial conditions. For this case, all the parameters are given as follows:

$$n = 2 \quad (\psi = \frac{\pi}{2}), \quad \alpha = \frac{\pi}{3}, \quad K_1 = K_2 = K \quad (\beta = 1), \quad M_1 = M_2 = M,$$

$$v_{(k,1)}(0) = 0, \ \dot{v}_{(k,1)}(0) = 0, \ v_{(k,2)}(0) = 0, \ \dot{v}_{(k,2)}(0) = 0,$$

$$k=1,2,\ldots,n \tag{3.3.45}$$

and the loading functions can be expressed as

$$F_{(2,1)} = PH(t), \quad F_{(1,1)} = F_{(3,1)} = 0, \quad F_{(1,2)} = F_{(2,2)} = 0, \tag{3.3.46}$$

where H(t) denotes the unit step function; *i.e.*,

$$H(t) = \begin{cases} 0, & t < 0, \\ 1, & t \geq 0. \end{cases} \tag{3.3.47}$$

Substituting equation (3.3.46) into equations (3.3.7) and noting that n=2 and $\psi = \pi/2$, yields

$$f_{m,1} = -i\sin\frac{m\pi}{2}PH(t), \quad f_{m,2} = 0, \qquad m=1,2,3. \tag{3.3.48}$$

Since the generalized forces are equal to zero for m=2, the dynamic equation (3.3.38) needs be considered only for the case of $m = 1, 3$ in this example. Natural frequencies may be obtained by inserting $\alpha = \pi/3$ and $\psi = \pi/2$ into equation (3.3.32), as

$$\omega_{m,1}^2 = \frac{6}{49}(11 - 6\sqrt{2})\Omega^2,$$

$$\omega_{m,2}^2 = \frac{6}{49}(11 + 6\sqrt{2})\Omega^2, \qquad m=1,3 \tag{3.3.49}$$

where the suffix m will be dropped from the symbol for natural frequencies in this case.

Finally, the generalized displacements $q_{m,2}(t)$ and $q_{m,4}(t)$ can be obtained as

$$q_{m,2}(t) = -i\sin\frac{m\pi}{2}\frac{P}{12K}\left[22 - (11 + 6\sqrt{2})\cos\omega_1 t - (11 - 6\sqrt{2})\cos\omega_2 t\right],$$

$$q_{m,4}(t) = (1 - i\sin\frac{m\pi}{2})\frac{P}{24K}\left[24 - (11\sqrt{2} + 12)\cos\omega_1 t + (11\sqrt{2} - 12)\cos\omega_2 t\right],$$

$$m=1,3 \tag{3.3.50}$$

and other generalized displacements $q_{m,1}(t)$ and $q_{m,3}(t)$ can be obtained from equations (3.3.18), where $f_{m,1}$ and $f_{m,2}$ need to be replaced by $(f_{m,1} - M_1\ddot{q}_{m,2})$ and $(f_{m,2} - M_2\ddot{q}_{m,4})$ respectively, as

$$q_{m,1}(t) = \frac{\sqrt{3}P}{12K}\left[2 - (1 + \sqrt{2})\cos\omega_1 t + (\sqrt{2} - 1)\cos\omega_2 t\right],$$

$$q_{m,3}(t) = (1 + i\sin\frac{m\pi}{2})\frac{\sqrt{3}P}{24K}\left[-4 + (2+\sqrt{2})\cos\omega_1 t + (2-\sqrt{2})\cos\omega_2 t\right],$$

$$m=1,3. \tag{3.3.51}$$

Substituting equations (3.3.50) and (3.3.51) into the **U**-transformation (3.3.3) yields

$$u_{(1,1)} = -u_{(3,1)} = \frac{\sqrt{3}P}{12K}\left[2-(1+\sqrt{2})\cos\omega_1 t + (\sqrt{2}-1)\cos\omega_2 t\right],$$

$$u_{(1,2)} = -u_{(2,2)} = \frac{\sqrt{3}P}{24K}\left[-4 + (2+\sqrt{2})\cos\omega_1 t + (2-\sqrt{2})\cos\omega_2 t\right],$$

$$u_{(2,1)} = 0,\ v_{(1,1)} = 0,\ v_{(3,1)} = 0,$$

$$v_{(2,1)} = \frac{P}{12K}\left[22-(11+6\sqrt{2})\cos\omega_1 t - (11-6\sqrt{2})\cos\omega_2 t\right],$$

$$v_{(1,2)} = v_{(2,2)} = \frac{P}{24K}\left[24-(11\sqrt{2}+12)\cos\omega_1 t + (11\sqrt{2}-12)\cos\omega_2 t\right].$$

$$\tag{3.3.52}$$

where ω_1 and ω_2 have been defined by equation (3.3.49). The constant terms on the right hand sides of equations (3.3.52) represent the static displacements of the truss shown in Figure 3.3.4b subjected to a concentrated load of magnitude P at the node (2,1).

Chapter 4

LINEAR PERIODIC STRUCTURES - CONTINUOUS SYSTEMS

Chapter 3 was devoted exclusively to the analysis of discrete systems with linear periodicity, whereas this chapter is devoted to continuous systems with linear periodicity. Many linear periodic structures can be analyzed by the U-transformation, if the extreme boundary conditions satisfy certain requirements. The following sections will give some examples of the application on the U-transformation method to continuous system.

4.1 Continuous Beams - Static Displacements

4.1.1 Continuous Beam with Extreme Ends Simply Supported

Consider a beam with uniform flexural rigidity EI that runs over n+1 number of roller supports as shown in Figure 4.1.1(a). Let the span length between any two adjacent supports be denoted by l. Firstly, we need to create a cyclic periodic system which is equivalent to the actual system. In order to form an equivalent system, it is necessary to extend the original beam by its symmetrical image and apply anti-symmetric loading on the corresponding extended part about the original loading as shown in Figure 4.1.1.(b). This is equivalent to extending the flexural deflection curve of the original beam anti-symmetrically at one end. The slopes at the extreme ends of the extended beam will have equal values and so do the curvatures which are zero at both ends. Hence the two ends can be imaginarily put together and treated as one point in mathematics. A structure having a flexural curve with such characteristics is a kind of cyclic periodic structure. The flexural deflection shape is one of the possible deflection shapes for a cyclic periodic structure and therefore can be expressed as a series of mode subspace coordinates given by equation (2.1.8b). It is thus possible to use the U-transformation to deal with the equivalent system as shown in Figure 4.1.1(b), instead of the original beam as shown in Figure 4.1.1(a). The boundary conditions at both extreme ends for the original beam are satisfied automatically in its equivalent system.

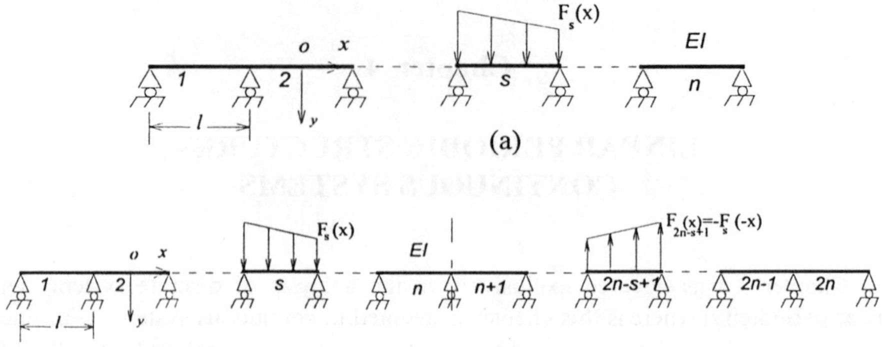

Figure 4.1.1 Continuous beam with simply supported extreme ends:
(a) Actual system; (b) Equivalent system

In each substructure a local coordinate system oxy with the origin o at its mid-span is established. The equilibrium equations for all substructures are of the same form, *i.e.*,

$$\mathrm{EI}\frac{d^4 w_k(x)}{d x^4} = F_k(x) \ , \qquad -\frac{l}{2} < x < \frac{l}{2}, \qquad k=1,2,\ldots,N \tag{4.1.1}$$

with N=2n, where $w_k(x)$ is the transverse displacement; and $F_k(x)$ the loading function for the k-th substructure. Continuity across the roller supports requires the following conditions to be met:

$$w_k\Big|_{x=-\frac{l}{2}} = 0, \qquad w_k\Big|_{x=\frac{l}{2}} = 0,$$

$$\left.\frac{d w_k}{dx}\right|_{x=\frac{l}{2}} = \left.\frac{d w_{k+1}}{dx}\right|_{x=-\frac{l}{2}}, \quad \left.\frac{d^2 w_k}{d x^2}\right|_{x=\frac{l}{2}} = \left.\frac{d^2 w_{k+1}}{d x^2}\right|_{x=-\frac{l}{2}}, \qquad k=1,2,\ldots, N \tag{4.1.2}$$

where w_{N+1} should be replaced by w_1.

Apply the U-transformation to equations (4.1.1) and (4.1.2), *i.e.*, let

$$w_k = \frac{1}{\sqrt{N}} \sum_{m=1}^{N} e^{i(k-1)m\psi} q_m , \qquad k=1,2,...,N \tag{4.1.3a}$$

or

$$q_m = \frac{1}{\sqrt{N}} \sum_{k=1}^{N} e^{-i(k-1)m\psi} w_k , \qquad m=1,2,..,N, \tag{4.1.3b}$$

multiplying the k-th equation shown in (4.1.1) and (4.1.2) by $e^{-i(k-1)m\psi}/\sqrt{N}$, respectively, and finding the sum of the equations for k=1,2,...,N yield

$$EI \frac{d^4 q_m}{d x^4} = f_m , \qquad m=1,2,...,N \tag{4.1.4}$$

and

$$q_m\big|_{x=-\frac{1}{2}} = 0, \qquad q_m\big|_{x=\frac{1}{2}} = 0 ,$$

$$q_m'\big|_{x=\frac{1}{2}} = e^{im\psi} q_m'\big|_{x=-\frac{1}{2}} , \quad q_m''\big|_{x=\frac{1}{2}} = e^{im\psi} q_m''\big|_{x=-\frac{1}{2}} , \qquad m=1,2,...,N \tag{4.1.5}$$

where a prime denotes differentiation with respect to x, and f_m is defined as

$$f_m = \frac{1}{\sqrt{N}} \sum_{k=1}^{N} e^{-i(k-1)m\psi} F_k , \qquad m=1,2,...,N \tag{4.1.6}$$

in which the loading functions F_k (k=1,2,...,N) must satisfy the anti-symmetric condition, *i.e.*,

$$F_{N-k+1}(x) = -F_k(-x) , \qquad k=1,2,...,n. \tag{4.1.7}$$

Equation (4.1.4) must be solved subject to the continuity condition (4.1.5) for each generalized displacement. This may be illustrated easily by considering a beam with a specific loading condition [7].

Firstly, let an uniformly distributed load p which is independent of time be applied on the middle span, say the s-th span of a continuous beam with n number of spans (n=2s-1). For this case, the loading functions can be expressed as

$$F_s(x) = p \quad \text{(constant)}, \quad F_{N-s+1}(x) = -p;$$

$$F_k(x) = 0,. \qquad k \neq s, N\text{-}s\text{+}1 \tag{4.1.8}$$

where N=2n. Substituting equation (4.1.8) into equation (4.1.6) and noting $\psi = 2\pi/N$, yields

$$f_m(x) = 2p\,e^{-i(s-1)m\psi}/\sqrt{N}\,, \qquad m=1,3,...,2n\text{-}1;$$

$$f_m(x) = 0\,, \qquad m=2,4,...,2n. \tag{4.1.9}$$

Substituting equation (4.1.9) into the right hand side of equation (4.1.4), yields

$$EI\frac{d^4 q_m}{d x^4} = \begin{cases} 2p\,e^{-i(s-1)m\psi}/\sqrt{N} & m = 1,3,\dots,2n-1, \\ 0 & m = 2,4,\dots,2n. \end{cases} \tag{4.1.10}$$

The solution for $q_m(x)$ of equations (4.1.10) and (4.1.5) can be expressed as

$$q_m(x) = \frac{2p}{EI\sqrt{2n}} e^{-i(s-1)m\psi}(C_{m0} + C_{m1}\,x + C_{m2}\,x^2 + C_{m3}\,x^3 + \frac{x^4}{24}), \quad -\frac{1}{2} \le x \le \frac{1}{2}\,,$$

$$m=1,3,..,2n\text{-}1 \tag{4.1.11a}$$

$$q_m(x) = 0, \qquad m=2,4,...,2n \tag{4.1.11b}$$

where

$$C_{m0} = \frac{l^4}{384}(\frac{4 - \cos m\psi}{2 + \cos m\psi}), \qquad C_{m1} = -i\frac{l^3}{96}(\frac{\sin m\psi}{2 + \cos m\psi}),$$

$$C_{m2} = -\frac{l^2}{16}(\frac{1}{2 + \cos m\psi}), \qquad C_{m3} = i\frac{1}{24}(\frac{\sin m\psi}{2 + \cos m\psi}). \tag{4.1.12}$$

Substituting equations (4.1.11) with (4.1.12) into the s-th equation in (4.1.3a) yields

$$w_s(x) = \frac{p l^4}{384EI}\frac{1}{n}\sum_{m=1,3,..}^{2n-1}\left[\frac{4-\cos m\psi}{2+\cos m\psi} - 24\left(\frac{1}{2+\cos m\psi}\right)\left(\frac{x}{l}\right)^2 + 16\left(\frac{x}{l}\right)^4\right],$$

$$-l/2 \le x \le l/2 \tag{4.1.13}$$

This deflection function is applicable to the loaded span only. For other spans the deflection functions can be obtained similarly without particular difficulty.

The maximum deflection occurs at mid-span and its magnitude is

$$w_{max} = w_s(0) = \frac{p l^4}{384EI}\frac{1}{n}\sum_{m=1,3,...}^{2n-1}\frac{4-\cos m\psi}{2+\cos m\psi}. \tag{4.1.14}$$

The maximum moment also occurs at mid-span with magnitude:

$$M_{max} = -EI\, w_s''(0) = \frac{p l^2}{8}\frac{1}{n}\sum_{m=1,3,...}^{2n-1}\frac{1}{2+\cos m\psi}. \tag{4.1.15}$$

Now consider the sum of terms in the series on the right hand side of equation (4.1.13). It has the general form:

$$In \equiv \frac{1}{n}\sum_{m=1,3,...}^{2n-1} f(m\psi) = \frac{1}{n}\sum_{m=1,3,...}^{2n-1} f\left(\frac{m\pi}{n}\right). \tag{4.1.16}$$

Let $\Delta\theta = 2\pi/n$. Then, equation (4.1.16) takes the form:

$$In = \frac{1}{2\pi}\sum_{m=1,3,...}^{2n-1} f\left(\frac{m}{2}\Delta\theta\right)\Delta\theta \tag{4.1.17}$$

By letting n approach infinity, the limit of the series summation becomes the definite integral:

$$\lim_{n\to\infty}\frac{1}{n}\sum_{m=1,3,...}^{2n-1} f(m\psi) = \frac{1}{2\pi}\int_0^{2\pi} f(\theta)d\theta \tag{4.1.18}$$

Applying this result to equation (4.1.13) yields the solution for the deflection $w_s^*(x)$ of the continuous beam with infinite number of spans:

$$w_s^*(x) = \lim_{n\to\infty} w_s(x) = \frac{p l^4}{384EI}\left[2\sqrt{3} - 1 - 8\sqrt{3}(\frac{x}{l})^2 + 16(\frac{x}{l})^4\right],$$

$$-l/2 \le x \le l/2 \tag{4.1.19}$$

The maximum deflection and moment occur at mid-span with magnitudes

$$w_{max}^* = w_s^*(0) = \frac{(2\sqrt{3} - 1)p l^4}{384EI} \tag{4.1.20}$$

and

$$M_{max}^* = -EI\, w_s^{*''}(0) = \frac{\sqrt{3}}{24} p l^2 \tag{4.1.21}$$

respectively. The numerical result are summarized in Table 4.1.1.

Table 4.1.1 Deflections and bending moments of continuous beams subjected to uniform load p over middle span

Number of spans	1	3	5	7	9	∞
$w_{max}/(p l^4/384EI)^a$	5.00000	2.60000	2.47368	2.46479	2.46415	2.46410162
$w_{max}/(p l^4/384EI)^b$	1.00000	2.33333	2.45455	2.46341	2.46405	2.46410162
$M_{max}/(p l^2/16)^a$	2.00000	1.20000	1.15789	1.15499	1.15472	1.1547005
$M_{max}/(p l^2/16)^b$	-1.3333c	1.11111	1.15152	1.15445	1.15468	1.1547005

a Case is of simply supported extreme end conditions.
b Case is of clamped extreme end conditions.
c Maximum moment for a uniformly loaded clamped beam occurs at the supports.

Secondly, let a concentrate load of magnitude P act at the midpoint of the middle span, say the s-th span. Equal but opposite concentrated loads must be applied to the s-th and (N-s+1)-th spans of the equivalent system with all other spans unloaded, *i.e.*,

$$F_s(x) = P\delta(x), \qquad F_{N-s+1}(x) = -P\delta(x)$$

$$F_k(x) = 0, \qquad k \neq s \text{ and } N\text{-}s\text{+}1 \tag{4.1.22}$$

where $\delta(x)$ = the Dirac delta function; and N=2n=2 (2s-1).

Substituting equation (4.1.22) into equation (4.1.6) yields

$$f_m(x) = 2Pe^{-i(s-1)m\psi}\delta(x)/\sqrt{N}, \qquad m = 1,3,\dots,2n-1;$$

$$f_m(x) = 0, \qquad m = 2,4,\dots,2n; \tag{4.1.23}$$

where $\psi = \pi/n$.

The solution for q_m of equation (4.1.4) with equation (4.1.23) subject to continuity condition (4.1.5) may be written as:

$$q_m(x) = \frac{2Pe^{-i(s-1)m\psi}}{EI\sqrt{2n}}(C_{m0} + C_{m1}x + C_{m2}x^2 + C_{m3}x^3 + \frac{1}{12}|x|^3),$$

$$-l/2 \le x \le l/2, \quad m=1,3,..,2n-1 \tag{4.1.24a}$$

$$q_m(x) = 0, \qquad m=2,4,\dots,2n \tag{4.1.24b}$$

where

$$C_{m0} = \frac{l^3}{384}(\frac{7-\cos m\psi}{2+\cos m\psi}),$$

$$C_{m1} = -i\frac{l^2}{64}(\frac{\sin m\psi}{2+\cos m\psi}),$$

$$C_{m2} = -\frac{l}{32}(\frac{5+\cos m\psi}{2+\cos m\psi}),$$

$$C_{m3} = i\frac{1}{16}(\frac{\sin m\psi}{2+\cos m\psi}). \tag{4.1.25}$$

The deflection in each substructure may now be found by substituting equation (4.1.24) and equation (4.1.25) into equation (4.1.3a). The deflection in the loaded span can be expressed as

$$w_s(x) = \frac{Pl^3}{384EI}\frac{1}{n}\sum_{m=1,3...}^{2n-1}[\frac{7-\cos m\psi}{2+\cos m\psi} - 12(\frac{5+\cos m\psi}{2+\cos m\psi})(\frac{x}{l})^2 + 32(\frac{|x|}{l})^3],$$

$$-l/2 \le x \le l/2 \qquad (4.1.26)$$

The maximum deflection occurs at midpoint of the loaded span and its magnitude is

$$w_{max} = w_s(0) = \frac{Pl^3}{384EI}\frac{1}{n}\sum_{m=1,3...}^{2n-1}\frac{7-\cos m\psi}{2+\cos m\psi}. \qquad (4.1.27)$$

The maximum moment also occurs at mid-span of the loaded span with magnitude:

$$M_{max} = -EIw_s''(0) = \frac{Pl}{16}\frac{1}{n}\sum_{m=1,3...}^{2n-1}\frac{5+\cos m\psi}{2+\cos m\psi}. \qquad (4.1.28)$$

Using the limiting process given by equation (4.1.18), the deflection curve $w_s^*(x)$ for the loaded span of the infinite continuous beam under a concentrated load P becomes:

$$w_s^*(x) = \frac{Pl^3}{384EI}[3\sqrt{3}-1-12(\sqrt{3}+1)(\frac{x}{l})^2 + 32(\frac{|x|}{l})^3],$$

$$-l/2 \le x \le l/2\,. \qquad (4.1.29)$$

Evaluation of the maximum deflection and moment in this case shows that both occur at mid-span:

$$w_{max}^* = w_s^*(0) = \frac{(3\sqrt{3}-1)Pl^3}{384EI}, \qquad (4.1.30)$$

$$M_{max}^* = -EIw_s^{*''}(0) = \frac{(\sqrt{3}+1)}{16}Pl\,. \qquad (4.1.31)$$

The numerical results are summarized in Table 4.1.2.

Table 4.1.2 Deflections and bending moments of continuous beams subjected to concentrated load P at center of middle span

Number of spans	1	3	5	7	9	∞
$w_{max}/(Pl^3/384EI)^a$	8.00000	4.40000	4.21053	4.19747	4.19623	4.19615242
$w_{max}/(Pl^3/384EI)^b$	2.00000	4.00000	4.18182	4.19512	4.19608	4.19615242
$M_{max}/(Pl/16)^a$	4.00000	2.80000	2.73684	2.73249	2.73208	2.7320508
$M_{max}/(Pl/16)^b$	2.00000	2.66667	2.72727	2.73171	2.73203	2.7320508

a Case is of simply supported extreme end conditions.
b Case is of clamped extreme end conditions.

4.1.2 Continuous Beam with Extreme Ends Clamped

Let us consider that both extreme ends of the continuous beam are clamped. The actual system is shown in Figure 4.1.2(a). Its equivalent system with cyclic periodicity can be obtained by producing a symmetrical image of the original beam and loading about one fixed end as shown in Figure 4.1.2(b). The curvatures at both ends are equal and so are the slopes which are equal to zero. Hence the system shown in Figure 4.1.2(b) may be regarded as a cyclically periodic system. The clamped end conditions for the original beam can be satisfied automatically in its equivalent system.

The loading functions for the equivalent system must satisfy the symmetric condition, *i.e.*,

$$F_{N-k+1}(x) = F_k(-x), \qquad k=1,2,...,n \tag{4.1.32}$$

where N=2n. Consider now a continuous beam with a specific loading condition.

Firstly, let us discuss the case of an uniformly distributed load p applied on the middle span, say the s-th span of the continuous beam with n number of spans (n=2s-1). For this case, the loading functions may be expressed as follows:

$$F_s(x) = p, \qquad F_{N-s+1}(x) = p,$$

$$F_k(x) = 0, \qquad k \neq s \text{ and } N\text{-}s\text{+}1 \tag{4.1.33}$$

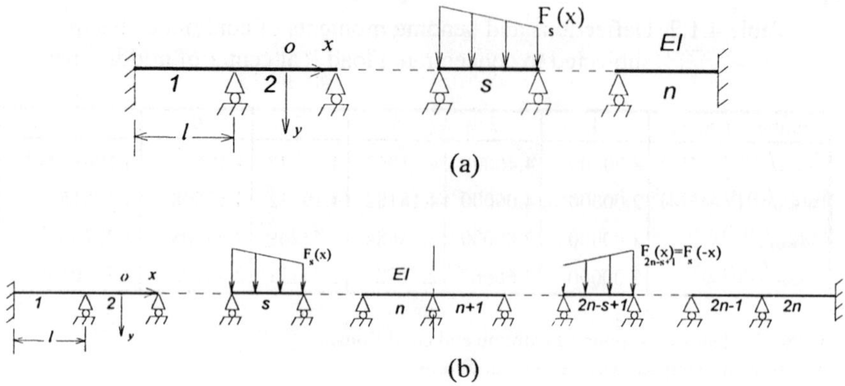

Figure 4.1.2 Continuous beam with clamped extreme ends:
(a) Actual system; (b) Equivalent system.

Substituting equation (4.1.33) into equation(4.1.6) yields

$$f_m(x) = 2pe^{-i(s-1)m\psi}/\sqrt{N}\,, \qquad m=2,4,....2n,$$

$$f_m(x) = 0\,, \qquad m=1,3,...,2n\text{-}1 \tag{4.1.34}$$

Both equations (4.1.9) and (4.1.34) have the same form except for the different parity of the subscript m. The solution to equation (4.1.4) with equation (4.1.34) subjected to the same continuity conditions as those given by equation (4.1.5) is:

$$q_m(x) = \frac{2pe^{-i(s-1)m\psi}}{EI\sqrt{2n}}(C_{m0} + C_{m1}x + C_{m2}x^2 + C_{m3}x^3 + \frac{x^4}{24}),$$

$$-l/2 \le x \le l/2\,, \qquad m=2,4,..,2n. \tag{4.1.35a}$$

$$q_m(x) = 0 \qquad m=1,3,..,2n\text{-}1 \tag{4.1.35b}$$

where $C_{m0}, C_{m1}, C_{m2},$ and C_{m3} have been defined by equation (4.1.12).

Substituting equations (4.1 35a-b) with equation (4.1.12) into the s-th equation in (4.1.3a) yields

$$w_s(x) = \frac{p\,l^4}{384EI}\frac{1}{n}\sum_{m=2,4,..}^{2n}\left[\frac{4-\cos m\psi}{2+\cos m\psi} - 24\left(\frac{1}{2+\cos m\psi}\right)\left(\frac{x}{l}\right)^2 + 16\left(\frac{x}{l}\right)^4\right],$$

$$-l/2 \le x \le l/2 \,. \tag{4.1.36}$$

The magnitude of the maximum deflection is

$$w_{max} = w_s(0) = \frac{p\,l^4}{384EI}\frac{1}{n}\sum_{m=2,4,..}^{2n}\frac{4-\cos m\psi}{2+\cos m\psi}\,. \tag{4.1.37}$$

The magnitudes of the moments at mid-span and supports of the loaded span are

$$M_s(0) = \frac{p\,l^2}{8}\frac{1}{n}\sum_{m=2,4...}^{2n}\frac{1}{2+\cos m\psi}\,, \tag{4.1.38a}$$

$$M_s\left(\frac{l}{2}\right) = \frac{p\,l^2}{8}\frac{1}{n}\sum_{m=2,4...}^{2n}\left(\frac{1}{2+\cos m\psi} - 1\right), \tag{4.1.38b}$$

respectively.

By letting n approach infinity, the limit of deflection function shown in equation (4.1.36) is the same as that shown in equation (4.1.19). Therefore, the magnitudes of the maximum deflection and bending moment are also the same as those shown in equations (4.1.20) and (4.1.21), respectively. A comparison of these results is summarized in Table 4.1.1.

Secondly, consider a concentrated load of magnitude P acting at the midpoint of the middle span, say the s-th span. It is plain enough to mention that this case is solved in exactly the same way as the case for uniform load in the previous example. The deflection function $w_s(x)$ of the loaded span may be obtained directly from equation (4.1.26) by assigning the even numbers instead of odd numbers for m in the series on the right hand side of equation (4.1.26), *i.e.*,

$$w_s(x) = \frac{P\,l^3}{384EI}\frac{1}{n}\sum_{m=2,4...}^{2n}\left[\frac{7-\cos m\psi}{2+\cos m\psi} - 12\left(\frac{5+\cos m\psi}{2+\cos m\psi}\right)\left(\frac{x}{l}\right)^2 + 32\left(\frac{|x|}{l}\right)^3\right],$$

$$-l/2 \le x \le l/2 \tag{4.1.39}$$

and then

$$w_{max} = \frac{P l^3}{384EI} \frac{1}{n} \sum_{m=2,4....}^{2n} \frac{7 - \cos m\psi}{2 + \cos m\psi}, \qquad (4.1.40a)$$

$$M_{max} = \frac{Pl}{16} \frac{1}{n} \sum_{m=2,4....}^{2n} \frac{5 + \cos m\psi}{2 + \cos m\psi}. \qquad (4.1.40b)$$

Applying the limiting process with n approaching infinity to equation (4.1.39) yields the same limit as equation (4.1.29). A comparison of these results is summarized in Table 4.1.2.

4.1.3 Continuous Beam with One Extreme End Simply Supported and the Other Clamped

Firstly, the original beam and loading shown in Figure 4.1.3(a) is extended at the clamped end by its symmetrical image as shown in Figure 4.1.3(b).Then the new structure is treated as a continuous beam with both ends simply supported as in example 4.1.1. An equivalent system with cyclic periodicity and 4n number of spans will be obtained as shown in Figure 4.1.3(c) and the U-transformation can be applied .

The loading functions for the equivalent system must satisfy the following conditions:

$$F_{2n-k+1}(x) = F_k(-x), \qquad F_{2n+k}(x) = -F_k(x),$$

$$F_{4n-k+1}(x) = -F_k(-x), \qquad k=1,2,...,n \qquad (4.1.41)$$

where x is the local coordinate of each span with the origin o at its mid-span.

To take just one example, let us consider the case of an uniformly distributed load p applied on the middle span, say the s-th span of the continuous beam with n number of spans (n=2s-1). For the present case, the loading functions may be expressed as follows:

$$F_s(x) = p, \qquad F_{2n-s+1}(x) = p,$$

$$F_{2n+s}(x) = -p, \qquad F_{N-s+1}(x) = -p. \qquad (4.1.42)$$

with other loading functions vanish.

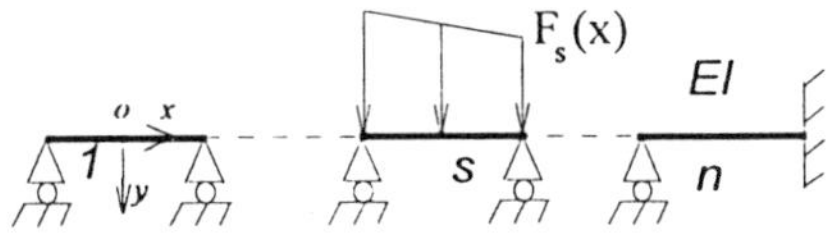

(a)

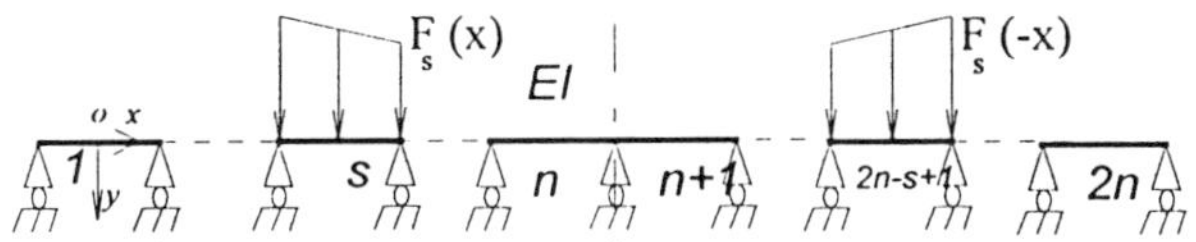

(b)

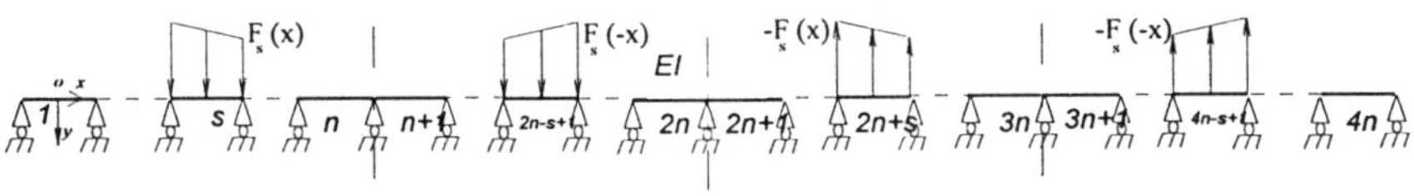

(c)

Figure 4.1.3. Continuous beam with simply supported and clamped ends; (a) Actual system; (b)Transition system; (c)Equivalent system.

Substituting equation (4.1.42) into equation (4.1.6) yields

$$f_m = 2(1 - i\sin\frac{m\pi}{2})p\,e^{-i(s-1)m\psi}\Big/\sqrt{N}\,, \qquad m=1,3,..,4n-1,$$

$$f_m(x) = 0\,, \qquad m=2,4,...,4n \tag{4.1.43}$$

where N=4n and $\psi = \pi/2n$.

The solution for $q_m(x)$ of equations (4.1.4) with (4.1.43) subjected to the continuity condition (4.1.5) may be expressed as

$$q_m(x) = \frac{2pe^{-i(s-1)m\psi}}{EI\sqrt{4n}}(1 - i\sin\frac{m\pi}{2})(C_{m0} + C_{m1}x + C_{m2}x^2 + C_{m3}x^3 + \frac{x^4}{24}),$$

$$-l/2 \le x \le l/2, \quad m=1,3,..,4n-1 \tag{4.1.44a}$$

$$q_m(x) = 0 \qquad m=2,4,...,4n \tag{4.1.44b}$$

where C_{m0} to C_{m3} have been defined by equation (4.1.12).

Substituting equations (4.1.44a-b) with (4.1.12) into the s-th equation in (4.1.3a) yields

$$w_s(x) = \frac{pl^4}{384EI}\frac{1}{n}\sum_{m=1,3,...}^{2n-1}[\frac{4-\cos m\psi}{2+\cos m\psi} - 24(\frac{1}{2+\cos m\psi})(\frac{x}{l})^2 + 16(\frac{x}{l})^4$$

$$-4\sin\frac{m\pi}{2}(\frac{\sin m\psi}{2+\cos m\psi})(\frac{x}{l}) + 16\sin\frac{m\pi}{2}(\frac{\sin m\psi}{2+\cos m\psi})(\frac{x}{l})^3],$$

$$-l/2 \le x \le l/2 . \tag{4.1.45}$$

By letting n approach infinity, the limit of deflection function $w_s(x)$ shown in equation (4.1.45) is the same as that shown in equation (4.1.19).

4.2 Continuous Beams - Dynamic Analysis

The dynamic equations for all substructures are of the same form, *i.e.*,

$$EI\frac{\partial^4 W_k(x,t)}{\partial x^4} + \rho A\frac{\partial^2 W_k(x,t)}{\partial t^2} = F_k(x,t),$$

$$-l/2 \le x \le l/2, \quad k=1,2,...,N \tag{4.2.1}$$

where ρA denotes the mass per unit length and N denotes the total number of spans for the equivalent system with cyclic periodicity. Continuity across the roller supports requires the following conditions to be satisfied:

$$W_k\big|_{x=-\frac{l}{2}} = 0, \qquad W_k\big|_{x=\frac{l}{2}} = 0,$$

$$\left.\frac{\partial W_k}{\partial x}\right|_{x=\frac{1}{2}} = \left.\frac{\partial W_{k+1}}{\partial x}\right|_{x=-\frac{1}{2}}, \qquad \left.\frac{\partial^2 W_k}{\partial x^2}\right|_{x=\frac{1}{2}} = \left.\frac{\partial^2 W_{k+1}}{\partial x^2}\right|_{x=-\frac{1}{2}},$$

$$k=1,2,\ldots,N. \tag{4.2.2}$$

with $W_{N+1} \equiv W_1$.

Apply the U transformation to equations (4.2.1) and (4.2.2), *i.e.,* let

$$W_k(x,t) = \frac{1}{\sqrt{N}} \sum_{m=1}^{N} e^{i(k-1)m\psi} q_m(x,t), \qquad k=1,2,\ldots,N \tag{4.2.3a}$$

or

$$q_m(x,t) = \frac{1}{\sqrt{N}} \sum_{k=1}^{N} e^{-i(k-1)m\psi} W_k(x,t), \qquad m=1,2,\ldots,N \tag{4.2.3b}$$

yield

$$EI \frac{\partial^4 q_m(x,t)}{\partial x^4} + \rho A \frac{\partial^2 q_m(x,t)}{\partial t^2} = f_m(x,t),$$

$$-1/2 \le x \le 1/2, \qquad m=1,2,\ldots,N \tag{4.2.4}$$

and

$$\left. q_m \right|_{x=-\frac{1}{2}} = 0, \qquad \left. q_m \right|_{x=\frac{1}{2}} = 0,$$

$$\left. q'_m \right|_{x=\frac{1}{2}} = e^{im\psi} \left. q'_m \right|_{x=-\frac{1}{2}}, \qquad \left. q''_m \right|_{x=\frac{1}{2}} = e^{im\psi} \left. q''_m \right|_{x=-\frac{1}{2}}, \qquad m=1,2,\ldots,N \tag{4.2.5}$$

where $f_m(x.t)$ is defined as

$$f_m(x,t) = \frac{1}{\sqrt{N}} \sum_{k=1}^{N} e^{-i(k-1)m\psi} F_k(x,t) \qquad m=1,2,\ldots,N \tag{4.2.6}$$

The initial condition for equation (4.2.4) can be written as

$$q_m\Big|_{t=0} = q_{m0}, \qquad \dot{q}_m\Big|_{t=0} = \dot{q}_{m0}, \tag{4.2.7a}$$

with

$$q_{m0} = \frac{1}{\sqrt{N}} \sum_{k=1}^{N} e^{-i(k-1)m\psi} W_k(x,0),$$

$$\dot{q}_{m0} = \frac{1}{\sqrt{N}} \sum_{k=1}^{N} e^{-i(k-1)m\psi} \dot{W}_k(x,0). \tag{4.2.7b}$$

4.2.1 Natural Vibration

Inserting $f_m = 0$ and $q_m = Q_m(x)\, e^{i\omega t}$ into equations (4.2.4) and (4.2.5) yields

$$\frac{d^4 Q_m}{d x^4} = \beta^4 Q_m \tag{4.2.8}$$

and

$$Q_m(-\frac{1}{2}) = 0, \qquad Q_m(\frac{1}{2}) = 0,$$

$$Q'_m(\frac{1}{2}) = e^{im\psi} Q'_m(-\frac{1}{2}), \qquad Q''_m(\frac{1}{2}) = e^{im\psi} Q''_m(-\frac{1}{2}), \tag{4.2.9}$$

where

$$\beta^4 = (\rho A/EI)\,\omega^2. \tag{4.2.10}$$

Equations (4.2.8) and (4.2.9) represent an eigenvalue problem of the self-adjoint differential equation. The eigenvalues are positive real numbers. The general solution for Q_m of equation (4.2.8) can be expressed as

$$Q_m(x) = A\sin\beta x + iB\cos\beta x + C\mathrm{sh}\beta x + iD\mathrm{ch}\beta x \tag{4.2.11}$$

where A, B, C, D are four constants of integration. Substituting equation (4.2.11) into equation (4.2.9) result in

$$-A\sin\alpha + iB\cos\alpha - C\mathrm{sh}\alpha + iD\mathrm{ch}\alpha = 0$$

$$A\sin\alpha + iB\cos\alpha + C\mathrm{sh}\alpha + iD\mathrm{ch}\alpha = 0$$

$$A\cos\alpha(1-e^{im\psi}) - iB\sin\alpha(1+e^{im\psi}) + C\mathrm{ch}\alpha(1-e^{im\psi}) + iD\mathrm{sh}\alpha(1+e^{im\psi}) = 0$$

$$-A\sin\alpha(1+e^{im\psi}) - iB\cos\alpha(1-e^{im\psi}) + C\mathrm{sh}\alpha(1+e^{im\psi}) + iD\mathrm{ch}\alpha(1-e^{im\psi}) = 0 \tag{4.2.12}$$

where

$$\alpha = \frac{\beta l}{2} \tag{4.2.13}$$

Let the coefficient determinant of the simultaneous equations (4.2.12) be equal to zero. Expanding this determinant yields the frequency equation

$$\cos m\psi(\mathrm{sh}2\alpha - \sin 2\alpha) + \mathrm{ch}2\alpha\sin 2\alpha - \mathrm{sh}2\alpha\cos 2\alpha = 0,$$

$$m=1,2,\ldots,N. \tag{4.2.14}$$

The natural frequency corresponding to the frequency parameter α_{mj} which denotes the j-th positive root of the m-th equation in equations (4.2.14), can be expressed as

$$\omega_{mj} = \frac{4}{l^2}\sqrt{\frac{EI}{\rho A}}\,\alpha_{mj}^2 \tag{4.2.15}$$

Upon substituting $\alpha = \alpha_{mj}$ into equation (4.2.12) the non-trivial solution for A, B, C and D can be obtained as

$$A(\alpha_{mj}) = \mathrm{sgn}\left[\cos\alpha_{mj}(\mathrm{ch}\,\alpha_{mj}\sin\alpha_{mj} + \mathrm{sh}\,\alpha_{mj}\cos\alpha_{mj})\right]\mathrm{sh}\,\alpha_{mj}$$

$$\times\sqrt{\mathrm{ch}\alpha_{mj}\left|\cos\alpha_{mj}(\mathrm{ch}\alpha_{mj}\sin\alpha_{mj} + \mathrm{sh}\alpha_{mj}\cos\alpha_{mj})\right|},$$

$$B(\alpha_{mj}) = -\mathrm{ch}\,\alpha_{mj}\sqrt{\mathrm{sh}\,\alpha_{mj}\left|\sin\alpha_{mj}(\mathrm{sh}\,\alpha_{mj}\cos\alpha_{mj} - \mathrm{ch}\,\alpha_{mj}\sin\alpha_{mj})\right|},$$

$$C(\alpha_{mj}) = -\,\mathrm{sgn}\left[\cos\alpha_{mj}(\mathrm{ch}\,\alpha_{mj}\sin\alpha_{mj} + \mathrm{sh}\,\alpha_{mj}\cos\alpha_{mj})\right]\sin\alpha_{mj}$$

$$\times\sqrt{\mathrm{ch}\,\alpha_{mj}\left|\cos\alpha_{mj}(\mathrm{ch}\,\alpha_{mj}\sin\alpha_{mj} + \mathrm{sh}\,\alpha_{mj}\cos\alpha_{mj})\right|},$$

$$D(\alpha_{mj}) = \cos\alpha_{mj}\sqrt{\mathrm{sh}\,\alpha_{mj}\left|\sin\alpha_{mj}(\mathrm{sh}\,\alpha_{mj}\cos\alpha_{mj} - \mathrm{ch}\,\alpha_{mj}\sin\alpha_{mj})\right|},$$

$$0 \le m\psi \le \pi \tag{4.2.16}$$

and the natural modes can be expressed as

$$Q_{m,j}(x) = Q_0 Q(\alpha_{mj}, x) \tag{4.2.17}$$

where Q_0 denotes a complex factor which is dependent on the extreme boundary conditions of the original continuous beam and

$$Q(\alpha_{mj}, x) = A(\alpha_{mj})\sin(\frac{2\alpha_{mj}}{l}x) + C(\alpha_{mj})\mathrm{sh}(\frac{2\alpha_{mj}}{l}x)$$

$$+ i[B(\alpha_{mj})\cos(\frac{2\alpha_{mj}}{l}x) + D(\alpha_{mj})\mathrm{ch}(\frac{2\alpha_{mj}}{l}x)] \tag{4.2.18}$$

Let us consider a continuous beam with n spans and the extreme ends simply supported. The natural modes of the equivalent system with cyclic periodicity must satisfy anti-symmetric condition, *i.e.*,

$$W_{N-k+1}(x) = -W_k(-x) \qquad k=1,2,\ldots,n \tag{4.2.19}$$

where N=2n. The corresponding generalized displacement can be obtained by substituting equation (4.2.19) into equation (4.2.3b), as

$$q_m = \frac{1}{\sqrt{2n}} e^{i\frac{1}{2}m\psi} \sum_{k=1}^{n} \{\cos(k - \frac{1}{2})m\psi[W_k(x) - W_k(-x)]$$

$$-i\sin(k-\frac{1}{2})m\psi\left[W_k(x)+W_k(-x)\right]\}$$

$$m=1,2,\ldots,2n. \tag{4.2.20}$$

The real part of each term in the series on the right hand side of equation (4.2.20) is an odd function of x and the imaginary part is an even function of x. Compare equations (4.2.17), (4.2.18) with equation (4.2.20), the complex factor Q_0 may be obtained as

$$Q_0=e^{i\frac{1}{2}m\psi}. \tag{4.2.21}$$

Let

$$Q_{2n-m,j}=\overline{Q}_{m,j} \tag{4.2.22}$$

with other generalized displacements vanish.

Substituting equations (4.2.17), (4.2.18), (4.2.21) and (4.2.22) into equation (4.2.3a), yields the natural mode:

$$W_k(x)=\cos(k-\frac{1}{2})m\psi\left[A(\alpha_{mj})\sin(\frac{2\alpha_{mj}}{l}x)+C(\alpha_{mj})sh(\frac{2\alpha_{mj}}{l}x)\right]$$

$$-\sin(k-\frac{1}{2})m\psi\left[B(\alpha_{mj})\cos(\frac{2\alpha_{mj}}{l}x)+D(\alpha_{mj})ch(\frac{2\alpha_{mj}}{l}x)\right],$$

$$k=1,2,\ldots,n \tag{4.2.23}$$

where a real constant factor has been neglected. It can be verified that the natural mode expressed by equation (4.2.23) satisfy the extreme boundary conditions, *i.e.*,

$$W_1(-\frac{1}{2})=0,\quad W_1''(-\frac{1}{2})=0,\quad W_n(\frac{1}{2})=0,\quad W_n''(\frac{1}{2})=0. \tag{4.2.24}$$

Moreover the continuous beam with clamped extreme ends is considered. The natural modes of its equivalent system must satisfy symmetric condition, *i.e.*,

$$W_{N-k+1}(x)=W_k(-x) \qquad k=1,2,\ldots,n \tag{4.2.25}$$

where N=2n. The corresponding generalized displacement may be expressed as

$$q_m = \frac{-i}{\sqrt{2n}} e^{i\frac{1}{2}m\psi} \sum_{k=1}^{n} \{\sin(k-\frac{1}{2})m\psi[W_k(x)-W_k(-x)]$$

$$+i\cos(k-\frac{1}{2})m\psi[W_k(x)+W_k(-x)]\}$$

$$m=1,2,...,2n. \tag{4.2.26}$$

By comparison with equation (4.2.17) and (4.2.18), it is obvious that the complex factor Q_0 must be equal to $i\,e^{i\frac{1}{2}m\psi}$. Let

$$Q_{m,j}(x) = -i\,e^{i\frac{1}{2}m\psi} Q(\alpha_{mj}, x), \quad Q_{2n-m,j}(x) = \overline{Q}_{m,j}(x),$$

$$Q_{k,j} = 0 \qquad k \neq m \text{ and } 2n\text{-}m. \tag{4.2.27}$$

The natural mode may be obtained by substituting equation (4.2.27) into equation (4.2.3a), as

$$W_k(x) = \cos(k-\frac{1}{2})m\psi[B(\alpha_{mj})\cos(\frac{2\alpha_{mj}}{l}x) + D(\alpha_{mj})\text{ch}(\frac{2\alpha_{mj}}{l}x)]$$

$$+\sin(k-\frac{1}{2})m\psi[A(\alpha_{mj})\sin(\frac{2\alpha_{mj}}{l}x) + C(\alpha_{mj})\text{sh}(\frac{2\alpha_{mj}}{l}x)]$$

$$k=1,2,...,n. \tag{4.2.28}$$

The natural modes shown in equation (4.2.28) satisfy the clamped extreme end conditions, *i.e.*,

$$W_1(-\frac{l}{2}) = 0, \quad W_1'(-\frac{l}{2}) = 0, \quad W_n(\frac{l}{2}) = 0, \quad W_n'(\frac{l}{2}) = 0. \tag{4.2.29}$$

For each frequency parameter α_{mj} ($m \neq N$, N/2) corresponding to the cyclic periodic system, there are two independent modes shown in equations (4.2.23) and (4.2.28), but for each frequency parameter α_{mj} (m=N, N/2, j=1,2,...) there is only one natural mode. Either the former solution shown in equation (4.2.23) or the latter solution shown in equation (4.2.28) must be trivial solution, *i.e.,* when m=N $(\sin(k-\frac{1}{2})m\psi=0)$, N/2 $(\cos(k-\frac{1}{2})m\psi=0)$, $B(\alpha_{mj})=D(\alpha_{mj})=0$ or $A(\alpha_{mj})=C(\alpha_{mj})=0$. This conclusion may be verified easily. For m=N, N/2 the frequency equation (4.2.14) becomes

$$4\sin\alpha\ \text{sh}\alpha(\text{ch}\alpha\sin\alpha+\text{sh}\alpha\cos\alpha)=0 \qquad (m=N) \tag{4.2.30a}$$

and

$$4\cos\alpha\ \text{ch}\alpha(\text{ch}\alpha\sin\alpha-\text{sh}\alpha\cos\alpha)=0 \qquad (m=N/2) \tag{4.2.30b}$$

respectively.

The frequency equation (4.2.30a) is equivalent to that

$$\sin\alpha=0 \qquad (B(\alpha)=D(\alpha)=0) \tag{4.2.31a}$$

or

$$\text{ch}\alpha\sin\alpha+\text{sh}\alpha\cos\alpha=0 \qquad (A(\alpha)=C(\alpha)=0) \tag{4.2.31b}$$

and the frequency equation (4.2.30b) is equivalent to that

$$\cos\alpha=0 \qquad (A(\alpha)=C(\alpha)=0) \tag{4.2.32a}$$

or

$$\text{ch}\alpha\sin\alpha-\text{sh}\alpha\cos\alpha=0 \qquad (B(\alpha)=D(\alpha)=0). \tag{4.2.32b}$$

4.2.2 Dynamic Response

The governing equation for generalized displacements may be expressed by equations (4.2.4) with (4.2.6). Equation (4.2.4) must be solved subject to boundary condition (4.2.5) and initial condition (4.2.7a) with (4.2.7b).

One can now solve equation (4.2.4) using the complex mode method. The frequency equation and natural modes are shown as equations (4.2.14) (4.2.17) and (4.2.18) where Q_0 is an undetermined multiplier, which can be determined by the normalization condition

$$\int_{-\frac{l}{2}}^{\frac{l}{2}} \rho A\, Q_{m,j}(x)\,\overline{Q}_{m,j}(x) = 1\,. \tag{4.2.33}$$

Substituting equations (4.2.17), (4.2.18) and (4.2.16) into equation (4.2.33) yields

$$Q_0 = 1\Big/\sqrt{\rho A l\, M(\alpha_{mj})} \tag{4.2.34}$$

where

$$M(\alpha_{mj}) = \frac{1}{2}\left[A^2(\alpha_{mj}) + B^2(\alpha_{mj}) - C^2(\alpha_{mj}) + D^2(\alpha_{mj})\right]. \tag{4.2.35}$$

The normalized mode can be obtained by inserting equations (4.2.34) and (4.2.35) into equation (4.2.17). This mode is applicable for $0 \le m\psi \le \pi$ $(\text{i.e.}, 0 \le m \le N/2)$.

Upon examining the mathematical characteristics of the eigenvalue equations (4.2.8) and (4.2.9), it is apparent that $\alpha_{(N-m)j} = \alpha_{mj}$ and $Q_{(N-m),j} = \overline{Q}_{m,j}$.

The normalized modes satisfy the orthonormality conditions:

$$\int_{-\frac{l}{2}}^{\frac{l}{2}} \rho A\, Q_{m,j}(x)\,\overline{Q}_{m,k}(x)dx = \delta_{jk} \tag{4.2.36a}$$

$$\int_{-\frac{l}{2}}^{\frac{l}{2}} EI \frac{d^4 Q_{m,j}(x)}{dx^4}\,\overline{Q}_{m,k}(x)dx = \omega_{mj}^2\,\delta_{jk} \tag{4.2.36b}$$

where δ_{jk} is the Kronecker symbol. The natural frequency ω_{mj} can be expressed as

$$\omega_{mj} = \Omega_0 \alpha_{mj}^2 \tag{4.2.37}$$

where

$$\Omega_0 = \frac{4}{l^2}\sqrt{\frac{EI}{\rho A}}\,. \tag{4.2.38}$$

According to the expansion theorem, the solution for q_m of equation (4.2.4) can be expressed as a series of natural modes, *i.e.*,

$$q_m(x,t) = \sum_{j=1}^{\infty} a_{mj}(t)\, Q_{m,j}(x). \tag{4.2.39}$$

Since it has satisfied the boundary conditions (4.2.5) previously, substituting equation (4.2.39) into equation(4.2.4) and carrying out the multiplication by $\overline{Q}_{m,j}(x)$, and finally integrating with respect to the space variable x between limits -l/2 and l/2 leads to the ordinary differential equation.

$$\omega_{mj}^2 a_{mj}(t) + \ddot{a}_{mj}(t) = \xi(m\psi, \alpha_{mj}, t) \tag{4.2.40}$$

where

$$\xi(m\psi, \alpha_{mj}, t) = \int_{-\frac{l}{2}}^{\frac{l}{2}} f_m(x,t)\, \overline{Q}_{m,j}(x) dx. \tag{4.2.41}$$

The solution for a_{mj} of equation (4.2.40) can be expressed as a Duhamel integral

$$a_{mj}(t) = \frac{1}{\omega_{mj}} \int_0^t \xi(m\psi,\ \alpha_{mj}, \tau) \sin\omega_{mj}(t-\tau) d\tau + a_{mj}(0) \cos\omega_{mj} t + \frac{\dot{a}_{mj}(0)}{\omega_{mj}} \sin\omega_{mj} t \tag{4.2.42}$$

where $a_{mj}(0)$, $\dot{a}_{mj}(0)$ denotes initial value of mode coordinate and its velocity, respectively. They may be expressed as

$$a_{mj}(0) = \int_{-\frac{l}{2}}^{\frac{l}{2}} \rho A\, \overline{Q}_{m,j}(x)\, q_{m0}\, dx, \quad \dot{a}_{mj}(0) = \int_{-\frac{l}{2}}^{\frac{l}{2}} \rho A\, \overline{Q}_{m,j}(x)\, \dot{q}_{m0}\, dx \tag{4.2.43}$$

in which q_{m0} and $\dot{q}_{m0}$ have been defined as equation (4.2.7b). The response of deflection can be obtained by substituting equations (4.2.39) and (4.2.42) into equation (4.2.3a).

To take an example [8], the response of an infinite prismatic beam on periodic rigid supports and subjected to a moving force at uniform velocity (see Figure 4.2.1) will be studied in the following section.

It is well known that, unlike a finite periodic structure, the frequencies of an infinite periodic beam are continuously distributed in every natural frequency band. Consequently, the mode methods for finite and infinite periodic structures are distinct in mathematical form. A new mathematical skill must be applied to determine the response of an infinite continuous beam under a moving load using the mode method.

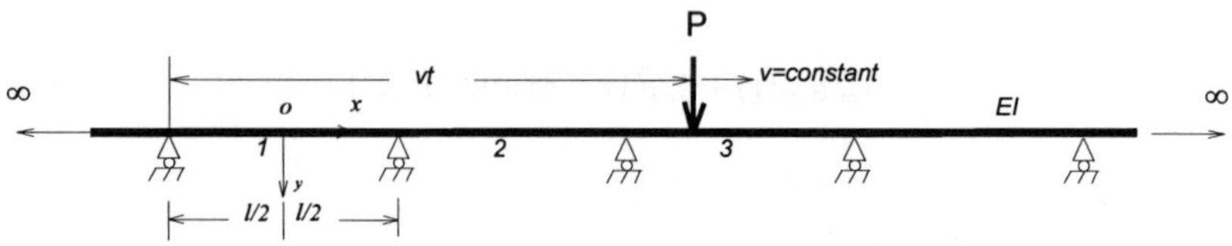

Figure 4.2.1 Infinite continuous beam subjected to a moving load

At the outset, the response of a rotationally periodic beam having a finite number of spans, say N, subjected to a moving load is studied by using the U-transformation and mode method as stated above. Then by a limiting process (with $N \to \infty$), the limiting solution becomes applicable for an infinite continuous beam under the same loading condition.

Consider the zero initial conditions, *i.e.*,

$$W_k(x,0) = 0, \quad \dot{W}_k(x,0) = 0 \quad \text{or} \quad q_{m0} = 0, \quad \dot{q}_{m0} = 0$$

$$\text{or} \quad a_{mj}(0) = 0, \quad \dot{a}_{mj}(0) = 0, \tag{4.2.44}$$

It is assumed that the concentrated load of magnitude P acts at the left end of the first span at the initial time and thenceforth moves toward the right at an uniform velocity V. For such a case, the loading functions can be expressed as

$$F_k(x,t)=\begin{cases} P\delta\left[x+l(k-\frac{1}{2}-\frac{t}{t_0})\right], & (k-1)t_0<t\le kt_0 \\ 0, & t\le(k-1)t_0 \text{ or } t>kt_0 \end{cases}$$

$$k=1,2,...,N \tag{4.2.45}$$

where

$$t_0=l/V \tag{4.2.46}$$

is the time to move through a span.

Substituting equation (4.2.45) into equation (4.2.6) , yields

$$f_m(x,t)=\frac{P}{\sqrt{N}}e^{-i(k-1)m\psi}\,\delta[x+l(k-\frac{1}{2}-\frac{t}{t_0})],\quad ,$$

$$(k-1)\,t_0<t\le k\,t_0 \qquad k=1,2,...,N. \tag{4.2.47}$$

Substituting equations (4.2.47), (4.2.17) into equation (4.2.41) will lead to the following result

$$\xi(m\psi,\alpha_{mj},t)=\frac{Q_0 P}{\sqrt{N}}e^{-i(k-1)m\psi}\,\overline{Q}\;(\alpha_{mj},x)\bigg|_{x=(\frac{t}{t_0}-k+\frac{1}{2})l},$$

$$(k-1)\,t_0<t\le k\,t_0,\qquad k=1,2,...,N \tag{4.2.48}$$

where Q_0 has been defined in equation (4.2.34) and $Q(\alpha_{mj},x)$ is as shown in equations (4.2.18) and (4.2.16).

Substituting equations (4.2.37), (4.2.48), (4.2.18) and (4.2.44) into equation (4.2.42) and integrating with respect to τ yields

$$a_{mj}(t)=\frac{P}{\sqrt{N}\sqrt{\rho AlM(\alpha_{mj})}}\eta(m\psi,\alpha_{mj},t) \tag{4.2.49}$$

where

$$\eta(m\psi,\alpha_{mj},t)=\frac{1}{\Omega_0^2\alpha_{mj}^2}\left[A(\alpha_{mj})I_1(m\psi,\alpha_{mj},t)-iB(\alpha_{mj})I_2(m\psi,\alpha_{mj},t)\right.$$
$$\left.+C(\alpha_{mj})I_3(m\psi,\alpha_{mj},t)-iD(\alpha_{mj})I_4(m\psi,\alpha_{mj},t)\right] \qquad (4.2.50)$$

in which

$$I_1(\theta,\alpha,t)=\frac{2}{\alpha_0^2-\alpha^2}\left[\frac{\alpha_0}{\alpha}\sin(\frac{\alpha^2}{\alpha_0})\cos\alpha-\sin\alpha\cos(\frac{\alpha^2}{\alpha_0})\right]$$
$$\times\sum_{k=1}^{n-1}\left\{e^{-i(k-1)\theta}\cos[(\frac{2t}{t_0}-2k+1)\frac{\alpha^2}{\alpha_0}]\right\}+\frac{e^{-i(n-1)\theta}}{\alpha_0^2-\alpha^2}$$
$$\times\left\{\frac{\alpha_0}{\alpha}\cos\alpha\sin[(\frac{t}{t_0}-n+1)\frac{2\alpha^2}{\alpha_0}]-\sin[(\frac{2t}{t_0}-2n+1)\alpha]-\sin\alpha\cos[(\frac{t}{t_0}-n+1)\frac{2\alpha^2}{\alpha_0}]\right\}$$

$$I_2(\theta,\alpha,t)=\frac{2}{\alpha_0^2-\alpha^2}\left[\frac{\alpha_0}{\alpha}\sin\alpha\cos(\frac{\alpha^2}{\alpha_0})-\sin(\frac{\alpha^2}{\alpha_0})\cos\alpha\right]$$
$$\times\sum_{k=1}^{n-1}\left\{e^{-i(k-1)\theta}\sin[(\frac{2t}{t_0}-2k+1)\frac{\alpha^2}{\alpha_0}]\right\}+\frac{e^{-i(n-1)\theta}}{\alpha_0^2-\alpha^2}$$
$$\times\left\{\frac{\alpha_0}{\alpha}\sin\alpha\sin[(\frac{t}{t_0}-n+1)\frac{2\alpha^2}{\alpha_0}]-\cos[(\frac{2t}{t_0}-2n+1)\alpha]+\cos\alpha\cos[(\frac{t}{t_0}-n+1)\frac{2\alpha^2}{\alpha_0}]\right\}$$

$$I_3(\theta,\alpha,t)=\frac{2}{\alpha_0^2+\alpha^2}\left[\mathrm{sh}\alpha\cos(\frac{\alpha^2}{\alpha_0})-\frac{\alpha_0}{\alpha}\mathrm{ch}\alpha\sin(\frac{\alpha^2}{\alpha_0})\right]$$
$$\times\sum_{k=1}^{n-1}\left\{e^{-i(k-1)\theta}\cos[(\frac{2t}{t_0}-2k+1)\frac{\alpha^2}{\alpha_0}]\right\}+\frac{e^{-i(n-1)\theta}}{\alpha_0^2+\alpha^2}$$
$$\times\left\{\mathrm{sh}[(\frac{2t}{t_0}-2n+1)\alpha]-\frac{\alpha_0}{\alpha}\mathrm{ch}\alpha\sin[(\frac{t}{t_0}-n+1)\frac{2\alpha^2}{\alpha_0}]+\mathrm{sh}\alpha\cos[(\frac{t}{t_0}-n+1)\frac{2\alpha^2}{\alpha_0}]\right\}$$

$$I_4(\theta,\alpha,t)=\frac{2}{\alpha_0^2+\alpha^2}\left[\frac{\alpha_0}{\alpha}\mathrm{sh}\alpha\cos(\frac{\alpha^2}{\alpha_0})+\mathrm{ch}\alpha\sin(\frac{\alpha^2}{\alpha_0})\right]$$

$$\times \sum_{k=1}^{n-1} \left\{ e^{-i(k-1)\theta} \sin[(\frac{2t}{t_0} - 2k + 1)\frac{\alpha^2}{\alpha_0}] \right\} + \frac{e^{-i(n-1)\theta}}{\alpha_0^2 + \alpha^2}$$

$$\times \left\{ \frac{\alpha_0}{\alpha} \mathrm{sh}\alpha \sin[(\frac{t}{t_0} - n + 1)\frac{2\alpha^2}{\alpha_0}] + \mathrm{ch}[(\frac{2t}{t_0} - 2n + 1)\alpha] - \mathrm{ch}\alpha \cos[(\frac{t}{t_0} - n + 1)\frac{2\alpha^2}{\alpha_0}] \right\}$$

$$(n-1)\, t_0 < t \le n\, t_0, \quad n=1,2,... \tag{4.2.51}$$

in which

$$\alpha_0 = \frac{lV}{2}\sqrt{\frac{\rho A}{EI}}\,. \tag{4.2.52}$$

α_0 is a non-dimensional parameter characterizing the velocity V. Substituting equations (4.2.39), (4.2.49), (4.2.17) and (4.2.34) into equation (4.2.3a) gives

$$W_k(x,t) = \frac{P}{N}\sum_{m=1}^{N} e^{i(k-1)m\psi} \sum_{j=1}^{\infty} \frac{\eta(m\psi, \alpha_{mj}, t)}{\rho A l M(\alpha_{mj})} Q(\alpha_{mj}, x)$$

$$= \frac{P}{\rho A l}\sum_{j=1}^{\infty}\{\frac{1}{N}\sum_{m=1}^{N} \mathrm{Re}\left[e^{i(k-1)m\psi}\, \eta(m\psi, \alpha_{mj}, t) Q(\alpha_{mj}, x) / M(\alpha_{mj})\right]\}\,. \tag{4.2.53}$$

By letting N approach infinity, the limit of the series summation in the curly brackets on the right hand side of equation (4.2.53) becomes

$$\lim_{N\to\infty}\frac{1}{N}\sum_{m=1}^{N} \mathrm{Re}\left[e^{i(k-1)m\psi}\, \eta(m\psi, \alpha_{mj}, t) Q(\alpha_{mj}, x) / M(\alpha_{mj})\right]$$

$$= \frac{1}{\pi}\int_0^{\pi} \mathrm{Re}\left[H(\theta, \alpha_j, x, t)\right] d\theta, \tag{4.2.54}$$

where

$$H(\theta, \alpha_j, x, t) = e^{i(k-1)\theta}\, \eta(\theta, \alpha_j, t) Q(\alpha_j, x) / M(\alpha_j)\,. \tag{4.2.55}$$

α_j denotes the j-th positive root of the following equation for a given θ:

$$\cos\theta\,(\mathrm{sh}2\alpha - \sin 2\alpha) + \mathrm{ch}2\alpha \sin 2\alpha - \mathrm{sh}2\alpha \cos 2\alpha = 0\,. \qquad (4.2.56)$$

Obviously, α_j is dependent on θ , but it cannot be expressed as an explicit function of θ. The value of α_j for an arbitrary θ is located in the j-th natural frequency band.

The dynamic deflection of the continuous beam with infinite spans can now be expressed as

$$W_k^*(x,t) = \lim_{N\to\infty} W_k(x,t) = \frac{P}{\rho A l \pi} \sum_{j=1}^{\infty} \int_0^{\pi} \mathrm{Re}\left[H(\theta,\alpha_j,x,t)\right] d\theta\,. \qquad (4.2.57)$$

Substituting equations (4.2.55), (4.2.50) into the right hand side of equation (4.2.57) and transforming the integral variable from θ to α yields

$$W_k^*(x,t) = \frac{P l^3}{16\pi EI} \sum_{j=1}^{\infty} \int_{\alpha_j(0)}^{\alpha_j(\pi)} \frac{\Phi(\alpha)}{\alpha^2 M(\alpha)} \mathrm{Re}\Big\{e^{i(k-1)\theta}\big[A(\alpha)\, I_1(\theta,\alpha,t)$$

$$- iB(\alpha)\, I_2(\theta,\alpha,t) + C(\alpha)\, I_3(\theta,\alpha,t) - iD(\alpha)\, I_4(\theta,\alpha,t)\big] Q(\alpha,x)\Big\} d\alpha$$

$$(4.2.58)$$

where

$$\theta = \arccos \frac{\mathrm{sh}2\alpha \cos 2\alpha - \mathrm{ch}2\alpha \sin 2\alpha}{\mathrm{sh}2\alpha - \sin 2\alpha}, \qquad 0 \le \theta \le \pi \qquad (4.2.59)$$

$$\Phi(\alpha) = \frac{d\theta}{d\alpha}$$

$$= \frac{2\left[(\mathrm{ch}2\alpha + \cos 2\alpha)(\mathrm{ch}2\alpha \sin 2\alpha + \mathrm{sh}2\alpha \cos 2\alpha) - 2(\mathrm{sh}2\alpha + \sin 2\alpha)\right]}{(\mathrm{sh}2\alpha - \sin 2\alpha)\sqrt{2\mathrm{sh}2\alpha \sin 2\alpha(\mathrm{ch}2\alpha \cos 2\alpha - 1)}}\,.$$

$$(4.2.60)$$

Functions $M(\alpha)$, $I_1(\theta,\alpha,t),\cdots,I_4(\theta,\alpha,t),A(\alpha),\cdots,D(\alpha)$ are defined by equations (4.2.35), (4.2.51) and (4.2.16) respectively. $\alpha_j(0)$ $(\alpha_j(\pi))$ is the upper (lower) limit or the lower (upper) limit of the j-th frequency band.

By using a numerical method, the roots for α can be determined from equation (4.2.56) in which θ takes the values $0, \pi$ respectively. The first six values are shown in Table 4.2.1

Table 4.2.1. The upper and lower limits of the first six natural frequency bands

$\alpha_1(0)$	$\alpha_2(0)$	$\alpha_3(0)$	$\alpha_4(0)$	$\alpha_5(0)$	$\alpha_6(0)$
2.36502	π	5.49780	2π	8.63938	3π
$\alpha_1(\pi)$	$\alpha_2(\pi)$	$\alpha_3(\pi)$	$\alpha_4(\pi)$	$\alpha_5(\pi)$	$\alpha_6(\pi)$
$\pi/2$	3.92660	$3\pi/2$	7.06858	$5\pi/2$	10.21017

The expression on the right hand side of equation (4.2.58) includes two independent parameters k and α_0. k denotes the ordinal number of the span considered and α_0 has been defined by equation (4.2.52). For given arbitrary values of k, α_0, x and t, the dynamic deflection $W_k^*(x,t)$ can be calculated from equation (4.2.58) by a numerical integration method.

As a numerical example, the response of the deflection at the mid-point of the 6th span is considered over a period in which the concentrated load moves from the left end of the first span to the right end of the 13th span (*i.e.*, k=6, x=0 and $0<t\leq 13t_0$). The resonant case is considered; $\alpha_0=\pi/2$, i.e., . $V=(\pi/l)\sqrt{EI/(\rho A)}$ The dynamic deflection curve of the 6th span at the mid-point versus time is shown in Figure 4.2.2. The first five terms of the series on the right hand side of equation (4.2.58) are taken into account and the remaining terms are neglected. (A very accurate result can already be obtained by taking the first term only.) The rectangular integration formula is adopted for calculating the definite integral. The value of the integrand in the rectangular integration formula is defined by the value of integrand at the mid-point of each sub-interval. When using the rectangular integration formula, each integration interval is divided into 200 sub-intervals. The curve of the response is calculated step by step with each time step being equal to $0.2t_0$.

The maximum dynamic deflection is $0.036\mathrm{P}\,l^3/\mathrm{EI}$. The ratio of the maximum dynamic deflection to the static deflection shown in equation (4.1.30) equals 3.3.

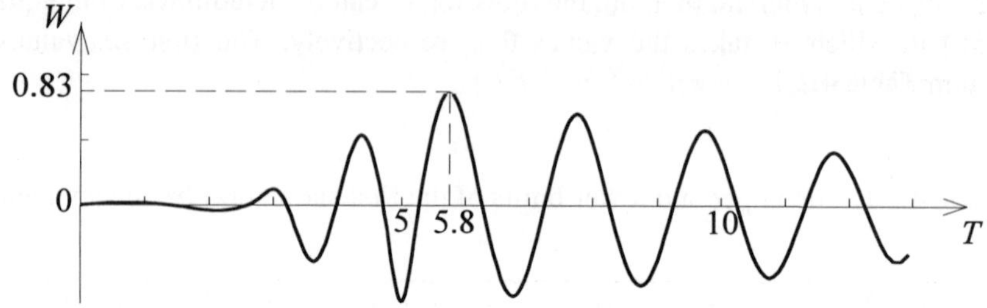

Figure 4.2.2 Non-dimensional deflection W versus non-dimensional time T

$$W_6(0,t) = (\mathrm{P}\,l^3/16\pi\mathrm{EI})\mathrm{W}, \qquad t = t_0\mathrm{T}$$

4.3 Continuous Beams with Periodic Elastic Supports

4.3.1 Static Problem

Consider a beam with uniform flexural rigidity EI and mass per unit length ρA that runs over infinite number of elastic supports as shown in Figure 4.3.1. Let the span length between any two adjacent supports be denoted by l. We shall consider the case that the displacements , slopes and moments at the two infinite extreme ends are equal to zero. Therefore it may be imagined that the two extreme ends are connected in the same way as with other adjacent spans. The structure considered can be regarded as a rotationally periodic structure with infinite number of substructures.

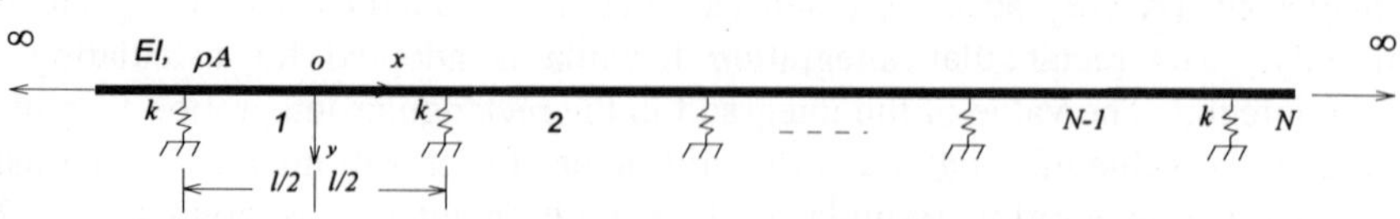

Figure 4.3.1 Infinite continuous beam on periodic elastic supports

At first, the static analysis[9] of a cyclic periodic beam having N number of spans subjected to arbitrary loads is studied. The N independent equilibrium equations for the N spans must be solved subject to the continuity requirements over all supports. Secondly, by a limiting process with N approaching infinity, the results can be generalized to render a solution for the case of an infinite continuous beam with periodic elastic supports and same loads.

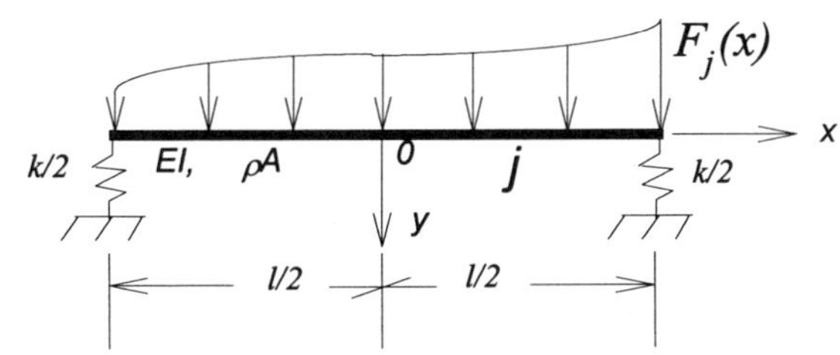

Figure 4.3.2. Substructure j with local co-ordinate system

In each span a local co-ordinate system oxy is established with the origin o at its mid-span as shown in Figure 4.3.2. The equilibrium equations for all substructures are of the same form, *i.e.*,

$$EI\frac{d^4 W_j(x)}{dx^4} = F_j(x), \qquad -\frac{l}{2} < x < \frac{l}{2}, \qquad j=1,2,..,N \tag{4.3.1}$$

where $W_j(x)$ is the transverse displacement and $F_j(x)$ the loading function for the j-th substructure. Continuity across the elastic supports requires the following conditions to be satisfied

$$W_j(\frac{l}{2}) = W_{j+1}(-\frac{l}{2}), \quad W_j'(\frac{l}{2}) = W_{j+1}'(-\frac{l}{2}), \quad W_j''(\frac{l}{2}) = W_{j+1}''(-\frac{l}{2}),$$

$$-EIW_j'''(\frac{l}{2}) + \frac{k}{2}W_j(\frac{l}{2}) = -EIW_{j+1}'''(-\frac{l}{2}) - \frac{k}{2}W_{j+1}(-\frac{l}{2}),$$

$$j=1,2,...,N \tag{4.3.2}$$

where $W_{N+1} \equiv W_1$ and k denotes the stiffness constant for each elastic support.

Applying the U-transformation to equations (4.3.1) and (4.3.2), let

$$W_j(x) = \frac{1}{\sqrt{N}} \sum_{m=1}^{N} e^{i(j-1)m\psi} q_m(x), \qquad j=1,2,\ldots,N \tag{4.3.3a}$$

or

$$q_m(x) = \frac{1}{\sqrt{N}} \sum_{j=1}^{N} e^{-i(j-1)m\psi} W_j(x), \qquad m=1,2,\ldots,N \tag{4.3.3b}$$

with $\psi = 2\pi/N$ and $i = \sqrt{-1}$. The equations (4.3.1) and (4.3.2) becomes

$$EI \frac{d^4 q_m(x)}{d x^4} = f_m(x), \qquad -\frac{1}{2} < x < \frac{1}{2} \qquad m=1,2,\ldots,N. \tag{4.3.4}$$

and

$$q_m(\frac{1}{2}) = e^{im\psi} q_m(-\frac{1}{2}), \quad q'_m(\frac{1}{2}) = e^{im\psi} q'_m(-\frac{1}{2}), \quad q''_m(\frac{1}{2}) = e^{im\psi} q''_m(-\frac{1}{2}),$$

$$q'''_m(\frac{1}{2}) - \frac{K}{2l^3} q_m(\frac{1}{2}) = e^{im\psi} \left[q'''_m(-\frac{1}{2}) + \frac{K}{2l^3} q_m(-\frac{1}{2}) \right],$$

$$m=1,2,\ldots,N. \tag{4.3.5}$$

with $K = kl^3/EI$ and

$$f_m(x) = \frac{1}{\sqrt{N}} \sum_{j=1}^{N} e^{-i(j-1)m\psi} F_j(x)\,. \tag{4.3.6}$$

In a general way , the solution for $q_m(x)$ of equations (4.3.4), (4.3.5) and (4.3.6) may be expressed as

$$q_m(x) = q(m\psi, x)/\sqrt{N}\,, \qquad q_{N-m}(x) = \bar{q}_m(x)\,. \tag{4.3.7}$$

Substituting equation (4.3.7) into equation (4.3.3a) gives

$$W_j(x) = \frac{1}{N}\sum_{m=1}^{N} e^{i(j-1)m\psi} q(m\psi, x) = \frac{1}{N}\sum_{m=1}^{N} \mathrm{Re}\left[e^{i(j-1)m\psi} q(m\psi, x)\right]. \tag{4.3.8}$$

In this form, $W_j(x)$ has the interpretation of the average value of the function $\mathrm{Re}\left[e^{i(j-1)\theta} q(\theta, x)\right]$ over the interval $0 \le \theta \le 2\pi$. By letting N approach infinity, the limit of the summation of the series on the right hand side of equation (4.3.8) becomes a definite integral, *i.e.*,

$$W_j(x) = \lim_{N\to\infty} W_j(x) = \frac{1}{2\pi}\int_0^{2\pi} \mathrm{Re}\left[e^{i(j-1)\theta} q(\theta, x)\right] d\theta\,, \qquad j=1,2,\ldots,N. \tag{4.3.9}$$

Several specific examples will be discussed in detail below:

Uniformly Distributed Load

Let us discuss the case of an uniformly distributed load p applied on a span, say the 1st span, *i.e.*,

$$F_1(x) = p, \qquad F_2(x) = F_3(x) = \cdots = F_N(x) = 0\,. \tag{4.3.10}$$

Substituting equation (4.3.10) into equation (4.3.6) yields

$$f_m(x) = p/\sqrt{N}\,, \qquad m=1,2,\ldots,N. \tag{4.3.11}$$

The solution for $q_m(x)$ of equations (4.3.4), (4.3.11) and (4.3.5) can be expressed as

$$q_m(x) = \frac{p l^4}{384 EI\sqrt{N}}\left(C_{m0} + C_{m1}\frac{x}{l} + C_{m2}\frac{x^2}{l^2} + C_{m3}\frac{x^3}{l^3} + 16\frac{x^4}{l^4}\right), \tag{4.3.12}$$

where

$$C_{m0} = 1 + [8(5 - K_1)\cos m\psi + 8(K_1 + 7)]/\Delta m\,,$$

$$C_{m1} = 8i\sin m\psi(\cos m\psi + 11 - 2K_1)/\Delta m\,,$$

$$C_{m2} = 24 - 96[(K_1 - 1)\cos m\psi + 1 + 3K_1]/\Delta m\,,$$

$$C_{m3} = 32i\sin m\psi(\cos m\psi + 2K_1 - 1)/\Delta m\,,$$

$$\Delta m = \cos^2 m\psi + (4K_1 - 2)\cos m\psi + 8K_1 + 1\,,$$

$$K_1 = K/48 \qquad (4.3.13)$$

The deflection functions of the infinite beam with periodic elastic supports under uniform load applying on the 1st span can be obtained by substituting $q(\theta, x) = \sqrt{N}\, q_m(x)\big|_{m\psi=\theta}$ into equation (4.3.9) for j=1,2, as

$$W_1(x) = \frac{pl^4}{384EI}\left\{1 + \frac{4}{3K_1K_3}\left(\frac{K_1+1}{K_2} + 6\right) + \left[24 - \frac{16}{K_1K_3}\left(\frac{K_1+1}{K_2} + 2K_1\right)\right]\frac{x^2}{l^2} + 16\frac{x^4}{l^4}\right\},$$

$$-\frac{l}{2} \le x \le \frac{l}{2}\,. \qquad (4.3.14a)$$

$$W_2(x) = \frac{pl^4}{48EI}\left\{5 - K_1 + \frac{1}{K_3}\left(K_1 - 5 + \frac{1}{K_1}\right) + \frac{1}{6K_2K_3}\left(2K_1 - 9 - \frac{11}{K_1}\right)\right.$$

$$-\left[6K_1 - 13 + \frac{14 - 6K_1}{K_3} + \frac{1}{K_2K_3}\left(2 - 2K_1 + \frac{4}{K_1}\right)\right]\frac{x}{l}$$

$$-12\left[K_1 - 1 - \frac{3K_1 - 4}{3K_3} - \frac{1}{6K_2K_3}\left(2K_1 + 1 - \frac{1}{K_1}\right)\right]\frac{x^2}{l^2}$$

$$\left.-\left[8K_1 - 4 + \frac{8(1 - K_1)}{K_3} - \frac{8(K_1 + 1)}{2K_2K_3}\right]\frac{x^3}{l^3}\right\}, \quad -\frac{l}{2} \le x \le \frac{l}{2} \qquad (4.3.14b)$$

where

$$K_2 = \sqrt{\frac{K_1+1}{3K_1}}, \qquad K_3 = \sqrt{\frac{4}{3}+2K_2}\,. \tag{4.3.15}$$

The bending moment for loaded span can be expressed as

$$M_1(x) = -EIW_1''(x) = -\frac{pl^2}{24}\left[3 - \frac{2}{K_1K_3}\left(\frac{K_1+1}{K_2}+2K_1\right)+12\frac{x^2}{l^2}\right] \tag{4.3.16}$$

The maximum deflection and moment occur at mid-span with magnitude:

$$W_{max} = W_1(0) = \frac{pl^4}{384EI}\left[1+\frac{4}{3K_1K_3}\left(\frac{K_1+1}{K_2}+6\right)\right], \tag{4.3.17}$$

and

$$M_{max} = -EIW_1''(0) = \frac{pl^2}{24}\left[\frac{2}{K_1K_3}\left(\frac{K_1+1}{K_2}+2K_1\right)-3\right]. \tag{4.3.18}$$

By letting K or K_1 approach infinity, equations (4.3.14) and (4.3.16) become

$$\lim_{K\to\infty} W_1(x) = \frac{pl^4}{384EI}\left(2\sqrt{3}-1-8\sqrt{3}\frac{x^2}{l^2}+16\frac{x^4}{l^4}\right), \qquad -\frac{1}{2}\le x\le\frac{1}{2},$$

$$\lim_{K\to\infty} W_2(x) = \frac{pl^4}{48EI}\left[\frac{3}{4}-\frac{\sqrt{3}}{2}+\left(1-\frac{\sqrt{3}}{2}\right)\frac{x}{l}+(2\sqrt{3}-3)\frac{x^2}{l^2}-(4-2\sqrt{3})\frac{x^3}{l^3}\right],$$

$$-\frac{1}{2}\le x\le\frac{1}{2} \tag{4.3.19}$$

and

$$\lim_{K\to\infty} M_1(x) = \frac{pl^2}{24}\left(\sqrt{3}-12\frac{x^2}{l^2}\right), \qquad -\frac{1}{2}\le x\le\frac{1}{2}. \tag{4.3.20}$$

The above results are applicable to the infinite continuous beam with periodic roller supports, which are the same as those shown in equations (4.1.19)-(4.1.21).

Concentrated Load

Let a concentrated load of magnitude P act at the midpoint of the 1st span. The loading functions can be expressed as

$$F_1(x) = P\delta(x), \qquad F_j(x) = 0, \qquad j=2,3,\ldots,N. \tag{4.3.21}$$

Substituting equation (4.3.21) into equation (4.3.6) yields

$$f_m(x) = \frac{P}{\sqrt{N}}\delta(x), \qquad m=1,2,\ldots,N. \tag{4.3.22}$$

The solution for $q_m(x)$ of equations (4.3.4), (4.3.22) and (4.3.5) may be obtained without any difficulty, then substituting $q(\theta,x) = \sqrt{N}\, q_m(x)\big|_{m\psi=\theta}$ into equation (4.3.9) and integrating the definite integral on the right hand side of equation (4.3.9) result in

$$W_1(x) = \frac{Pl^3}{12EI}\left[\frac{1}{12K_1K_3}(3+\frac{3}{4}K_1+\frac{K_1+1}{K_2}) - \frac{1}{4K_1K_3}(3K_1+\frac{2K_1+2}{K_2})\frac{x^2}{l^2}+\frac{|x|^3}{l^3}\right],$$

$$-\frac{l}{2} \le x \le \frac{l}{2} \tag{4.3.23a}$$

$$W_2(x) = \frac{Pl^3}{12EI}\left\{1-\frac{K_1}{8}+\frac{1}{16K_3}(2K_1-15+\frac{4}{K_1})+\frac{1}{24K_2K_3}(K_1-9-\frac{10}{K_1})\right.$$

$$+3\left[1-\frac{K_1}{4}-\frac{4-K_1}{4K_3}+\frac{1}{12K_2K_3}(K_1-3-\frac{4}{K_1})\right]\frac{x}{l}$$

$$+3\left[1-\frac{K_1}{2}+\frac{2K_1-5}{4K_3}+\frac{1}{6K_2K_3}(K_1-\frac{1}{K_1})\right]\frac{x^2}{l^2}$$

$$+(1-K_1+\frac{K_1-2}{K_3}+\frac{K_1+1}{3K_2K_3})\frac{x^3}{l^3}\Big\}, \quad -\frac{l}{2}\leq x\leq\frac{l}{2}. \tag{4.3.23b}$$

The bending moment for the loaded span may be expressed as

$$M_1(x)=-EI\,W_1''(x)=\frac{Pl}{12}\left[\frac{1}{2K_1K_3}(3K_1+\frac{2K_1+2}{K_2})-\frac{6|x|}{l}\right],$$

$$-\frac{l}{2}\leq x\leq\frac{l}{2}. \tag{4.3.24}$$

The maximum deflection and moment occur at mid-span with magnitude

$$W_{max}=W_1(0)=\frac{Pl^3}{144EI\,K_1K_3}(3+\frac{3}{4}K_1+\frac{K_1+1}{K_2}) \tag{4.3.25}$$

and

$$M_{max}=M_1(0)=\frac{Pl}{24\,K_1K_3}(3K_1+\frac{2K_1+2}{K_2}). \tag{4.3.26}$$

By letting K approach infinity, the limits of equations (4.3.23a), (4.3.23b) and (4.3.24) become

$$\lim_{K\to\infty}W_1(x)=\frac{Pl^3}{384EI}\left[3\sqrt{3}-1-12(\sqrt{3}+1)\frac{x^2}{l^2}+32\frac{|x|^3}{l^3}\right],$$

$$-\frac{l}{2}\leq x\leq\frac{l}{2} \tag{4.3.27a}$$

$$\lim_{K\to\infty}W_2(x)=\frac{Pl^3}{128EI}\left[3-2\sqrt{3}+2(2-\sqrt{3})\frac{x}{l}+4(2\sqrt{3}-3)\frac{x^2}{l^2}-8(2-\sqrt{3})\frac{x^3}{l^3}\right],$$

$$-\frac{l}{2}\leq x\leq\frac{l}{2} \tag{4.3.27b}$$

and

$$\lim_{K\to\infty} M_1(x) = \frac{Pl}{16}(\sqrt{3}+1-8\frac{|x|}{l}), \qquad -\frac{l}{2} \le x \le \frac{l}{2}. \tag{4.3.28}$$

The limits of maximum deflection and moment are

$$\lim_{K\to\infty} W_1(0) = \frac{(3\sqrt{3}-1)Pl^3}{384EI} \tag{4.3.29}$$

and

$$\lim_{K\to\infty} M_1(0) = \frac{(\sqrt{3}+1)Pl}{16}. \tag{4.3.30}$$

Equations (4.3.27a), (4.3.29) and (4.3.30) are the same as equations (4.1.29), (4.1.30) and (4.1.31), respectively.

It is well known that a model of the beam on elastic foundation can be used instead of the beam supported by discrete springs in engineering. We discuss now the relative error due to the use of the approximate model.

Firstly, we consider the case of an uniformly distributed load p applied on the section with length l of the infinite beam on elastic foundation.

The maximum deflection occurs at the midpoint of loaded section and its magnitude is

$$\overline{W}_{max} = \frac{p}{\overline{K}}(1-e^{-\beta l/2}\cos\frac{\beta l}{2}) \tag{4.3.31}$$

where

$$\overline{K} = \frac{k}{l} = \frac{KEI}{l^4}, \qquad \beta = \sqrt[4]{\overline{K}/4EI} = \sqrt[4]{K/4}\Big/l\ . \tag{4.3.32}$$

The maximum moment also occurs at midpoint of loaded section with magnitude:

$$\overline{M}_{max} = \frac{p}{2\beta^2}e^{-\beta l/2}\sin\frac{\beta l}{2}. \tag{4.3.33}$$

Substituting equation (4.3.32) into equations (4.3.31) and (4.3.33) yields

$$\overline{W}_{max} = \frac{p l^4}{EIK}[1 - e^{-\sqrt[4]{K/64}} \cos(\sqrt[4]{K/64})], \tag{4.3.34}$$

and

$$\overline{M}_{max} = \frac{p l^2}{\sqrt{K}} e^{-\sqrt[4]{K/64}} \sin(\sqrt[4]{K/64}), \tag{4.3.35}$$

respectively.

The exact values of the maximum deflection and moment are shown in equations (4.3.17) and (4.3.18). The relative error of the maximum deflection and bending moment may be denoted by e_1 and e_2, respectively, *i.e.*,

$$e_1 = (W_{max} - \overline{W}_{max})/W_{max} , \tag{4.3.36}$$

$$e_2 = (M_{max} - \overline{M}_{max})/M_{max} . \tag{4.3.37}$$

Substituting equations (4.3.17), (4.3.34) and equations (4.3.18), (4.3.35) into equations (4.3.36) and (4.3.37), yields

$$e_1 = \frac{1 + \dfrac{4}{3K_1K_3}(\dfrac{K_1+1}{K_2} + 6) - \dfrac{384}{K}[1 - e^{-\sqrt[4]{K/64}} \cos\ (\sqrt[4]{K/64})]}{1 + \dfrac{4}{3K_1K_3}(\dfrac{K_1+1}{K_2} + 6)} \tag{4.3.38}$$

and

$$e_2 = \frac{\dfrac{2}{K_1K_3}(\dfrac{K_1+1}{K_2} + 2K_1) - 3 - \dfrac{24}{\sqrt{K}} e^{-\sqrt[4]{K/64}} \sin(\sqrt[4]{K/64})}{\dfrac{2}{K_1K_3}(\dfrac{K_1+1}{K_2} + 2K_1) - 3} . \tag{4.3.39}$$

It is obvious that the relative error is the function of the non-dimensional spring constant K only. By letting K approach zero , the limits of e_1 and e_2 become zero, *i.e.*,

$$\lim_{K\to 0} e_1 = 0, \qquad \lim_{K\to 0} e_2 = 0 . \tag{4.3.40}$$

The relationships of e_1 and e_2 to K are shown in Figure 4.3.3 which shows e_1 and e_2 are always positive, *i.e.*, the approximate values shown in equations (4.3.34) and (4.3.35) are less than the exact values given in equations (4.3.17) and (4.3.18), respectively.

Secondly, let a concentrated load of magnitude P act at the beam with elastic foundation. The maximum deflection and bending moment occur at the acting point of the concentrated load and their magnitudes are

$$\overline{W}_{max} = \frac{P}{8\beta^3 EI} = \frac{\sqrt{2}Pl^3}{4EI}K^{-3/4}, \tag{4.3.41}$$

and

$$\overline{M}_{max} = \frac{P}{4\beta} = \frac{\sqrt{2}}{4}PlK^{-1/4}. \tag{4.3.42}$$

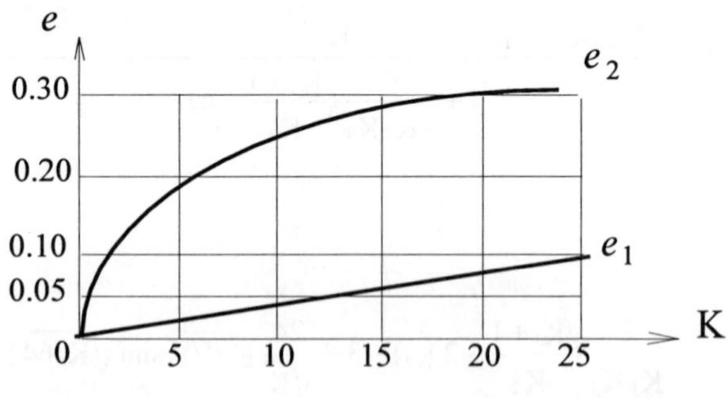

Figure 4.3.3 Relative error versus K for the maximum deflection, moment obtained by the approximate model with uniform load

The exact values of the maximum deflection and moment are given in equations (4.3.25) and (4.3.26). Substituting equations (4.3.25) and (4.3.41) and equations (4.3.26) and (4.3.42) into equations (4.3.36) and (4.3.37), yields

$$e_1 = (3 + \frac{3}{4}K_1 + \frac{K_1+1}{K_2} - 36\sqrt{2}\,K_1 K_3 K^{-3/4}) \Big/ (3 + \frac{3}{4}K_1 + \frac{K_1+1}{K_2}), \quad (4.3.43)$$

and

$$e_2 = (3K_1 + \frac{2K_1+2}{K_2} - 6\sqrt{2}\,K_1 K_3 K^{-1/4}) \Big/ (3K_1 + \frac{2K_1+2}{K_2}). \quad (4.3.44)$$

It can be proved that

$$\lim_{K\to 0} e_1 = 0, \qquad \lim_{K\to 0} e_2 = 0 \ . \quad (4.3.45)$$

The relative errors e_1 and e_2 are only dependent on the value of the nondimensional spring constant K which are shown in Figure 4.3.4.

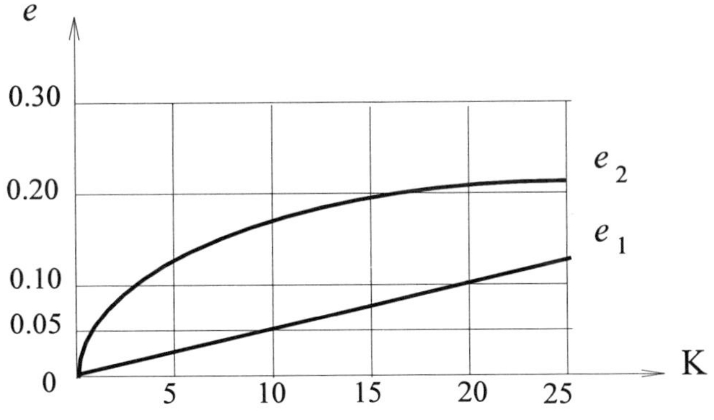

Figure 4.3.4 e_1 and e_2 versus K for the maximum deflection, moment obtained by the approximate model with concentrated load

4.3.2. Natural frequency

The dynamic equation can be obtained from equation (4.3.1) with the addition of the inertia term, *i.e.*,

$$EI\frac{\partial^4 W_j}{\partial x^4}+\rho A\frac{\partial^2 W_j}{\partial t^2}=F_j \qquad -\frac{1}{2}\le x\le\frac{1}{2}, \quad j=1,2,...,N \tag{4.3.46}$$

where ρA denotes the mass per unit length of the beam.

Applying the U transformation (4.3.3) to equations (4.3.46) and (4.3.2) and letting $F_j=0$ and $q_m=Q_m(x)e^{i\omega t}$, yields

$$\frac{d^4Q_m}{dx^4}=\beta^4 Q_m \tag{4.3.47}$$

and

$$Q_m(\frac{1}{2})=e^{im\psi}Q_m(-\frac{1}{2}),\quad Q'_m(\frac{1}{2})=e^{im\psi}Q'_m(-\frac{1}{2}),$$

$$Q''_m(\frac{1}{2})=e^{im\psi}Q''_m(-\frac{1}{2}),$$

$$Q'''_m(\frac{1}{2})-\frac{K}{2l^3}Q_m(\frac{1}{2})=e^{im\psi}\left[Q'''_m(-\frac{1}{2})+\frac{K}{2l^3}Q_m(-\frac{1}{2})\right], \tag{4.3.48}$$

with $K=kl^3/EI$ and m=1,2,...,N. β has been defined by equation (4.2.10).

The solution for Q_m of equation (4.3.47) can be generally expressed as [10]

$$Q_m(x)=A\sin\beta x+B\cos\beta x+C\mathrm{sh}\beta x+D\mathrm{ch}\beta x\,. \tag{4.3.49}$$

Substituting equation (4.3.49) into equation (4.3.48), yields

$$\sin\alpha(1+e^{im\psi})A+\cos\alpha(1-e^{im\psi})B+\mathrm{sh}\alpha(1+e^{im\psi})C+\mathrm{ch}\alpha(1-e^{im\psi})D=0\,,$$

$$\cos\alpha(1-e^{im\psi})A-\sin\alpha(1+e^{im\psi})B+\mathrm{ch}\alpha(1-e^{im\psi})C+\mathrm{sh}\alpha(1+e^{im\psi})D=0\,,$$

$$-\sin\alpha(1+e^{im\psi})A-\cos\alpha(1-e^{im\psi})B+\text{sh}\alpha(1+e^{im\psi})C+\text{ch}\alpha(1-e^{im\psi})D=0\,,$$

$$-(\cos\alpha+\frac{K}{16\alpha^3}\sin\alpha)(1-e^{im\psi})A+(\sin\alpha-\frac{K}{16\alpha^3}\cos\alpha)(1+e^{im\psi})B$$

$$+(\text{ch}\alpha-\frac{K}{16\alpha^3}\text{sh}\alpha)(1-e^{im\psi})C+(\text{sh}\alpha-\frac{K}{16\alpha^3}\text{ch}\alpha)(1+e^{im\psi})D=0\,,$$

(4.3.50)

with $\alpha=\beta l/2$.

Let the coefficient determinant of the simultaneous equations (4.3.50) be equal to zero. Expanding this determinant yields the frequency equation, *i.e.*,

$$\frac{32\alpha^3}{K}\cos^2 m\psi+\left[\text{sh}2\alpha-\sin 2\alpha-\frac{32\alpha^3}{K}(\text{ch}2\alpha+\cos 2\alpha)\right]\cos m\psi$$

$$+\sin 2\alpha\;\text{ch}2\alpha-\cos 2\alpha\;\text{sh}2\alpha+\frac{32\alpha^3}{K}\text{ch}2\alpha\;\cos 2\alpha=0\,,$$

$$m=1,2,\ldots,N. \qquad (4.3.51)$$

When K approaches infinity, the limit of the frequency equation (4.3.51) becomes equation (4.2.14). The frequency bands are dependent on the non-dimensional spring constant K. The upper and lower limit of the frequency bands can be determined by substituting $m\psi=0$ and π into equation (4.3.51), respectively. For instance, while K=20000, the first five positive roots of equation (4.3.51), in which $m\psi=0$ and π respectively, are shown in Table 4.3.1 where $\alpha_j(\theta)$ denotes the j-th positive root of equation (4.3.51) for $m\psi=\theta$.

Table 4.3.1 The upper and lower limits of the natural frequency bands while K=20000

$\alpha_1(0)$	$\alpha_2(0)$	$\alpha_3(0)$	$\alpha_4(0)$	$\alpha_5(0)$
2.35483	3.14159	5.35837	6.28319	8.02529
$\alpha_1(\pi)$	$\alpha_2(\pi)$	$\alpha_3(\pi)$	$\alpha_4(\pi)$	$\alpha_5(\pi)$
1.57080	3.87763	4.71239	6.75268	7.85398

4.4 Continuous Deep Beams

Consider a deep beam with uniform flexural rigidity D and shearing rigidity C running over n+1 number of roller supports as shown in Figure 4.4.1(a). Let the span length between any two adjacent supports be denoted by l. In order to create an equivalent system with cyclic periodicity, it is necessary to extend the original beam by its symmetrical image and apply anti-symmetric loading on the corresponding extended part about the original loading as shown in Figure 4.4.1(b).

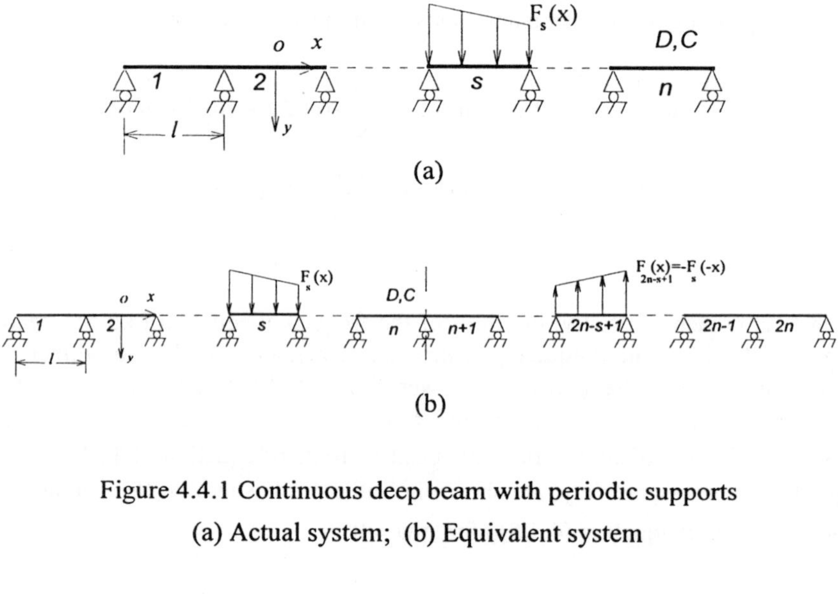

Figure 4.4.1 Continuous deep beam with periodic supports
(a) Actual system; (b) Equivalent system

In each substructure a local coordinate system oxy with origin o at its mid span is established . The equilibrium equations for all substructures are of the same form[11], *i.e.*,

$$-\frac{d}{dx}\left[C\left(\frac{dW_k(x)}{dx}-\theta_k(x)\right)\right]=p_k(x)\,,$$

$$-\frac{d}{dx}(D\frac{d\theta_k(x)}{dx}) - C(\frac{dW_k(x)}{dx} - \theta_k(x)) = m_k(x),$$

$$-\frac{l}{2} \leq x \leq \frac{l}{2}, \qquad k=1,2,\ldots,N \tag{4.4.1}$$

where N=2n, $W_k(x), \theta_k(x)$ denote the deflection and rotational angle functions for the k-th substructure, respectively and $p_x(x), m_k(x)$ denote the transverse loading and distributed moment functions, respectively. Continuity across the roller supports requires the following conditions to be satisfied:

$$W_k(\frac{l}{2}) = W_k(-\frac{l}{2}) = 0,$$

$$\theta_k(\frac{l}{2}) = \theta_{k+1}(-\frac{l}{2}), \quad \theta'_k(\frac{l}{2}) = \theta'_{k+1}(-\frac{l}{2}), \qquad k=1,2,\ldots,N \tag{4.4.2}$$

where $\theta_{N+1} \equiv \theta_1$. The last equation in equations (4.4.2) represents the continuity condition of the bending moment.

Applying the U-transformation to equations (4.4.1) and (4.4.2), *i.e.*, letting

$$W_k(x) = \frac{1}{\sqrt{N}} \sum_{m=1}^{N} e^{i(k-1)m\psi} q_m(x), \qquad k=1,2,\ldots,N \tag{4.4.3}$$

$$\theta_k(x) = \frac{1}{\sqrt{N}} \sum_{m=1}^{N} e^{i(k-1)m\psi} Q_m(x), \qquad k=1,2,\ldots,N. \tag{4.4.4}$$

with $\psi = 2\pi/N$ and $i = \sqrt{-1}$, equations (4.4.1) and (4.4.2) becomes

$$-\frac{d}{dx}\left[C(\frac{dq_m}{dx} - Q_m)\right] = f_m,$$

$$-\frac{d}{dx}(D\frac{dQ_m}{dx}) - C(\frac{dq_m}{dx} - Q_m) = F_m, \qquad m=1,2,\ldots,N \tag{4.4.5}$$

with $-\frac{1}{2} \le x \le \frac{1}{2}$ and

$$q_m(\frac{1}{2}) = q_m(-\frac{1}{2}) = 0\,,$$

$$Q_m(\frac{1}{2}) = e^{im\psi} Q_m(-\frac{1}{2}), \quad Q'_m(\frac{1}{2}) = e^{im\psi} Q'_m(-\frac{1}{2})\,,$$

$$m=1,2,\ldots,N \tag{4.4.6}$$

where

$$f_m(x) = \frac{1}{\sqrt{N}} \sum_{k=1}^{N} e^{-i(k-1)m\psi} p_k(x)\,,$$

$$F_m(x) = \frac{1}{\sqrt{N}} \sum_{k=1}^{N} e^{-i(k-1)m\psi} m_k(x)\,, \qquad m=1,2\ldots,N. \tag{4.4.7}$$

The equilibrium equation (4.4.1) and continuity condition (4.4.2) have been uncoupled in the domain of single substructure by the U-transformation, as shown in equations (4.4.5) and (4.4.6).

Two specific examples will be discussed in detail.

4.4.1. Uniformly Distributed Load

Let us discuss the case of an uniformly distributed load p applied on the middle span, say the s-th span of a continuous deep beam with n number of spans (n=2s-1). For this case, N=2n=2(2s-1) and

$$p_s(x) = p, \qquad p_{N-s+1}(x) = -p,$$

$$p_j(x) = 0, \quad j \neq s \text{ and } N-s+1,$$

$$m_j(x) = 0, \qquad j=1,2,\ldots,N. \tag{4.4.8}$$

Substituting equation (4.4.8) into equation (4.4.7), yields

$$f_m(x) = \frac{2p}{\sqrt{N}} e^{-i(s-1)m\psi} , \qquad m=1,3,...,2n-1 \qquad (4.4.9a)$$

$$f_m(x) = 0 , \qquad m=2,4,...,2n \qquad (4.4.9b)$$

and

$$F_m(x) = 0 , \qquad m=1,2,...,2n. \qquad (4.4.10)$$

The solutions for q_m and Q_m of equations (4.4.5) and (4.4.6) with equations (4.4.9a-b) and (4.4.10) may be expressed as

$$q_m(x) = \frac{2p}{D\sqrt{N}} e^{-i(s-1)m\psi} (C_{m0} + C_{m1} x + C_{m2} x^2 + C_{m3} x^3 + \frac{x^4}{24}) ,$$

$$Q_m(x) = \frac{2p}{D\sqrt{N}} e^{-i(s-1)m\psi} (D_{m0} + D_{m1} x + D_{m2} x^2 + \frac{x^3}{6}) ,$$

$$m=1,3,...,2n-1 \qquad (4.4.11a)$$

and

$$q_m = Q_m = 0, \qquad m=2,4,...,2n \qquad (4.4.11b)$$

where

$$C_{m0} = \frac{l^4}{384}(\frac{a + b\cos m\psi}{e + f\cos m\psi}), \qquad C_{m1} = -i\frac{l^3}{96}(\frac{\sin m\psi}{e + f\cos m\psi}),$$

$$C_{m2} = -\frac{l^2}{16}(\frac{c + d\cos m\psi}{e + f\cos m\psi}), \qquad C_{m3} = -\frac{4}{l^2} C_{m1} ,$$

$$D_{m0} = (1 - 24\eta) C_{m1} , \qquad D_{m1} = 2 C_{m2} + \frac{D}{C} , \qquad D_{m2} = -\frac{12}{l^2} C_{m1} ,$$

$$m=1,3,...,2n-1 \qquad (4.4.12)$$

with

$$\eta = D/C\,l^2,$$

$$a = 126\eta + 288\eta^2 + 4, \quad b = 18\eta - 288\eta^2 - 1, \quad c = 22\eta + 48\eta^2 + 1,$$

$$d = 2\eta - 48\eta^2, \quad e = 2 + 6\eta, \quad f = 1 - 6\eta. \tag{4.4.13}$$

The transverse displacements and rotational angles for all substructures can be obtained by substituting equations (4.4.11a-b),(4.4.12) and (4.4.13) into equations (4.4.3) and (4.4.4). The transverse displacement for the loaded span can be expressed as

$$W_s(x) = \frac{p\,l^4}{384D}\frac{1}{n}\sum_{m=1,3...}^{2n-1}\left[\frac{a + b\cos m\psi}{e + f\cos m\psi} - 24\left(\frac{c + d\cos m\psi}{e + f\cos m\psi}\right)\left(\frac{x}{l}\right)^2 + 16\left(\frac{x}{l}\right)^4\right],$$

$$-\frac{l}{2} \le x \le \frac{l}{2} \tag{4.4.14}$$

and the rotational angle is

$$\theta_s(x) = \frac{p\,l^3}{48D}\frac{1}{n}\sum_{m=1,3...}^{2n-1}\left\{\left[48\eta - 6\left(\frac{c + d\cos m\psi}{e + f\cos m\psi}\right)\right]\left(\frac{x}{l}\right) + 8\left(\frac{x}{l}\right)^3\right\},$$

$$-\frac{l}{2} \le x \le \frac{l}{2} \tag{4.4.15}$$

The maximum deflection occurs at mid-span and its magnitude is

$$W_{max} = W_s(0) = \frac{p\,l^4}{384D}\frac{1}{n}\sum_{m=1,3,...}^{2n-1}\frac{a + b\cos m\psi}{e + f\cos m\psi} \tag{4.4.16}$$

and the maximum rotational angle occur at supports of the loaded span with magnitude

$$\theta_{max} = \theta_s\left(\pm\frac{l}{2}\right) = \pm\frac{p\,l^3}{48D}\frac{1}{n}\sum_{m=1,3,...}^{2n-1}\left[24\eta + 1 - 3\left(\frac{c + d\cos m\psi}{e + f\cos m\psi}\right)\right]. \tag{4.4.17}$$

The bending moment and shear force for the loaded span can be expressed as

$$M_s(x) = -D\theta_s'(x) = \frac{pl^2}{8}\frac{1}{n}\sum_{m=1,3...}^{2n-1}[\frac{c+d\cos m\psi}{e+f\cos m\psi} - 8\eta - 4(\frac{x}{l})^2],$$

$$-\frac{l}{2} \le x \le \frac{l}{2} \qquad (4.4.18)$$

and

$$V_s(x) = C(W_s'(x) - \theta_s(x)) = \frac{p}{n}\sum_{m=1,3...}^{2n-1}(-x) = -px,$$

$$-\frac{l}{2} \le x \le \frac{l}{2} \qquad (4.4.19)$$

The maximum moment occurs at midspan with magnitude

$$M_{max} = \frac{pl^2}{8}\frac{1}{n}\sum_{m=1,3...}^{2n-1}(\frac{c+d\cos m\psi}{e+f\cos m\psi} - 8\eta), \qquad (4.4.20)$$

and the maximum shear force occurs at the supports with magnitude:

$$V_{max} = V_s(\pm\frac{l}{2}) = \mp\frac{pl}{2}. \qquad (4.4.21)$$

Consider now the limiting process with n approaching infinity. Applying the result shown in equation (4.1.18) to equations (4.4.14) and (4.4.15) yield

$$W_s^*(x) = \lim_{n\to\infty} W_s(x)$$

$$= \frac{pl^4}{384D}\left\{\frac{af - be + b\sqrt{3(e-f)}}{f\sqrt{3(e-f)}} - 24\left[\frac{cf - de + d\sqrt{3(e-f)}}{f\sqrt{3(e-f)}}\right](\frac{x}{l})^2 + 16(\frac{x}{l})^4\right\},$$

$$-l/2 \le x \le l/2 \qquad (4.4.22)$$

and

$$\theta_s^*(x) = \lim_{n\to\infty}\theta_s(x) = \frac{pl^3}{48D}\left\{[48\eta - 6\frac{cf - de + d\sqrt{3(e-f)}}{f\sqrt{3(e-f)}}](\frac{x}{l}) + 8(\frac{x}{l})^3\right\},$$

$$-l/2 \le x \le l/2 . \tag{4.4.23}$$

Table 4.4.1 Deflections and bending moments of continuous deep beams subjected to uniform load p over middle span

Relative stiffness D/cl^2	0					
Number of spans n	1	3	5	7	9	∞
$W_{max}/(pl^4/384D)$	5.00000	2.60000	2.47368	2.46479	2.46415	2.46410162
$M_{max}/(pl^2/16)$	2.00000	1.20000	1.15789	1.15499	1.15472	1.1547005
Relative stiffness D/cl^2	0.5					
Number of spans n	1	3	5	7	9	∞
$W_{max}/(pl^4/384D)$	29.00000	27.50000	27.42105	27.41772	27.41743	27.41739861
$M_{max}/(pl^2/16)$	2.00000	1.50000	1.47368	1.47258	1.47248	1.4724662
Relative stiffness D/cl^2	1.0					
Number of spans n	1	3	5	7	9	∞
$W_{max}/(pl^4/384D)$	53.00000	51.90909	51.72847	51.70540	51.70241	51.70197760
$M_{max}/(pl^2/16)$	2.00000	1.63636	1.57616	1.56846	1.56747	1.5673432

The maximum deflection and bending moment occur at mid-span with magnitudes

$$W_{max}^* = W_s^*(0) = \frac{pl^4}{384D}\left[\frac{af - be + b\sqrt{3(e-f)}}{f\sqrt{3(e-f)}}\right], \tag{4.4.24}$$

and

$$M^*_{max} = -D\theta_s^{*'}(0) = \frac{p l^2}{8}\left[\frac{cf - de + d\sqrt{3(e-f)}}{f\sqrt{3(e-f)}} - 8\eta\right]. \qquad (4.4.25)$$

A comparison of above results is summarized in Table 4.4.1.

4.4.2 Concentrated Load

Let a concentrated load of magnitude P act at the midpoint of the middle span, say the s-th span of a continuous deep beam with n number of spans (n=2s-1). For this case, the loading functions for the equivalent system with cyclic periodicity may be expressed as

$$p_s(x) = P\delta(x), \qquad p_{N-s+1}(x) = -P\delta(x),$$

$$p_j(x) = 0, \quad j \neq s \text{ and } N-s+1, \quad m_j(x) = 0, \quad j=1,2,...,N \qquad (4.4.26)$$

where $\delta(x)$ denotes the Dirac delta function and N=2n.

Substituting equation (4.4.26) into equation (4.4.7) yields

$$f_m(x) = \frac{2P}{\sqrt{N}} e^{-i(s-1)m\psi}\delta(x), \quad m=1,3,...,N-1 \qquad (4.4.27a)$$

$$f_m(x) = 0, \quad m=2,4,...,N, \qquad (4.4.27b)$$

$$F_m(x) = 0, \quad m=1,2,...,N. \qquad (4.4.28)$$

The solutions for q_m and Q_m of equations (4.4.5), (4.4.6), (4.4.27a-b) and (4.4.28) may be expressed as

$$q_m(x) = \frac{2P}{D\sqrt{N}} e^{-i(s-1)m\psi}(C_{m0} + C_{m1}x + C_{m2}x^2 + C_{m3}x^3 - \frac{D}{2C}|x| + \frac{|x|^3}{12})$$

$$Q_m(x) = \frac{2P}{D\sqrt{N}} e^{-i(s-1)m\psi}(D_{m0} + D_{m1}x + D_{m2}x^2 + \mathrm{sgn}(x)\frac{x^2}{4}),$$

$$-l/2 \le x \le l/2, \quad m=1,3,..,2n-1 \qquad (4.4.29a)$$

and

$$q_m = Q_m = 0, \qquad m=2,4,..,2n \tag{4.4.29b}$$

where sgn denotes the signum function and

$$C_{m0} = \frac{l^3}{384}\left(\frac{\bar{a} - \bar{b}\cos m\psi}{e + f\cos m\psi} + 96\eta\right), \qquad C_{m1} = -i\frac{l^2}{64}\left(\frac{\sin m\psi}{e + f\cos m\psi}\right),$$

$$C_{m2} = -\frac{1}{32}\left(\frac{\bar{c} + \bar{d}\cos m\psi}{e + f\cos m\psi}\right), \qquad C_{m3} = -\frac{4}{l^4}C_{m1},$$

$$D_{m0} = (1 - 24\eta)C_{m1}, \qquad D_{m1} = 2C_{m2}, \qquad D_{m2} = -\frac{12}{l^2}C_{m1} \tag{4.4.30a}$$

with

$$\bar{a} = 7 + 48\eta, \ \bar{b} = 1 + 48\eta, \ \bar{c} = 5 + 24\eta, \ \bar{d} = 1 - 24\eta,$$

$$e = 2 + 6\eta, \ f = 1 - 6\eta \text{ and } \eta = D/Cl^2. \tag{4.4.30b}$$

Substituting equations (4.4.29a-b) and (4.4.30a-b) into equations (4.4.3) and (4.4.4) for k=s, yields

$$W_s(x) = \frac{Pl^3}{384D}\frac{1}{n}\sum_{m=1,3...}^{2n-1}\left[\frac{\bar{a} - \bar{b}\cos m\psi}{e + f\cos m\psi} + 96\eta - 192\eta\left(\frac{|x|}{l}\right) - 12\left(\frac{\bar{c} + \bar{d}\cos m\psi}{e + f\cos m\psi}\right)\left(\frac{x}{l}\right)^2 + 32\left(\frac{|x|}{l}\right)^3\right]$$

$$-l/2 \le x \le l/2 \tag{4.4.31}$$

$$\theta_s(x) = \frac{Pl^2}{16D}\frac{1}{n}\sum_{m=1,3...}^{2n-1}\left[-\frac{\bar{c} + \bar{d}\cos m\psi}{e + f\cos m\psi}\left(\frac{x}{l}\right) + 4\,\mathrm{sgn}(x)\left(\frac{x}{l}\right)^2\right],$$

$$-l/2 \le x \le l/2 \tag{4.4.32}$$

The maximum deflection occurs at mid-span and its magnitude is

$$W_{max} = W_s(0) = \frac{P\,l^3}{384D}\frac{1}{n}\sum_{m=1,3...}^{2n-1}\left(\frac{\bar{a} - \bar{b}\cos m\psi}{e + f\cos m\psi} + 96\eta\right). \qquad (4.4.33)$$

The maximum rotational angle occurs at supports of the loaded span with magnitude

$$\theta_{max} = \theta_s(\pm\frac{l}{2}) = \mp\frac{P\,l^2}{32D}\frac{1}{n}\sum_{m=1,3...}^{2n-1}\left(\frac{\bar{c} + \bar{d}\cos m\psi}{e + f\cos m\psi} - 2\right). \qquad (4.4.34)$$

The bending moment and shear force for the loaded span can be obtained as

$$M_s(x) = -D\theta_s'(x) = \frac{Pl}{16}\frac{1}{n}\sum_{m=1,3...}^{2n-1}\left[\frac{\bar{c} + \bar{d}\cos m\psi}{e + f\cos m\psi} - 8\,\mathrm{sgn}(x)(\frac{x}{l})\right], \qquad (4.4.35)$$

and

$$V_s(x) = C\left[W_s'(x) - \theta_s(x)\right] = -\mathrm{sgn}(x)\frac{P}{2}. \qquad (4.4.36)$$

The maximum moment occurs at mid-span with magnitude

$$M_{max} = M_s(0) = \frac{Pl}{16}\frac{1}{n}\sum_{m=1,3,...}^{2n-1}\frac{\bar{c} + \bar{d}\cos m\psi}{e + f\cos m\psi}. \qquad (4.4.37)$$

Consider now the limiting process with n approaching infinity. Applying that result shown in equation (4.1.18) to equations (4.4.31) and (4.4.32) yield

$$W_s^*(x) = \lim_{n\to\infty} W_s(x) = \frac{P\,l^3}{384D}\left\{\frac{\bar{a}f + \bar{b}e - \bar{b}\sqrt{3(e-f)}}{f\sqrt{3(e-f)}} + 96\eta - 192\eta(\frac{|x|}{l})\right.$$

$$\left. - 12\left[\frac{\bar{c}f - \bar{d}e + \bar{d}\sqrt{3(e-f)}}{f\sqrt{3(e-f)}}\right](\frac{x}{l})^2 + 32(\frac{|x|}{l})^3\right\}, \qquad (4.4.38)$$

and

$$\theta_s^*(x) = \lim_{n\to\infty} \theta_s(x) = \frac{P\,l^2}{16D}\left\{ -\frac{\bar{c}f - \bar{d}e + \bar{d}\sqrt{3(e-f)}}{f\sqrt{3(e-f)}}\left(\frac{x}{l}\right) + 4\,\mathrm{sgn}(x)\left(\frac{x}{l}\right)^2 \right\}.$$

(4.4.39)

The maximum deflection and moment occur at mid-span and their magnitudes are

$$W_{max}^* = W_s^*(0) = \frac{P\,l^3}{384D}\left[\frac{\bar{a}f + \bar{b}e - \bar{b}\sqrt{3(e-f)}}{f\sqrt{3(e-f)}} + 96\eta\right] \tag{4.4.40}$$

and

$$M_{max}^* = -D\theta_s^{*\prime}(0) = \frac{Pl}{16}\left[\frac{\bar{c}f - \bar{d}e + \bar{d}\sqrt{3(e-f)}}{f\sqrt{3(e-f)}}\right], \tag{4.4.41}$$

respectively. A comparison of these results is summarized in Table 4.4.2.

Table 4.4.2 Deflections and bending moments of continuous deep beams subjected to concentrated load P at center of middle span

Relative stiffness D/cl^2	0					
Number of spans n	1	3	5	7	9	∞
$W_{max}/(Pl^3/384D)$	8.00000	4.40000	4.21053	4.19747	4.19623	4.19615242
$M_{max}/(Pl/16)$	4.00000	2.80000	2.73684	2.73249	2.73208	2.7320508
Relative stiffness D/cl^2	0.5					
Number of spans n	1	3	5	7	9	∞
$W_{max}/(Pl^3/384D)$	56.00000	53.75000	53.63157	53.62659	53.62615	53.62613645
$M_{max}/(Pl/16)$	4.00000	3.25000	3.21052	3.20886	3.20872	3.2087122
Relative stiffness D/cl^2	1.0					
Number of spans n	1	3	5	7	9	∞
$W_{max}/(Pl^3/384D)$	104.00000	102.36363	102.09271	102.05810	102.05361	102.0530012
$M_{max}/(Pl/16)$	4.00000	3.45455	3.36424	3.35270	3.35120	3.3510004

4.5 Continuous Plates

Mathematically two-dimensional periodic structures can be analyzed by the same procedures used for beams. As an illustration consider an elastic, homogeneous isotropic rectangular plate spanning over a finite (or infinite) number of roller supports. Let "a" be the distance between adjacent supports and b the plate's width. D denotes the flexural rigidity and μ the Poisson's ratio. In each substructure, establish a local cartesian coordinates oxy with the origin o at the midpoint of the edge as shown in Figure 4.5.1(a). Let the plate's edges y=0, b be simply supported and other extreme edges of the continuous plate be either simply supported or clamped. These extreme edge conditions can be satisfied automatically by the method of images used for beams. We replace the actual continuous plate with n spans by an equivalent continuous plate with 2n spans upon which anti-symmetric or symmetric loads are acting as shown in Figure 4.5.1b and 4.5.2b. Such an equivalent system can be regarded as a cyclic periodic system mathematically.

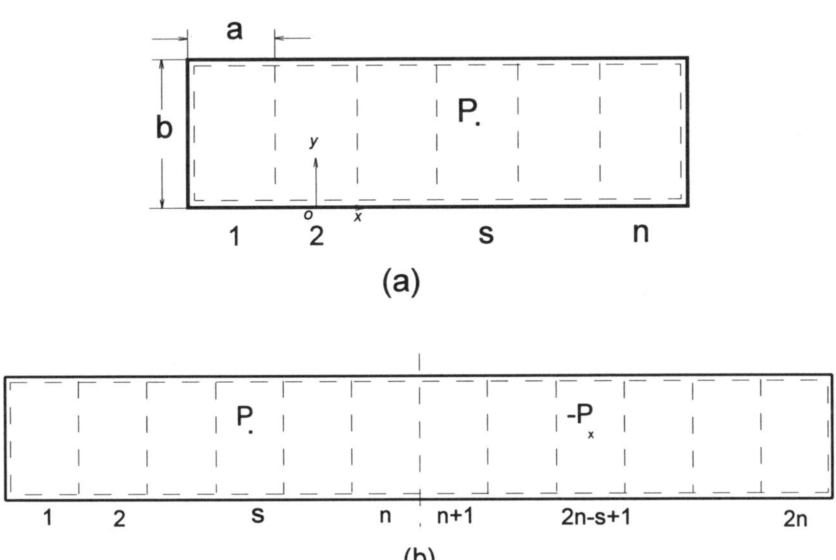

Figure 4.5.1 Continuous plate with simply supported continuous edges and extreme edges (a) Actual system; (b) Equivalent system

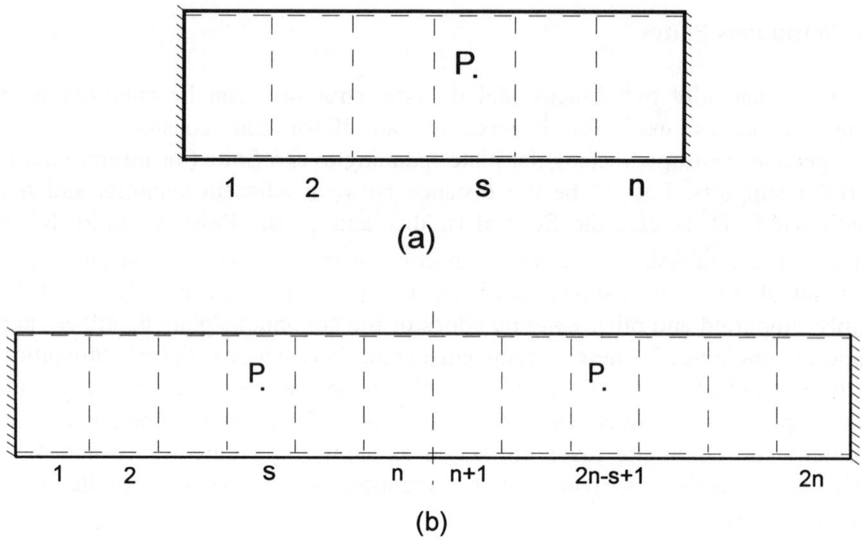

Figure 4.5.2 Continuous plate with simply supported continuous edges and clamped extreme edges (a) Actual system; (b) Equivalent system

The governing equation for the transverse displacement $W_k(x,y)$ of the k-th plate substructure is

$$D\left(\frac{\partial^4 W_k}{\partial x^4} + 2\frac{\partial^2 W_k}{\partial x^2 \partial y^2} + \frac{\partial^4 W_k}{\partial y^4}\right) = F_k(x,y), \quad k=1,2,\ldots,N \tag{4.5.1}$$

where F_k denotes the transverse load and N=2n. The boundary conditions at y=0,b and continuity conditions over the roller supports can be summarized as follows:

$$W_k(x,0) = 0, \qquad W_{k,yy}(x,0) = 0,$$

$$W_k(x,b) = 0, \qquad W_{k,yy}(x,b) = 0, \qquad k=1,2,\ldots,N \tag{4.5.2}$$

$$W_k(-\frac{a}{2},y) = 0, \qquad W_k(\frac{a}{2},y) = 0,$$

$$W_{k,x}(\frac{a}{2},y) = W_{k+1,x}(-\frac{a}{2},y),$$

$$W_{k,xx}(\frac{a}{2},y) = W_{k+1,xx}(-\frac{a}{2},y), \qquad k=1,2,...,N \tag{4.5.3}$$

where W_{N+1} should be replaced by W_1 and the subscripts x and y denote the partial differentiation with respect to x and y, respectively.

The transverse displacement $W_k(x,y)$ in terms of the generalized coordinates q_m (m=1,2,...,N) is similar in form to equation (4.1.3) except that now they are functions of two spatial variables, *i.e.*, $q_m = q_m(x,y)$.The governing equation on $q_m(x,y)$ has the form:

$$D(\frac{\partial^4 q_m}{\partial x^4} + 2\frac{\partial^2 q_m}{\partial x^2 \partial y^2} + \frac{\partial^4 q_m}{\partial y^4}) = f_m(x,y), \qquad m=1,2,...,N \tag{4.5.4}$$

where $f_m(x,y)$ is the two-dimensional analog of equation (4.1.6), i.e.,

$$f_m(x,y) = \frac{1}{\sqrt{N}}\sum_{k=1}^{N} e^{-i(k-1)m\psi} F_k(x,y). \tag{4.5.5}$$

Boundary and continuity conditions shown in equations (4.5.2-3) expressed in terms of q_m become:

$$q_m(x,0) = 0, \qquad q_{m,yy}(x,0) = 0,$$

$$q_m(x,b) = 0, \qquad q_{m,yy}(x,b) = 0, \qquad m=1,2,...,N \tag{4.5.6}$$

$$q_m(-\frac{a}{2},y) = 0, \qquad q_m(\frac{a}{2},y) = 0,$$

$$q_{m,x}(\frac{a}{2},y) = e^{im\psi} q_{m,x}(-\frac{a}{2},y),$$

$$q_{m,xx}(\frac{a}{2},y) = e^{im\psi} q_{m,xx}(-\frac{a}{2},y), \qquad m=1,2,...,N \tag{4.5.7}$$

with $\psi = 2\pi/N$.

In view of simply supported conditions at y=0,b. Levy's method of solution may be applied with the solution form in the y-direction taken as a Fourier series expansion. The Fourier series expression for q_m has the form:

$$q_m(x,y) = \sum_{k=1}^{\infty} X_{mk}(x) \sin\frac{k\pi y}{b}, \quad m=1,2,\ldots,N, \tag{4.5.8}$$

Substituting equation (4.5.8) into equations (4.5.4) and (4.5.7) yield

$$D[\frac{d^4 X_{mk}}{dx^4} - 2\left(\frac{k\pi}{b}\right)^2 \frac{d^2 X_{mk}}{dx^2} + \left(\frac{k\pi}{b}\right)^4 X_{mk}] = g_{mk}(x),$$

$$m=1,2,..,N,\ k=1,2,3,..,\infty \tag{4.5.9}$$

and

$$X_{mk}(-\frac{a}{2}) = 0, \qquad X_{mk}(\frac{a}{2}) = 0,$$

$$X'_{mk}(\frac{a}{2}) = e^{im\psi} X'_{mk}(-\frac{a}{2}),$$

$$X''_{mk}(\frac{a}{2}) = e^{im\psi} X''_{mk}(-\frac{a}{2}), \qquad m=1,2,..,N \quad k=1,2,\ldots,\infty \tag{4.5.10}$$

where g_{mk} is the Fourier coefficient of function f_m, *i.e.*,

$$g_{mk}(x) = \frac{2}{b}\int_0^b f_m(x,y) \sin\frac{k\pi y}{b} dy . \tag{4.5.11}$$

The simply supported edge conditions (4.5.6) have been satisfied by equation (4.5.8). The general solution for $X_{mk}(x)$ of equation (4.5.9) takes the form

$$X_{mk}(x) = C_{mk1} \operatorname{sh}\frac{k\pi x}{b} + C_{mk2} \operatorname{ch}\frac{k\pi x}{b} + C_{mk3}\, x \operatorname{sh}\frac{k\pi x}{b} + C_{mk4}\, x \operatorname{ch}\frac{k\pi x}{b} + X^*_{mk}(x) \tag{4.5.12}$$

where $X^{*}_{mk}(x)$ denotes an arbitrary particular solution of equation (4.5.9); and C_{mk} to C_{mk4} are four constants of integration that will be determined by equation (4.5.10).

The deflection surface $W_k(x,y)$ can be found by substituting equation (4.5.8) with the solution for $X_{mk}(x)$ of equations (4.5.9) and (4.5.10) into equation (4.1.3a). Two loading conditions will be discussed: (1) uniform load over the middle span and (2) a concentrated load at the center of the middle span. In order to simplify the expressions, let a=b in the following examples.

4.5.1 Uniformly Distributed Load

Let us consider a rectangular continuous plate with n (n=2s-1) number of spans subjected to an uniformly distributed load p over the middle (s-th) span [7].

(a) With Both Extreme Edges Simply Supported

Firstly we discuss the case of simply supported edge conditions at two extreme edges. By analogy with continuous beam (see section 4.1.1), equation (4.5.5) takes the following form for this case:

$$f_m(x,y) = 2p\,e^{-i(s-1)m\psi}\big/\sqrt{N}\,, \qquad m=1,3,...,2n-1;$$

$$f_m(x,y) = 0\,, \qquad m=2,4,...,2n \tag{4.5.13}$$

where N=2n and $\psi = \pi/n$.

Substituting equation (4.5.13) into equation (4.5.11), yields

$$g_{mk}(x) = \frac{8p}{k\pi\sqrt{N}}e^{-i(s-1)m\psi}\,, \qquad m=1,3,...,2n-1, \quad k=1,3,5,..,\infty;$$

$$g_{mk}(x) = 0\,, \qquad m=2,4..,2n \quad \text{or} \quad k=2,4,6,..,\infty. \tag{4.5.14}$$

Integrating equation (4.5.9) with equation (4.5.14) subjected to continuity conditions (4.5.10) gives:

$$X_{mk}(x)=\frac{8pa^4e^{-i(s-1)m\psi}}{D\pi^5\sqrt{2n}\,k^5}(C_{mk1}\,\text{sh}\frac{k\pi x}{a}+C_{mk2}\,\text{ch}\frac{k\pi x}{a}$$

$$+C_{mk3}\frac{x}{a}\,\text{sh}\frac{k\pi x}{a}+C_{mk4}\frac{x}{a}\,\text{ch}\frac{k\pi x}{a}+1)$$

$$m=1,3,\dots,2n-1,\quad k=1,3,5,\dots,\infty$$

$$X_{mk}(x)=0,\qquad m=2,4,\dots,2n,\quad \text{or}\quad k=2,4,\dots,\infty \tag{4.5.15}$$

where

$$C_{mk1}=-i\sin m\psi\,\text{ch}\,\alpha_k(\text{sh}\,\alpha_k\,\text{ch}\,\alpha_k-\alpha_k)\alpha_k/\Delta_{mk},$$

$$C_{mk2}=-(1+\frac{1}{2}\text{sh}\,\alpha_k C_{mk3})\Big/\text{ch}\,\alpha_k,$$

$$C_{mk3}=\frac{2\alpha_k}{\Delta_{mk}}[\cos m\psi(2\text{sh}^3\alpha_k-\text{ch}^2\alpha_k\,\text{sh}\,\alpha_k+\alpha_k\,\text{ch}\,\alpha_k)$$

$$+2\text{sh}^3\alpha_k+\text{ch}^2\alpha_k\,\text{sh}\,\alpha_k-\alpha_k\,\text{ch}\,\alpha_k\,]$$

$$C_{mk4}=-\frac{2\text{sh}\,\alpha_k}{\text{ch}\,\alpha_k}C_{mk1},$$

$$\Delta_{mk}=\cos m\psi(2\alpha_k\,\text{ch}2\alpha_k-\text{sh}2\alpha_k)+(\text{ch}2\alpha_k\,\text{sh}2\alpha_k-2\alpha_k), \tag{4.5.16}$$

with $\alpha_k=k\pi/2$, and $\psi=\pi/n$.

Substituting equations (4.5.8) (4.5.15) and (4.5.16) into equation (4.1.3a) yields

$$W_s(x,y)=\frac{4pa^4}{D\pi^5}\frac{1}{n}\sum_{k=1,3\dots}^{\infty}\left[\sum_{m=1,3\dots}^{2n-1}(C_{mk2}\,\text{ch}\frac{k\pi x}{a}+C_{mk3}\frac{x}{a}\,\text{sh}\frac{k\pi x}{a}+1)\right]\frac{1}{k^5}\sin\frac{k\pi y}{a} \tag{4.5.17}$$

The maximum deflection occurs at the center of the loaded span and its magnitude is

$$W_{max} = W_s(0,\frac{a}{2}) = \frac{4pa^4}{D\pi^5}\sum_{k=1,3...}^{\infty}\frac{(-1)^{(k-1)/2}}{k^5}(1+\frac{1}{n}\sum_{m=1,3...}^{2n-1}C_{mk2}) \ . \tag{4.5.18}$$

Using the limiting process given by equation (4.1.18), the deflection surface $W_s^*(x,y)$ for the loaded span of the continuous plate with infinite number of spans becomes:

$$W_s^*(x,y) = \lim_{n\to\infty} W_s(x,y) = \frac{4pa^4}{D\pi^5}\sum_{k=1,3...}^{\infty}[\frac{C_{k2}\,ch\frac{k\pi x}{a} + C_{k3}\frac{x}{a}sh\frac{k\pi x}{a}+1}{k^5}]\sin\frac{k\pi y}{a}\ , \tag{4.5.19}$$

where

$$C_{k2} = \frac{-(1+\frac{1}{2}C_{k3}\,sh\,\alpha_k)}{ch\,\alpha_k}\ ,$$

$$C_{k3} = \frac{2\alpha_k}{2\alpha_k ch2\alpha_k - sh2\alpha_k}\left[2sh^3\alpha_k - ch^2\alpha_k\,sh\alpha_k + \alpha_k ch\alpha_k - \frac{sh\alpha_k(sh2\alpha_k - 2\alpha_k)^2}{2\sqrt{sh^2 2\alpha_k - 4\alpha_k^2}}\right] \tag{4.5.20}$$

It is easy to prove that the series on the right-hand side of equations (4.5.17) and (4.5.19) are uniformly convergent.

The deflection and bending moments at the center of the loaded span can be expressed respectively as:

$$W_s^*(0,\frac{a}{2}) = \frac{4pa^4}{D\pi^5}\sum_{k=1,3...}^{\infty}\frac{(-1)^{(k-1)/2}}{k^5}(1+C_{k2}) = 0.002887\frac{pa^4}{D} \tag{4.5.21}$$

and

$$M_x^*(0,\frac{a}{2}) = -D\left[W_{s,xx}^*(0,\frac{a}{2}) + \mu\,W_{s,yy}^*(0,\frac{a}{2})\right]$$

$$=\frac{4pa^2}{\pi^3}\sum_{k=1,3,...}^{\infty}\frac{(-1)^{(k-1)/2}}{k^3}[\mu-(1-\mu)C_{k2}-\frac{C_{k3}}{\alpha_k}]=0.03985pa^2$$

$$\text{(when } \mu=0.3\text{)}, \tag{4.5.22}$$

$$M_y^*(0,\frac{a}{2})=-D\left[W_{s,yy}^*(0,\frac{a}{2})+\mu W_{s,xx}^*(0,\frac{a}{2})\right]$$

$$=\frac{4pa^2}{\pi^3}\sum_{k=1,3,...}^{\infty}\frac{(-1)^{(k-1)/2}}{k^3}[1+(1-\mu)C_{k2}-\frac{\mu C_{k3}}{\alpha_k}]=0.03500pa^2$$

$$\text{(when } \mu=0.3\text{)}. \tag{4.5.23}$$

(b) With Both Extreme Edges Clamped

Let us discuss now the case of clamped conditions at both extreme edges. By analogy with equation (4.1.34), equation (4.5.5) takes the following form for this case:

$$f_m(x,y)=2pe^{-i(s-1)m\psi}/\sqrt{N}, \quad m=2,4,...2n$$

$$f_m(x,y)=0, \quad m=1,3,...,2n\text{-}1. \tag{4.5.24}$$

In the same way as before, the deflection surface $W_s(x,y)$ of the loaded substructure has the form:

$$W_s(x,y)=\frac{4pa^4}{D\pi^5}\frac{1}{n}\sum_{k=1,3...}^{\infty}\left[\sum_{m=2,4...}^{2n}(C_{mk2}\,ch\frac{k\pi x}{a}+C_{mk3}\frac{x}{a}sh\frac{k\pi x}{a}+1)\right]\frac{1}{k^5}\sin\frac{k\pi y}{a} \tag{4.5.25}$$

where C_{mk2} and C_{mk3} have been defined by equation (4.5.16).

Applying a limiting process with n approaching infinity to equation (4.5.25), yields the same equation as equation (4.5.19) with equation (4.5.20).

A comparison of some numerical results is summarized in Table 4.5.1.

Table 4.5.1 Deflections and bending moments of continuous plates subjected to uniform load p on middle span [a]

Number of spans n	1	3	5	7	∞
$W_{max}/(pa^4/D)^b$	0.004062	0.002896	0.002887	0.002887	0.002887
$W_{max}/(pa^4/D)^c$	0.001917	0.002877	0.002887	0.002887	0.002887
$M_x/(pa^2)^{b,d}$	0.04788	0.03993	0.03986	0.03986	0.03985
$M_x/(pa^2)^{c,d}$	0.03324	0.03980	0.03986	0.03986	0.03985
$M_y/(pa^2)^{b,d}$	0.04788	0.03510	0.03500	0.03500	0.03500
$M_y/(pa^2)^{c,d}$	0.02438	0.03489	0.03500	0.03500	0.03500

[a] The numerical results in the table are obtained by evaluating the sum of the first N terms (N=1,2,..) in the infinite series, say $\sum_{k=1}^{N} \alpha_k$, until $|\alpha_{N+1} / \sum_{k=1}^{N} \alpha_k| < 10^{-4}$.

[b] Case is of simply supported extreme edges.

[c] Case is of clamped extreme edges.

[d] M_x and M_y denote the bending moments at the center of the loaded span of the continuous plate with $\mu = 0.3$.

4.5.2 Concentrated Load

Let a concentrated load of magnitude P act at the center of the middle (s-th) span of a continuous plate with n (n=2s-1) number of spans. The solutions for two extreme edge conditions are outlined as follows.

(a) With Both Extreme Edges Simply Supported

For this case, equal but opposite concentrated loads must be applied to the s-th and (2n-s+1)-th spans with all other spans unloaded, *i.e.*,

$$F_s(x,y) = P\delta(x)\delta(y-\frac{a}{2}), \qquad F_{2n-s+1}(x,y) = -P\delta(x)\delta(y-\frac{a}{2}),$$

$$F_k(x,y) = 0, \qquad \text{when } k \neq s,\ 2n-s+1. \tag{4.5.26}$$

Substituting equation (4.5.26) into equation (4.5.5) yields

$$f_m(x,y) = 2P e^{-i(s-1)m\psi}\,\delta(x)\delta(y-\frac{a}{2})\Big/\sqrt{2n}\,, \qquad m=1,3,...,2n\text{-}1$$

$$f_m(x,y) = 0\,, \qquad m=2,4,...,2n \tag{4.5.27}$$

with $\psi = \pi/n$.

Substituting equation (4.5.27) into equation (4.5.11) yields

$$g_{mk} = (-1)^{(k-1)/2}\frac{4P}{a\sqrt{2n}}e^{-i(s-1)m\psi}\,\delta(x)\,, \quad m=1,3,...,2n\text{-}1, \quad k=1,3,5,...,\infty$$

$$g_{mk} = 0\,, \qquad m=2,4,...,2n \ \text{ or } \ k=2,4,6,...,\infty\,. \tag{4.5.28}$$

Integrating equation (4.5.9) taking into account of equation (4.5.28) and subject to condition (4.5.10) gives:

$$X_{mk}(x) = (-1)^{(k-1)/2}\frac{4Pa^2}{D\pi^3k^3\sqrt{2n}}e^{-i(s-1)m\psi}\Big(C_{mk1}\,\mathrm{sh}\frac{k\pi x}{a} + C_{mk2}\,\mathrm{ch}\frac{k\pi x}{a} + C_{mk3}\frac{x}{a}\,\mathrm{sh}\frac{k\pi x}{a}$$

$$+ C_{mk4}\frac{x}{a}\,\mathrm{ch}\frac{k\pi x}{a} - \frac{1}{4}\,\mathrm{sh}\frac{k\pi|x|}{a} + \frac{\alpha_k|x|}{2a}\,\mathrm{ch}\frac{k\pi x}{a}\Big)\,,$$

$$m=1,3,...,2n\text{-}1, \qquad k=1,3,5,...,\infty$$

$$X_{mk}(x) = 0\,, \quad m=2,4,...,2n \ \text{ or } \ k=2,4,6,...,\infty \tag{4.5.29}$$

where

$$C_{mk1} = -\frac{\mathrm{ch}\,\alpha_k}{2\mathrm{sh}\,\alpha_k}C_{mk4}\,, \qquad C_{mk2} = \frac{\mathrm{sh}\,\alpha_k - \alpha_k\,\mathrm{ch}\,\alpha_k}{4\mathrm{ch}\,\alpha_k} - \frac{\alpha_k\,\mathrm{sh}^2\,\alpha_k}{2\mathrm{ch}\,\alpha_k}\eta_{mk}\,,$$

$$C_{mk3} = \alpha_k\,\mathrm{sh}\,\alpha_k\cdot\eta_{mk}\,, \qquad C_{mk4} = \frac{i\alpha_k(1-\cos m\psi)}{\sin m\psi}\Big(\mathrm{ch}\,\alpha_k\cdot\eta_{mk} + \frac{1}{2}\Big)\,,$$

$$\eta_{mk} = \frac{\cos m\psi(\mathrm{sh}\,\alpha_k - \alpha_k\,\mathrm{ch}\,\alpha_k) + (\alpha_k\,\mathrm{ch}\,\alpha_k - \mathrm{sh}\,\alpha_k\,\mathrm{ch}2\,\alpha_k)}{\cos m\psi(2\,\alpha_k\,\mathrm{ch}2\,\alpha_k - \mathrm{sh}2\,\alpha_k) + (\mathrm{sh}2\,\alpha_k\,\mathrm{ch}2\,\alpha_k - 2\,\alpha_k)}$$

$$\alpha_k = \frac{k\pi}{2} \tag{4.5.30}$$

Substituting equations (4.5.8) and (4.5.29) with equation (4.5.30) into equation (4.1.3a) yields

$$W_s(x,y) = \frac{2Pa^2}{D\pi^3}\frac{1}{n}\sum_{k=1,3...}^{\infty}\left[(-1)^{(k-1)/2}k^{-3}\sin\frac{k\pi y}{a}\sum_{m=1,3...}^{2n-1}(C_{mk2}\,\mathrm{ch}\frac{k\pi x}{a}\right.$$

$$\left.+C_{mk3}\frac{x}{a}\,\mathrm{sh}\frac{k\pi x}{a}-\frac{1}{4}\mathrm{sh}\frac{k\pi|x|}{a}+\frac{\alpha_k|x|}{2a}\mathrm{ch}\frac{k\pi x}{a})\right],$$

$$-a/2 \le x \le a/2,\ 0 \le y \le a. \tag{4.5.31}$$

The maximum deflection occurs at the center of the loaded span and its magnitude is :

$$W_{max} = W_s(0,\frac{a}{2}) = \frac{2Pa^2}{D\pi^3}\frac{1}{n}\sum_{k=1,3...}^{\infty}\left(k^{-3}\sum_{m=1,3...}^{2n-1}C_{mk2}\right). \tag{4.5.32}$$

Applying the limiting process given by equation (4.1.18) to equation (4.5.31), the deflection surface $W_s^*(x,y)$ for the loaded span of the continuous plate with infinite number of spans under a concentrated load P becomes:

$$W_s^*(x,y) = \lim_{n\to\infty} W_s(x,y)$$

$$= \frac{2Pa^2}{D\pi^3}\sum_{k=1,3...}^{\infty}[(-1)^{(k-1)/2}k^{-3}\sin\frac{k\pi y}{a}(C_{k2}\,\mathrm{ch}\frac{k\pi x}{a}+C_{k3}\frac{x}{a}\mathrm{sh}\frac{k\pi x}{a},$$

$$-\frac{1}{4}\mathrm{sh}\frac{k\pi|x|}{a}+\frac{\alpha_k|x|}{2a}\mathrm{ch}\frac{k\pi x}{a})\ ]$$

$$-a/2 \le x \le a/2,\ 0 \le y \le a \tag{4.5.33}$$

where

$$C_{k2}=\frac{1}{4\text{ch}\,\alpha_k}(\text{sh}\,\alpha_k-\alpha_k\,\text{ch}\,\alpha_k-2\text{sh}\,\alpha_k\,C_{k3}),$$

$$C_{k3}=\frac{\alpha_k\,\text{sh}\,\alpha_k}{2\alpha_k\,\text{ch}2\alpha_k-\text{sh}2\alpha_k}\left[\text{sh}\,\alpha_k-\alpha_k\,\text{ch}\,\alpha_k+\frac{\alpha_k\,\text{sh}\,\alpha_k(2\alpha_k-\text{sh}2\alpha_k)}{\sqrt{\text{sh}^2 2\alpha_k-4\alpha_k^2}}\right].$$

(4.5.34)

The maximum deflection occurs at the center of the loaded span with magnitude:

$$W^*_{max}=W^*_s(0,\frac{a}{2})=\frac{2Pa^2}{D\pi^3}\sum_{k=1,3...}^{\infty}k^{-3}C_{k2}=0.009081\frac{Pa^2}{D}. \tag{4.5.35}$$

It can be proved that the series on the right-hand side of equations (4.5.31) and (4.5.33) are uniformly convergent.

(b) With Both Extreme Edges Clamped

In the same way as before the deflection surface $W_s(x,y)$ for the loaded (s-th) span of the continuous plate with n (n=2s-1) number of spans takes the following form for present case:

$$W_s(x,y)=\frac{2Pa^2}{D\pi^3}\frac{1}{n}\sum_{k=1,3...}^{\infty}\left[(-1)^{(k-1)/2}k^{-3}\sin\frac{k\pi y}{a}\sum_{m=2,4...}^{2n}(C_{mk2}\,\text{ch}\frac{k\pi x}{a}\right.$$

$$\left.+C_{mk3}\frac{x}{a}\text{sh}\frac{k\pi x}{a}-\frac{1}{4}\text{sh}\frac{k\pi|x|}{a}+\frac{\alpha_k|x|}{2a}\text{ch}\frac{k\pi x}{a})\right],$$

$$-\frac{a}{2}\le x\le\frac{a}{2},0\le y\le a \tag{4.5.36}$$

where C_{mk2} and C_{mk3} are defined by equation (4.5.30). The maximum deflection also occurs at center of the loaded span and its magnitude is

$$W_{max}=W_s(0,\frac{a}{2})=\frac{2Pa^2}{D\pi^3}\frac{1}{n}\sum_{k=1,3...}^{\infty}\left(k^{-3}\sum_{m=2,4....}^{2n}C_{mk2}\right). \tag{4.5.37}$$

Letting n approach infinity, the limit of equation (4.5.36) takes the same form as that shown in equations (4.5.33) and (4.5.34). A comparison of these results obtained in this example is summarized in Table 4.5.2.

Table 4.5.2 Maximum deflection of continuous plates subjected to concentrated load P at center of middle span [a]

Number of spans n	1	3	5	7	∞
$W_{max} / (Pa^2/D)^b$	0.011600	0.009102	0.009081	0.009081	0.009081
$W_{max} / (Pa^2/D)^c$	0.007024	0.009060	0.009081	0.009081	0.009081

[a] The numerical results in the table are obtained by evaluating the sum of the first N terms (N=1,2,..) in the infinite series, say $\sum_{k=1}^{N} a_k$, until $\left| a_{N+1} / \sum_{k=1}^{N} a_k \right| < 10^{-4}$.

[b] Case is of simply supported extreme edges.

[c] Case is of clamped extreme edges.

4.6 Stiffened Plates

Several approximate methods are usually adopted for the analysis of stiffened plates, such as the energy method and the orthotropic plate model. More accurate numerical results may be obtained by means of the finite element method, but it will normally involve high computational cost. However exact analytical solutions for general stiffened plates have not been found except for some particular cases.

A rectangular plate with equidistant stiffeners and four edges simply supported may be regarded as a periodic structure and then the U-transformation method can be applied to analysis this problem.

4.6.1 Static Solution of Stiffened Plates

The rectangular stiffened plate to be considered is regarded as a periodic structure with n substructures as shown in Figure 4.6.1(a). The plate is simply supported at four edges and has of flexural rigidity D, Poisson's ratio μ and width b. The stiffening ribs are all of the same flexural rigidity EI, length b and are spaced at "a" apart.

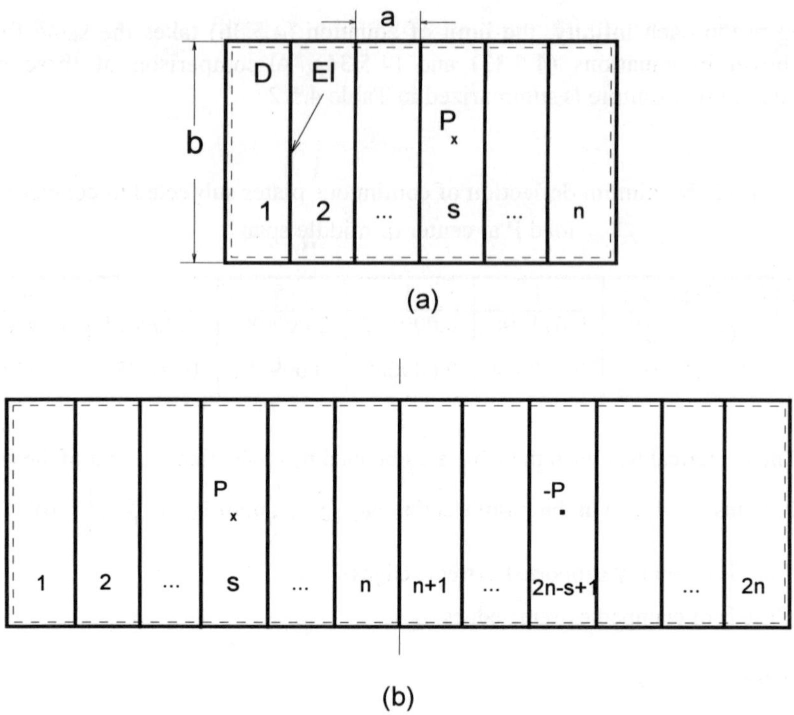

Figure 4.6.1 Stiffened plate with four simply supported edges
(a) Actual system; (b) Equivalent system.

At the outset , let us consider the equivalent system with cyclic periodicity , whose length is twice the actual size as shown in Figure 4.6.1 (b). The loading pattern is anti-symmetric with respect to the symmetric plane of the equivalent system so that the deformation is also anti-symmetric. The left edge of the first substructure and the right edge of the last (2n-th) substructure may be connected imaginarily by using the continuity condition . The simply supported boundary condition for the left and right edges of the actual system may be satisfied automatically in its equivalent system so that the corresponding restriction may be relaxed.

Consider now the equivalent system with 2n substructures and cyclic periodicity instead of the actual original structure. A typical substructure is shown in Figure 4.6.2. In each substructure a local coordinate system oxy is established as shown in Figure 4.6.2. The equilibrium equations for all substructures are of the same form, *i.e.*,

$$D\left(\frac{\partial^4 W_k}{\partial x^4}+2\frac{\partial^4 W_k}{\partial x^2 \partial y^2}+\frac{\partial^4 W_k}{\partial y^4}\right)=F_k(x,y), \quad k=1,2,\ldots,N \tag{4.6.1}$$

with (-a/2<x<a/2, 0<y<b) and N= 2n , where W_k is the transverse deflection and F_k the loading function for the k-th substructure.

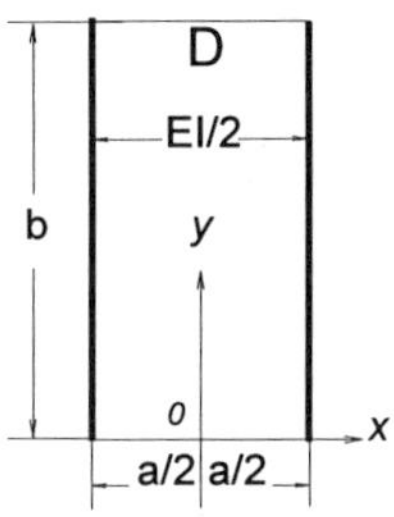

Figure 4.6.2 Substructure with local coordinate system

Continuity across the interface of adjacent substructures requires the following condition[12] to be satisfied:

$$W_k\left(\frac{a}{2},y\right)=W_{k+1}\left(-\frac{a}{2},y\right), \qquad W_{k,x}\left(\frac{a}{2},y\right)=W_{k+1,x}\left(-\frac{a}{2},y\right),$$

$$W_{k,xx}\left(\frac{a}{2},y\right)=W_{k+1,xx}\left(-\frac{a}{2},y\right),$$

$$\left\{-D\left[W_{k,xxx}+(2-\mu)W_{k,xyy}\right]+\frac{EI}{2}W_{k,yyyy}\right\}_{x=a/2}$$

$$=\left\{-D\left[W_{k+1,xxx}+(2-\mu)W_{k+1,xyy}\right]-\frac{EI}{2}W_{k+1,yyyy}\right\}_{x=-a/2} \tag{4.6.2}$$

with k=1,2, ..., N and $W_{N+1} \equiv W_1$. It must be noted that the middle terms on the left and right hand sides of the last equation can be eliminated by using the second equation in equation (4.6.2). The simply supported boundary condition at two longitudinal edges may be expressed as

$$W_k(x,0) = 0, \qquad W_{k,yy}(x,0) = 0,$$

$$W_k(x,b) = 0, \qquad W_{k,yy}(x,b) = 0, \qquad k=1,2,...,N. \tag{4.6.3}$$

Now apply the U- transformation shown in equation (4.1.3) to equations (4.6.1-3), yielding N uncoupled equations as

$$D\left(\frac{\partial^4 q_m(x,y)}{\partial x^4} + 2\frac{\partial^4 q_m(x,y)}{\partial x^2 \partial y^2} + \frac{\partial^4 q_m(x,y)}{\partial y^4}\right) = f_m(x,y),$$

$$m=1,2,...,N \tag{4.6.4}$$

and

$$q_m(\frac{a}{2},y) = e^{im\psi} q_m(-\frac{a}{2},y), \qquad q_{m,x}(\frac{a}{2},y) = e^{im\psi} q_{m,x}(-\frac{a}{2},y),$$

$$q_{m,xx}(\frac{a}{2},y) = e^{im\psi} q_{m,xx}(-\frac{a}{2},y),$$

$$q_{m,xxx}(\frac{a}{2},y) - \frac{Ka}{2} q_{m,yyyy}(\frac{a}{2},y) = e^{im\psi}\left[q_{m,xxx}(-\frac{a}{2},y) + \frac{Ka}{2} q_{m,yyyy}(-\frac{a}{2},y)\right],$$

$$0 \le y \le b, \quad m=1,2,...,N \tag{4.6.5}$$

and

$$q_m(x,0) = 0, \qquad q_{m,yy}(x,0) = 0,$$

$$q_m(x,b) = 0, \qquad q_{m,yy}(x,b) = 0,$$

$$-a/2 \le x \le a/2, \quad m=1,2,...,N \tag{4.6.6}$$

where

$$f_m(x,y) = \frac{1}{\sqrt{N}} \sum_{k=1}^{N} e^{-i(k-1)m\psi} F_k(x,y) , \tag{4.6.7}$$

$$K = \frac{EI}{Da}, \qquad \psi = \frac{\pi}{n} . \tag{4.6.8}$$

Noting the simply supported boundary condition (4.6.6) , the Levy's form of solution may be used, *i.e.,* let

$$q_m(x,y) = \sum_{k=1}^{\infty} X_{mk}(x) \sin\frac{k\pi y}{b} , \qquad m=1,2,...,N \tag{4.6.9}$$

The boundary condition (4.6.6) is readily satisfied.

Substituting equation (4.6.9) into equations (4.6.4) and (4.6.5) yields

$$D[\frac{d^4 X_{mk}}{d x^4} - 2(\frac{k\pi}{b})^2 \frac{d^2 X_{mk}}{d x^2} + (\frac{k\pi}{b})^4 X_{mk}] = g_{mk}(x) ,$$

$$m=1,2,...,N, \quad k=1,2,.., \infty \tag{4.6.10}$$

and

$$X_{mk}(\frac{a}{2}) = e^{im\psi} X_{mk}(-\frac{a}{2}), \;\; X'_{mk}(\frac{a}{2}) = e^{im\psi} X'_{mk}(-\frac{a}{2}), \;\; X''_{mk}(\frac{a}{2}) = e^{im\psi} X''_{mk}(-\frac{a}{2}),$$

$$X'''_{mk}(\frac{a}{2}) - \frac{Ka}{2}\left(\frac{k\pi}{b}\right)^4 X_{mk}(\frac{a}{2}) = e^{im\psi}\left[X'''_{mk}(-\frac{a}{2}) + \frac{Ka}{2}\left(\frac{k\pi}{b}\right)^4 X_{mk}(-\frac{a}{2})\right],$$

$$m=1,2,...,N, \quad k=1,2,3,.., \infty , \tag{4.6.11}$$

where

$$g_{mk}(x) = \frac{2}{b}\int_0^b f_m(x,y) \sin\frac{k\pi y}{b} dy . \tag{4.6.12}$$

The solution for X_{mk} of equations (4.6.10) and (4.6.11) can be derived by considering the stiffened plate with a specific loading condition. Let the uniform load with magnitude p be imposed on all substructures of the actual system. The actual and fictitious loads for the equivalent system may be expressed as

$$F_1 = F_2 \cdots = F_n = p \ ,$$

$$F_{n+1} = F_{n+2} \cdots = F_N = -p \ . \tag{4.6.13}$$

Substituting equation (4.6.13) into equation (4.6.7) yields

$$f_m(x,y) = \frac{4p}{\sqrt{N}(1-e^{-im\psi})}\,, \qquad m=1,3,\ldots,2n-1,$$

$$f_m(x,y) = 0\,, \qquad m=2,4,\ldots,2n. \tag{4.6.14}$$

Inserting equation (4.6.14) in equation (4.6.12) and then integrating with respect to y yields

$$g_{mk}(x) = \frac{16p}{\sqrt{N}k\pi(1-e^{-im\psi})}\,, \qquad m=1,3,\ldots,2n-1, \qquad k=1,3,\ldots,\infty$$

$$g_{mk}(x) = 0\,, \qquad m=2,4,\ldots,2n \ \text{ or } \ k=2,4,6,\ldots,\infty \tag{4.6.15}$$

Finally , the solution for $X_{mk}(x)$ of equations (4.6.10) and (4.6.11) may be expressed as

$$X_{mk}(x) = X_{mko}\left(1 + C_{mk1}\,\mathrm{sh}\frac{k\pi x}{b} + C_{mk2}\,\mathrm{ch}\frac{k\pi x}{b} + C_{mk3}\frac{2x}{a}\,\mathrm{sh}\frac{k\pi x}{b} + C_{mk4}\frac{2x}{a}\,\mathrm{ch}\frac{k\pi x}{b}\right) \tag{4.6.16}$$

where

$$X_{mk0} = \begin{cases} \dfrac{16p\,b^4}{\sqrt{N}D(k\pi)^5(1-e^{-im\psi})}, & m = 1,3,..,2n-1, k = 1,3,5,..,\infty \\ 0, & m = 2,4,..,2n \text{ or } k = 2,4,6,..,\infty \end{cases} \tag{4.6.17}$$

and

$$C_{mk1} = i\sin m\psi\big[K\alpha_k^2\,\mathrm{ch}\,\alpha_k(\alpha_k - \mathrm{sh}\,\alpha_k\,\mathrm{ch}\,\alpha_k) + \alpha_k\,\mathrm{ch}\,\alpha_k(\mathrm{ch}2\,\alpha_k - 2)$$

$$+2\mathrm{sh}\,\alpha_k\,\mathrm{ch}2\,\alpha_k + \cos m\psi(\alpha_k\,\mathrm{ch}\,\alpha_k - 2\mathrm{sh}\,\alpha_k)\big]/\Delta_{mk}$$

$$C_{mk2} = -\{K\alpha_k[\mathrm{sh}\,\alpha_k(2\mathrm{ch}2\,\alpha_k - \alpha_k^2) + \alpha_k\,\mathrm{ch}\,\alpha_k(3\mathrm{sh}^2\,\alpha_k - 2)$$

$$+\cos m\psi(\alpha_k^2\,\mathrm{sh}\,\alpha_k - 2\mathrm{sh}\,\alpha_k + \alpha_k\,\mathrm{sh}^2\,\alpha_k\,\mathrm{ch}\,\alpha_k + 2\,\alpha_k\,\mathrm{ch}\,\alpha_k)]$$

$$+(1-\cos m\psi)[2\mathrm{ch}\,\alpha_k\,\mathrm{ch}2\,\alpha_k + \alpha_k\,\mathrm{sh}\,\alpha_k + \alpha_k\,\mathrm{sh}2\,\alpha_k\,\mathrm{ch}\,\alpha_k$$

$$+\cos m\psi(\alpha_k\,\mathrm{sh}\,\alpha_k - 2\mathrm{ch}\,\alpha_k)]\}\,/\,\Delta_{mk}$$

$$C_{mk3} = -\alpha_k\{K\alpha_k[\alpha_k\,\mathrm{ch}\,\alpha_k - \mathrm{sh}\,\alpha_k(\mathrm{ch}2\,\alpha_k + \mathrm{sh}^2\,\alpha_k)$$

$$-\cos m\psi(\alpha_k\,\mathrm{ch}\,\alpha_k - 2\mathrm{sh}\,\alpha_k + \mathrm{sh}\,\alpha_k\,\mathrm{ch}^2\,\alpha_k)]$$

$$-(1-\cos m\psi)\mathrm{ch}\,\alpha_k(\mathrm{ch}2\,\alpha_k - \cos m\psi)\}\,/\,\Delta_{mk}$$

$$C_{mk4} = -i\,\alpha_k\sin m\psi\big[K\alpha_k^2\,\mathrm{sh}\,\alpha_k(\alpha_k - \mathrm{sh}\,\alpha_k\,\mathrm{ch}\,\alpha_k)$$

$$+\alpha_k\,\mathrm{sh}\,\alpha_k(\mathrm{ch}2\,\alpha_k - \cos m\psi)]\,/\,\Delta_{mk}$$

$$\Delta_{mk} = K\alpha_k[\mathrm{sh}2\,\alpha_k\,\mathrm{ch}2\,\alpha_k - 2\,\alpha_k + \cos m\psi(2\,\alpha_k\,\mathrm{ch}2\,\alpha_k - \mathrm{sh}2\,\alpha_k)]$$

$$+2(\mathrm{ch}2\,\alpha_k - \cos m\psi)^2$$

$$\alpha_k = \frac{k\pi a}{2b}\,. \tag{4. 6.18}$$

The solution for deflection of substructures may be found by substituting equations (4.6.9) and (4.6.16) - (4.6.18) into equation (4.1.3a) . Consider now the actual stiffened plate with an odd number of substructures, say n= 2s-1. The deflection surface of the middle (s-th) substructure of the actual system can be expressed as

$$W_s(x,y)=\frac{4pb^4}{D\pi^5 n}\sum_{k=1,3...}^{\infty}\left\{\frac{1}{k^5}\sum_{m=1,3...}^{2n-1}\left[\frac{\sin\frac{m\pi}{2}}{\sin\frac{m\psi}{2}}(1+C_{mk2}\,\mathrm{ch}\frac{k\pi x}{b}+C_{mk3}\frac{2x}{a}\mathrm{sh}\frac{k\pi x}{b})\right]\sin\frac{k\pi y}{b}\right\}$$

(4.6.19)

The maximum deflection occurs at the center of the s-th substructure and its magnitude is

$$W_{max}=W_s(0,\frac{b}{2})=\frac{4pb^4}{D\pi^5 n}\sum_{k=1,3,...}^{\infty}\left\{\frac{\sin\frac{k\pi}{2}}{k^5}\sum_{m=1,3,...}^{2n-1}\left[\frac{\sin\frac{m\pi}{2}}{\sin\frac{m\psi}{2}}(1+C_{mk2})\right]\right\} \quad (4.6.20)$$

The magnitudes of the bending moments at the center of the middle substructure are

$$M_x=-D\left[W_{s,xx}(0,\frac{b}{2})+\mu\,W_{s,yy}(0,\frac{b}{2})\right]$$

$$=\frac{4pb^2}{\pi^3 n}\sum_{k=1,3...}^{\infty}\left\{\frac{\sin\frac{k\pi}{2}}{k^3}\sum_{m=1,3,...}^{2n-1}\frac{\sin\frac{m\pi}{2}}{\sin\frac{m\psi}{2}}\left[\mu-(1-\mu)C_{mk2}-\frac{2}{\alpha_k}C_{mk3}\right]\right\},$$

$$M_y=-D\left[W_{s,yy}(0,\frac{b}{2})+\mu\,W_{s,xx}(0,\frac{b}{2})\right]$$

$$=\frac{4pb^2}{\pi^3 n}\sum_{k=1,3...}^{\infty}\left\{\frac{\sin\frac{k\pi}{2}}{k^3}\sum_{m=1,3...}^{2n-1}\frac{\sin\frac{m\pi}{2}}{\sin\frac{m\psi}{2}}\left[1+(1-\mu)C_{mk2}-\frac{2\mu}{\alpha_k}C_{mk3}\right]\right\} \quad (4.6.21)$$

As a numerical example , the deflection and bending moment M_x at the center of a square stiffened plate , *i.e.,* a = b/n , with all edges simply supported and under uniform load are calculated. The solutions shown in equations (4.6.20) and (4.6.21) include two independent parameters n and K except load p. The numerical results obtained for different values of n and K are shown in Tables 4.6.1 and 4.6.2 .

By using an orthotropic plate model with D_x=H=D and D_y=D(1+K), the corresponding results are obtained and also tabulated for comparison , where D_x , D_y and H are three independent stiffness constants in the governing differential equation for deflection of orthotropic plate. The results obtained by the orthotropic plate model increase in relative error with K. When the values of K and n are located at the area under and at the left side of the broken line shown in Table 4.6.2, the relative error for M_x is greater than 5%. Hence the present model is better than the orthotropic plate model.

Table 4.6.1 Deflections at the center point of square stiffened plates under uniform loading

K \ n	3	7	11	15	19	Orthotropic plate
0.0	4062	4062	4062	4062	4062	4062
0.2	3872	3871	3871	3871	3871	3865
0.4	3698	3696	3696	3696	3696	3686
0.6	3539	3536	3536	3536	3536	3523
0.8	3393	3389	3389	3389	3389	3373
1.0	3258	3253	3253	3253	3253	3235
2.0	2717	2707	2707	2707	2707	2684
3.0	2330	2316	2315	2315	2315	2292
4.0	2039	2021	2021	2021	2021	1998
5.0	1814	1793	1792	1792	1792	1769
6.0	1633	1609	1609	1609	1609	1587
7.0	1485	1460	1459	1459	1459	1438
8.0	1362	1335	1334	1334	1334	1314
9.0	1258	1229	1229	1228	1228	1209
10.0	1169	1139	1138	1138	1138	1119
15.0	863.4	829.6	828.8	828.7	828.6	813.7
20.0	685.4	649.9	649.0	648.9	648.9	636.4
30.0	486.6	450.3	449.4	449.2	449.2	439.8
40.0	378.2	342.5	341.5	341.4	341.3	333.9
50.0	309.9	275.3	274.3	274.1	274.1	267.9
Multiplier $(Pb^4/D)\times 10^{-6}$						

Table 4.6.2 Moments (M_x) at the center point of square stiffened plate under uniform load (μ =0.3)

K \ n	3	7	11	15	19	Orthotropic Plate
0.0	4789	4789	4789	4789	4789	4789
0.2	4578	4560	4558	4558	4557	4538
0.4	4385	4349	4345	4344	4344	4311
0.6	4208	4155	4149	4147	4146	4104
0.8	4046	3975	3968	3965	3964	3914
1.0	3897	3809	3800	3796	3795	3740
2.0	3297	3140	3122	3115	3113	3045
3.0	2869	2658	2633	2626	2622	2551
4.0	2548	2296	2266	2257	2253	2182
5.0	2298	2015	1981	1971	1966	1898
6.0	2098	1790	1753	1742	1736	1671
7.0	1935	1607	1567	1555	1549	1487
8.0	1799	1455	1413	1400	1394	1334
9.0	1683	1326	1283	1269	1263	1205
10.0	1584	1216	1171	1157	1151	1096
15.0	1247	845.5	795.5	780.3	773.6	727.8
20.0	1050	634.8	582.6	566.2	559.1	520.0
30.0	830.3	409.4	355.2	338.4	331.1	299.4
40.0	710.5	295.2	240.5	223.2	215.9	188.6
50.0	635.0	228.7	174.0	156.9	149.3	124.8
Multiplier	$Pb^2 \times 10^{-5}$					

4.6.2 Stability Analysis of Stiffened Plates

The rectangular stiffened plate with four edges simply supported, subjected to an uniformly distributed compression in y direction is considered in this section. The actual system and its equivalent system with cyclic periodicity are shown in Figure 4.6.3(a) and (b), respectively . In each substructure a local coordinate system oxy is established as shown in Figure 4.6.4.

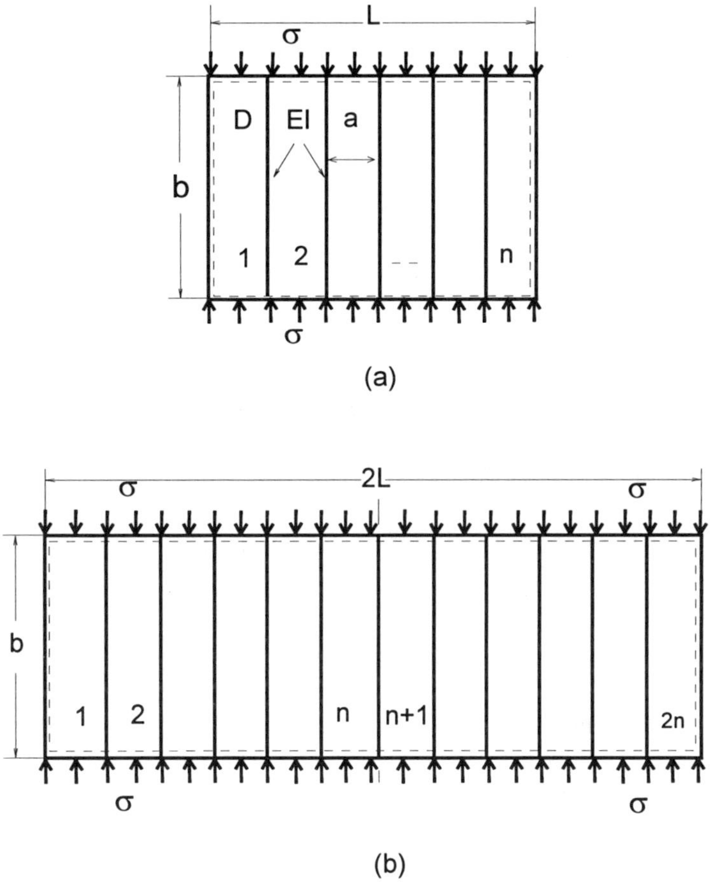

Figure 4.6.3 Stiffened plate under unidirectional compression
(a) Actual system; (b) Equivalent system

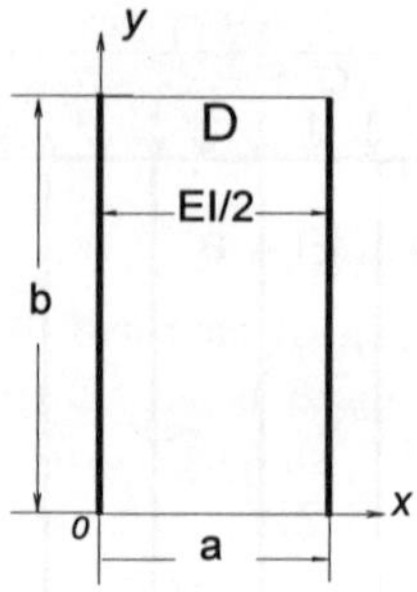

Figure 4.6.4 Substructure and local coordinate system

The equilibrium equation for each plate substructure under a uniform compression in *y*-direction takes the form

$$D\left(\frac{\partial^4 W_k}{\partial x^4} + 2\frac{\partial^4 W_k}{\partial x^2 \partial y^2} + \frac{\partial^4 W_k}{\partial y^4}\right) + \sigma \frac{\partial^2 W_k}{\partial y^2} = 0,$$

$$0<x<a, \quad 0<y<b, \quad k=1,2,...,N \tag{4.6.22}$$

where N=2n, σ denotes the pressure per unit length and $W_k(x,y)$ the transverse displacement for the k-th plate substructure . The simply supported boundary conditions at y=0, b and continuity conditions across the ribs may be summarized as follows :

$$W_k(x,0) = 0, \qquad W_{k,yy}(x,0) = 0,$$

$$W_k(x,b) = 0, \qquad W_{k,yy}(x,b) = 0, \qquad k=1,2,...,N \tag{4.6.23}$$

$$W_k(a,y) = W_{k+1}(0,y), \qquad W_{k,x}(a,y) = W_{k+1,x}(0,y)$$

$$W_{k,xx}(a,y) = W_{k+1,xx}(0,y),$$

$$\left\{-D\left[W_{k,xxx} + (2-\mu)W_{k,xyy}\right] + \frac{EI}{2}W_{k,yyyy} + \frac{\sigma_1}{2}W_{k,yy}\right\}_{x=a}$$

$$= \left\{ - D\left[W_{k+1,xxx} + (2-\mu)\, W_{k+1,xyy} \right] - \frac{EI}{2} W_{k+1,yyyy} - \frac{\sigma_1}{2} W_{k+1,yy} \right\}_{x=0} \qquad (4.6.24)$$

with k = 1, 2 ,..., N and $W_{N+1} \equiv W_1$ where μ, σ_1 denote the Poisson's ratio and pressure acting on the cross section of each rib. According to the property of the rotationally periodic structure, there is an integer number of waves along the periodic direction for each buckling shape. The anti-symmetric shapes with respect to the symmetric plane of the equivalent system automatically satisfy the simply supported boundary conditions for the extreme edges of the actual system.

Apply the U-transformation (4.1.3) to equations (4.6.22) to (4.6.24) result in

$$D\left(\frac{\partial^4 q_m}{\partial x^4} + 2\frac{\partial^4 q_m}{\partial x^2 \partial y^2} + \frac{\partial^4 q_m}{\partial y^4}\right) + \sigma \frac{\partial^2 q_m}{\partial y^2} = 0, \quad m=1,2,...,N \qquad (4.6.25)$$

$$\begin{aligned} &q_m(x,0) = 0, \qquad q_{m,yy}(x,0) = 0, \\ &q_m(x,b) = 0, \qquad q_{m,yy}(x,b) = 0, \end{aligned} \qquad (4.6.26)$$

$$\begin{aligned} &q_m(a,y) = e^{im\psi} q_m(0,y), \\ &q_{m,x}(a,y) = e^{im\psi} q_{m,x}(0,y), \\ &q_{m,xx}(a,y) = e^{im\psi} q_{m,xx}(0,y), \\ &-D q_{m,xxx}(a,y) + \frac{EI}{2} q_{m,yyyy}(a,y) + \frac{\sigma_1}{2} q_{m,yy}(a,y) \\ &= e^{im\psi}\left[-D q_{m,xxx}(0,y) - \frac{EI}{2} q_{m,yyyy}(0,y) - \frac{\sigma_1}{2} q_{m,yy}(0,y) \right] \end{aligned} \qquad (4.6.27)$$

In view of the simply supported edge conditions (4.6.26), Levy's method of solution may be applied, *i.e.*, let

$$q_m(x,y) = \sum_{k=1}^{\infty} X_{mk}(x) \sin\frac{k\pi y}{b} \qquad (4.6.28)$$

Substituting equation (4.6.28) into equations (4.6.25) and (4.6.27) yields

$$D\left[X''''_{mk}(x)-2(\frac{k\pi}{b})^2X''_{mk}(x)+(\frac{k\pi}{b})^4X_{mk}(x)\right]-\sigma(\frac{k\pi}{b})^2X_{mk}(x)=0\,, \tag{4.6.29a}$$

and

$$X_{mk}(a)=e^{im\psi}X_{mk}(0),\qquad X'_{mk}(a)=e^{im\psi}X'_{mk}(0)\,,$$

$$X''_{mk}(a)=e^{im\psi}X''_{mk}(0)\,,$$

$$-DX'''_{mk}(a)+\frac{EI}{2}\left(\frac{k\pi}{b}\right)^4X_{mk}(a)-\frac{\sigma_1}{2}\left(\frac{k\pi}{b}\right)^2X_{mk}(a)$$

$$=e^{im\psi}\left[-DX'''_{mk}(0)-\frac{EI}{2}\left(\frac{k\pi}{b}\right)^4X_{mk}(0)+\frac{\sigma_1}{2}\left(\frac{k\pi}{b}\right)^2X_{mk}(0)\right]. \tag{4.6.29b}$$

The general solution for X_{mk} of equation (4.6.29a) can be expressed as

$$X_{mk}=A_{mk}\,\mathrm{sh}\,\lambda_k x+B_{mk}\,\mathrm{ch}\,\lambda_k x+C_{mk}\sin\mu_k x+D_{mk}\cos\mu_k x \tag{4.6.30}$$

where

$$\lambda_k^2=\frac{k^2\pi^2}{b^2}+\frac{k\pi}{b}\sqrt{\frac{\sigma}{D}};\qquad \mu_k^2=-\frac{k^2\pi^2}{b^2}+\frac{k\pi}{b}\sqrt{\frac{\sigma}{D}} \tag{4.6.31}$$

and $A_{mk}, B_{mk}, C_{mk}, D_{mk}$ denote four constants of integration.

Substituting equation (4.6.30) into equation (4.6.29b) yields a set of simultaneous equations

$$\mathrm{sh}(\lambda_k a)A_{mk}+(\mathrm{ch}\,\lambda_k a-e^{im\psi})B_{mk}+\sin(\mu_k a)C_{mk}+(\cos\mu_k a-e^{im\psi})D_{mk}=0\,,$$

$$\lambda_k(\mathrm{ch}\,\lambda_k a-e^{im\psi})A_{mk}+\lambda_k\,\mathrm{sh}(\lambda_k a)B_{mk}+\mu_k(\cos\mu_k a-e^{im\psi})C_{mk}$$
$$-\mu_k\sin(\mu_k a)D_{mk}=0\,,$$

$$\lambda_k^2\,\mathrm{sh}(\lambda_k a)A_{mk}+\lambda_k^2(\mathrm{ch}\,\lambda_k a-e^{im\psi})B_{mk}-\mu_k^2\sin(\mu_k a)C_{mk}$$

$$-\mu_k^2(\cos\mu_k a - e^{im\psi})\,D_{mk} = 0\,,$$

$$\left[\lambda_k^3(\mathrm{ch}\,\lambda_k a - e^{im\psi}) - 2\eta_k\left(\frac{k\pi}{b}\right)^4 \mathrm{sh}\,\lambda_k a\right]A_{mk} + \left[\lambda_k^3\,\mathrm{sh}\,\lambda_k a - 2\eta_k\left(\frac{k\pi}{b}\right)^4 \mathrm{ch}\,\lambda_k a\right]B_{mk}$$

$$-\left[\mu_k^3(\cos\mu_k a - e^{im\psi}) + 2\eta_k\left(\frac{k\pi}{b}\right)^4 \sin\mu_k a\right]C_{mk}$$

$$+\left[\mu_k^3(\sin\mu_k a - 2\eta_k\left(\frac{k\pi}{b}\right)^4 \cos\mu_k a\right]D_{mk} = 0\,. \tag{4.6.32}$$

where

$$\eta_k = \frac{EI}{2D} - \frac{\sigma_1}{2D}\left(\frac{b}{k\pi}\right)^2 \tag{4.6.33}$$

Let the coefficient determinant of the simultaneous equations (4.6.32) be equal to zero. Expanding this determinant results in[13]

$$\lambda_k\mu_k(\lambda_k^2+\mu_k^2)(\mathrm{ch}\,\lambda_k a - \cos m\psi)(\cos\mu_k a - \cos m\psi)$$

$$+\eta_k\left(\frac{k\pi}{b}\right)^4\left[\lambda_k\sin\mu_k a(\mathrm{ch}\,\lambda_k a - \cos m\psi) - \mu_k\,\mathrm{sh}\,\lambda_k a(\cos\mu_k a - \cos m\psi)\right] = 0$$

$$m=1,2,\ldots,N \quad k=1,2,3,\ldots,\infty \tag{4.6.34}$$

This is the buckling load equation which includes two loading parameters σ and σ_1. If the pressures per unit area for both plate and ribs are the same, then

$$\sigma_1 = \frac{\sigma A}{h} \tag{4.6.35}$$

in which A and h denote the area of cross section of a rib and the thickness of plate respectively.

The critical load σ_{cr} is the minimum root for σ of characteristic equation (4.6.34) with equation (5.6.35). It is assumed that σ_{cr} is the root of equation

(4.6.34) when m=r and k=j. Then four constants of integration in equation (4.6.30) where m=r and k=j, can be determined except an arbitrary complex factor. Upon substituting $\sigma = \sigma_{cr}$ into equation (4.6.32) where m=r and k=j, the non-trivial solution for $A_{rj}, B_{rj}, C_{rj}, D_{rj}$ can be obtained as

$$A_{rj} = E_{rj}\mu_j(\text{ch}\,\lambda_j a - e^{ir\psi})(\cos r\psi - \cos\mu_j a),$$

$$B_{rj} = -E_{rj}\mu_j \text{sh}\,\lambda_j a(\cos r\psi - \cos\mu_j a),$$

$$C_{rj} = E_{rj}\lambda_j(\cos\mu_j a - e^{ir\psi})(\text{ch}\,\lambda_j a - \cos r\psi),$$

$$D_{rj} = -E_{rj}\lambda_j \sin\mu_j a(\text{ch}\,\lambda_j a - \cos r\psi), \qquad r \neq n, 2n \qquad (4.6.36)$$

where E_{rj} denotes an arbitrary complex factor. Substituting equation (4.6.36) into equation (4.6.30) yields

$$X_{rj} = E_{rj}\left\{\left[\text{sh}\,\lambda_j(x-a) - \text{sh}\,\lambda_j x\, e^{ir\psi}\right]\mu_j(\cos r\psi - \cos\mu_j a)\right.$$

$$\left.+\left[\sin\mu_j(x-a) - \sin\mu_j x\, e^{ir\psi}\right]\lambda_j(\text{ch}\,\lambda_j a - \cos r\psi)\right\},$$

$$X_{2n-r,j} = \overline{X}_{r,j},$$

$$X_{mk} = 0, \qquad m \neq r,\ 2n\text{-}r \ \text{ or } \ k \neq j. \qquad (4.6.37)$$

Substituting equation (4.6.37) into equation (4.6.28) yields

$$q_r = E_{rj}\left\{\left[\text{sh}\,\lambda_j(x-a) - \text{sh}\,\lambda_j x\, e^{ir\psi}\right]\mu_j(\cos r\psi - \cos\mu_j a)\right.$$

$$\left.+\left[\sin\mu_j(x-a) - \text{sh}\mu_j x\, e^{ir\psi}\right]\lambda_j(\text{ch}\,\lambda_j a - \cos r\psi)\right\}\sin\frac{j\pi y}{b}$$

$$q_{2n-r} = \overline{q}_r\,,$$

$$q_m = 0, \qquad m \neq r,\ 2n\text{-}r \qquad (4.6.38)$$

Substituting equation (4.6.38) into equation (4.1.3a) yields two independent buckling shapes (E_{rj}= 1,i respectively) for the equivalent system. Their linear combination is also a buckling shape. But there is only one shape which will satisfy the actual boundary conditions.

when E_{rj} = i (*i.e.*, $\sqrt{-1}$), neglecting an arbitrary real factor, the buckling shape can be expressed as

$$W_k(x,y) = \sin\frac{j\pi y}{b}\{[\mu_j \mathrm{sh}\lambda_j x(\cos\mu_j a - \cos r\psi) + \lambda_j \sin\mu_j x(\cos r\psi - \mathrm{ch}\lambda_j a)]\sin kr\psi$$

$$-[\mu_j(\cos\mu_j a - \cos r\psi)\mathrm{sh}\lambda_j(x-a) + \lambda_j(\cos r\psi - \mathrm{ch}\lambda_j a)\sin\mu_j(x-a)]\sin(k-1)r\psi\}$$

$$k=1,2,...,N, \quad r \neq n, 2n \tag{4.6.39}$$

It can be verified that the deformed shape shown in equation (4.6.39) satisfy the extreme edges simply supported conditions of the actual stiffened plate, *i.e.*,

$$W_1(0,y) = 0, \qquad W_{1,xx}(0,y) = 0,$$

$$W_n(a,y) = 0, \qquad W_{n,xx}(a,y) = 0. \tag{4.6.40}$$

For buckling shape shown in equation (4.6.39), r and j denote the half wave numbers in x and y directions.

Now considering the case of r=n, the equation (4.6.34) in which m=n and k=j, can be decomposed as

$$\cos\frac{\mu_j a}{2} = 0, \tag{4.6.41a}$$

or

$$\lambda_j\mu_j(\lambda_j^2+\mu_j^2)\mathrm{ch}\frac{\lambda_j a}{2}\cos\frac{\mu_j a}{2} + \eta_j\left(\frac{j\pi}{b}\right)^4(\lambda_j\sin\frac{\mu_j a}{2}\mathrm{ch}\frac{\lambda_j a}{2} - \mu_j\mathrm{sh}\frac{\lambda_j a}{2}\cos\frac{\mu_j a}{2}) = 0. \tag{4.6.41b}$$

The minimum positive root is

$$\mu_j a = \pi \qquad \text{or} \qquad \sigma = \frac{\pi^2 D}{a^2}\left(\frac{ja}{b} + \frac{b}{ja}\right)^2 \tag{4.6.42}$$

The non-trivial solution of equation (4.6.32) for present case can be expressed as

$$A_{nj} = B_{nj} = D_{nj} = 0, \qquad C_{nj} = E_{nj} \tag{4.6.43}$$

Substituting equation (4.6.43) into (4.6.30) yields

$$X_{nj} = E_{nj} \sin \mu_j x = E_{nj} \sin \frac{\pi x}{a},$$

$$X_{m,k} = 0, \qquad m \neq n \text{ or } k \neq j \tag{4.6.44}$$

and then

$$q_n(x, y) = E_{nj} \sin \frac{\pi x}{a} \sin \frac{j\pi y}{b},$$

$$q_m = 0, \qquad m \neq n \tag{4.6.45}$$

Substituting equation (4.6.45) into equation (4.1.3a) results in

$$W_k(x, y) = (-1)^{k-1} \sin \frac{\pi x}{a} \sin \frac{j\pi y}{b} \qquad k=1,2,\ldots,n \tag{4.6.46}$$

where the arbitrary constant factor has been neglected. It is obvious that each rib coincides with a nodal line of the buckling shape shown in equation (4.6.46). The critical load and buckling shape shown in equations (4.6.42) and (4.6.46) are the same with those for the single plate substructure with four simply supported edges.

Incidentally, it should be noted that it is impossible for the critical load σ_{cr} to satisfy equation (4.6.34) for k=2n.

Some numerical results of the critical load are given in Tables 4.6.3 and 4.6.4 where L=na, $\gamma = EI/LD$, $\beta = b/L$ and $\delta = A/Lh$.

Both Tables 4.6.3 and 4.6.4 show that when the flexural rigidity of the ribs is on the increase, the critical buckling load is also on the increase or invariant. The latter case is corresponding to that each rib is located at the nodal line of the buckling shape.

Table 4.6.3 Critical load parameter $\sigma_{cr} L^2/\pi^2 D$ (n=2)

	γ =5			γ =10			γ =15			γ =20		
	δ			δ			δ			δ		
β	0.05	0.10	0.20	0.05	0.10	0.20	0.05	0.10	0.20	0.05	0.10	0.20
0.6	16.5	16.5	16.5	16.5	16.5	16.5	16.5	16.5	16.5	16.5	16.5	16.5
*	16.5	16.5	16.5	16.5	16.5	16.5	16.5	16.5	16.5	16.5	16.5	16.5
0.8	15.1	14.3	12.8	16.8	16.8	16.8	16.8	16.8	16.8	16.8	16.8	16.8
*	15.4	14.6	13.0	16.8	16.8	16.8	16.8	16.8	16.8	16.8	16.8	16.8
1.0	11.9	11.1	9.67	16.0	15.5	16.0	16.0	16.0	16.0	16.0	16.0	16.0
*	12.0	11.1	9.72	16.0	16.0	15.8	16.0	16.0	16.0	16.0	16.0	16.0
1.2	9.78	9.03	7.82	15.2	14.1	12.4	16.5	16.5	16.5	16.5	16.5	16.5
*	9.83	9.06	7.88	15.3	14.2	12.4	16.5	16.5	16.5	16.5	16.5	16.5
1.6	8.00	7.35	6.32	11.4	10.5	9.05	14.6	13.5	11.7	16.1	16.1	14.3
*	8.01	7.38	6.32	11.4	10.5	9.05	14.7	13.6	11.8	16.1	16.1	14.4
2.0	7.95	7.29	6.25	10.2	9.35	8.03	12.4	11.4	9.79	14.5	13.4	11.5
*	7.96	7.29	6.24	10.2	9.35	8.03	12.4	11.4	9.80	14.6	13.4	11.6
2.4	8.79	8.06	6.91	10.4	9.50	8.15	11.9	10.9	9.38	13.5	12.4	10.6
*	8.79	8.06	6.91	10.4	9.49	8.15	11.9	10.9	9.37	13.5	12.4	10.6
2.8	8.59	7.91	6.81	11.4	10.4	8.94	12.5	11.5	9.85	13.7	12.5	10.8
*	8.62	7.91	6.81	11.4	10.4	8.94	12.5	11.5	9.85	13.7	12.6	10.8
3.2	8.00	7.35	6.32	11.4	10.5	9.05	13.9	12.7	10.9	14.8	13.5	11.6
*	8.01	7.38	6.32	11.4	10.5	9.05	13.9	12.7	10.9	14.8	13.5	11.6
3.6	7.83	7.19	6.17	10.6	9.70	8.35	13.2	12.2	10.5	15.8	14.6	12.6
*	7.84	7.19	6.16	10.6	9.70	8.35	13.2	12.2	10.5	15.9	14.7	12.6
4.0	7.95	7.29	6.25	10.2	9.35	8.03	12.4	11.4	9.79	14.5	13.4	11.5
*	7.96	7.29	6.24	10.2	9.35	8.03	12.4	11.4	9.8	14.6	13.4	11.6

* The numerical results are obtained by energy method

In order to design the ribs of stiffened plate with given value of the critical buckling load σ_{cr} , the minimum flexural rigidity EI_{min} can be determined by using the buckling load equation (4.6.34) with equations (4.6.33) and (4.6.35), *i.e.*,

$$\gamma_{min} = \underset{\substack{m=1,2,...,n-1 \\ k=1,2,3,...}}{\text{Max}} \left[\frac{2\lambda_k \mu_k (\lambda_k^2 + \mu_k^2)(\text{ch}\,\lambda_k a - \cos m\psi)(\cos m\psi - \cos \mu_k a)}{\lambda_k \sin \mu_k a(\text{ch}\,\lambda_k a - \cos m\psi) - \mu_k \,\text{sh}\,\lambda_k a(\cos \mu_k a - \cos m\psi)} \frac{b^4}{L k^4 \pi^4} \right.$$

$$+\frac{\delta\,\sigma_{cr}\,b^2}{D\,k^2\,\pi^2}] \qquad \sigma_{cr} \le \sigma_{cr}^* \tag{4.6.47}$$

in which σ_{cr}^* denotes the buckling load determined by equation (4.6.42) and $\gamma_{min} = EI_{min}/(LD)$. When $\sigma_{cr} < \sigma_{cr}^*$, substituting the EI_{min} defined by equation (4.6.47) into equations (4.6.34), the minimum root for σ of each equation in equations (4.6.34) is greater than or equal to σ_{cr}, *i.e.*, σ_{cr} is the buckling load. In order to keep every rib as a rigid one, *i.e.*, $\sigma_{cr} = \sigma_{cr}^*$, the minimum flexural rigidity EI_{min} can be obtained by using σ_{cr}^* instead of σ_{cr} in equation (4.6.47). Some numerical results are given in Tables 4.6.5 and 4.6.6.

Table 4.6.4 Critical load parameter $\sigma_{cr} L^2/\pi^2 D$ (n=3)

		γ=10/3		γ=5		γ=20/3		γ=10	
β		δ=0.05	0.10	δ=0.05	0.10	δ=0.05	0.10	δ=0.05	0.10
0.6		26.3	23.7	35.4	32.3	36.4	36.4	36.4	36.4
	*	26.8	24.1	36.4	33.2	36.4	36.4	36.4	36.4
0.8		16.8	15.0	23.0	20.6	29.0	26.0	37.2	36.4
	*	16.9	15.0	23.3	20.7	29.4	26.3	37.2	37.1
1.0		12.1	10.7	16.2	14.4	20.4	18.1	28.4	25.4
	*	12.1	10.7	16.3	14.5	20.5	18.2	28.7	25.6
1.2		9.60	8.50	12.6	11.1	15.5	13.7	21.3	19.0
	*	9.61	8.51	12.6	11.2	15.5	13.8	21.4	19.0
1.4		8.31	7.36	10.5	9.31	12.7	11.3	17.0	15.1
	*	8.32	7.36	10.5	9.32	12.7	11.3	17.2	15.2
1.6		7.70	6.81	9.39	8.31	11.1	9.80	14.4	12.8
	*	7.70	6.81	9.40	8.31	11.1	9.82	14.5	12.8
1.8		7.51	6.64	8.85	7.83	10.2	9.01	12.8	11.4
	*	7.51	6.64	8.85	7.83	10.2	9:02	12.9	11.4
2.0		7.61	6.73	8.69	7.69	9.73	8.65	11.9	10.6
	*	7.61	6.73	8.70	7.69	9.78	8.65	11.9	10.6

* The numerical results are obtained by energy method

Table 4.6.5 γ_{min} for the case that the rib is a nodal line of the buckling shape the stiffened plates with one rib

δ / β	0.05	0.10	02.0
0.6	3.7553 (3.2932)*	4.0530 (3.6312)	4.6484 (4.3072)
0.8	6.0933 (5.3372)	6.6312 (5.3872)	7.7071 (7.0232)
1.0	8.0254 (7.8500)	8.8254 (8.7000)	10.425 (10.400)
1.2	11.384 (10.754)	12.757 (11.947)	14.956 (14.351)
1.4	14.013 (13.917)	15.588 (15.535)	18.739 (18.771)
1.6	17.241 (17.241)	19.298 (19.339)	23.411 (23.535)
1.8	20.704 (20.573)	23.325 (23.215)	28.566 (28.499)
2.0	23.418 (23.750)	26.618 (27.000)	33.018 (33.500)
2.2	26.548 (26.589)	30.456 (30.511)	38.270 (38.355)
2.4	28.412 (28.889)	33.028 (33.547)	42.259 (42.862)
2.6	29.847 (30.429)	35.263 (35.887)	46.095 (46.803)
2.8	30.524 (30.97)	36.826 (37.291)	49.430 (49.935)
3.0	29.376 (30.97)	36.576 (37.68)	50.976 (52.00)
3.2	27.421 (30.97)	35.647 (37.68)	52.100 (52.72)
3.4	22.978 (30.97)	32.234 (37.68)	50.745 (52.72)
3.6	20.381 (30.97)	27.081 (37.68)	47.833 (52.72)
3.8	22.008 (30.97)	24.904 (37.68)	42.958 (52.72)
4.0	23.418 (30.97)	26.618 (37.68)	34.731 (52.72)

* The numerical results in the round brackets are quoted from reference [14]

4.6.3 Natural Vibration of Stiffened Plates

Considering the stiffened plate shown in Figure 4.6.3(a) with $\sigma = 0$, the natural vibration equation for each plate substructure can be expressed as

$$D\left(\frac{\partial^4 W_k}{\partial x^4} + 2\frac{\partial^4 W_k}{\partial x^2 \partial y^2} + \frac{\partial^4 W_k}{\partial y^4}\right) - \rho h \omega^2 W_k = 0 \tag{4.6.48}$$

where $W_k(x, y)$, ω denote the natural mode, frequency and ρ density of plate.

The boundary condition and continuity condition require the following equations to be satisfied:

$$W_k(x,0) = W_k(x,b) = 0\,, \qquad W_{k,yy}(x,0) = W_{k,yy}(x,b) = 0\,,$$
$$k=1,2,\ldots,N \tag{4.6.49}$$

and

$$W_k(a,y) = W_{k+1}(0,y), \qquad W_{k,x}(a,y) = W_{k+1,x}(0,y)\,,$$
$$W_{k,xx}(a,y) = W_{k+1,xx}(0,y)\,,$$
$$\left\{-D\left[W_{k,xxx} + (2-\mu)\,W_{k,xyy}\right] + \frac{EI}{2}W_{k,yyyy} - \frac{\rho_1 A}{2}\omega^2 W_k\right\}_{x=a}$$
$$= \left\{-D\left[W_{k+1,xxx} + (2-\mu)\,W_{k+1,xyy}\right] - \frac{EI}{2}W_{k+1,yyyy} + \frac{\rho_1 A}{2}\omega^2 W_{k+1}\right\}_{x=0} \tag{4.6.50}$$

with k=1,2,...,N and $W_{N+1} \equiv W_1$, where ρ_1 denotes the density of ribs.

Table 4.6.6 γ_{min} for the case that the ribs are nodal lines of the buckling shape of the stiffened plates with two ribs

δ / β	0.05	0.10	02.0
0.6	5.2000 (5.0681)*	5.8552 (5.7661)	7.1657 (7.4021)
0.8	9.0772 (8.6521)	10.268 (9.8541)	12.649 (12.685)
1.0	13.236 (13.183)	15.036 (15.033)	18.636 (19.200)
1.2	18.880 (18.604)	21.501 (21.246)	26.743 (27.490)
1.4	24.925 (24.844)	28.462 (28.422)	35.535 (36.885)
1.6	31.868 (31.820)	36.483 (36.478)	45.716 (47.501)
1.8	39.668 (39.436)	45.535 (45.318)	57.268 (59.242)
2.0	47.508 (47.583)	54.708 (54.833)	69.108 (72.000)

* see the footnote of Table 4.6.5

Applying the U-transformation (4.1.3) to equations (4.6.48) to (4.6.50) result in

$$D\left(\frac{\partial^4 q_m}{\partial x^4}+2\frac{\partial^4 q_m}{\partial x^2 \partial y^2}+\frac{\partial^4 q_m}{\partial y^4}\right)-\rho h \omega^2 q_m=0, \qquad m=1,2,...,N \quad (4.6.51)$$

$$q_m(x,0)=q_m(x,b)=0, \quad q_{m,yy}(x,0)=q_{m,yy}(x,b)=0$$
$$m=1,2,...,N \quad (4.6.52)$$

and

$$q_m(a,y)=e^{im\psi}q_m(0,y), \qquad q_{m,x}(a,y)=e^{im\psi}q_{m,x}(0,y),$$
$$q_{m,xx}(a,y)=e^{im\psi}q_{m,xx}(0,y),$$
$$\left\{-D\left[q_{m,xxx}+(2-\mu)q_{m,xyy}\right]+\frac{EI}{2}q_{m,yyyy}-\frac{\rho_1 A}{2}\omega^2 q_m\right\}_{x=a}$$
$$=e^{im\psi}\left\{-D\left[q_{m,xxx}+(2-\mu)q_{m,xyy}\right]-\frac{EI}{2}q_{m,yyyy}+\frac{\rho_1 A}{2}\omega^2 q_m\right\}_{x=0},$$
$$m=1,2,...,N \quad (4.6.53)$$

In view of the simply supported edge condition (4.6.52), Levy's method of solution can be applied. Substituting equation (4.6.28) into equations (4.6.51) and (4.6.53), yield

$$D[X''''_{mk}(x)-2\left(\frac{k\pi}{b}\right)^2X''_{mk}(x)+\left(\frac{k\pi}{b}\right)^4X_{mk}(x)]-\rho h\omega^2X_{mk}(x)=0,$$
$$k=1,2,...,\infty, \quad m=1,2,...,N \quad (4.6.54)$$

and

$$X_{mk}(a)=e^{im\psi}X_{mk}(0), \quad X'_{mk}(a)=e^{im\psi}X'_{mk}(0),$$
$$X''_{mk}(a)=e^{im\psi}X''_{mk}(0),$$

$$-DX'''_{mk}(a)+\frac{EI}{2}\left(\frac{k\pi}{b}\right)^4 X_{mk}(a)-\frac{\rho_1 A}{2}\omega^2 X_{mk}(a)$$

$$=e^{im\psi}\left[-DX'''_{mk}(0)-\frac{EI}{2}\left(\frac{k\pi}{b}\right)^4 X_{mk}(0)+\frac{\rho_1 A}{2}\omega^2 X_{mk}(0)\right],$$

$$k=1,2,\ldots,\infty, \quad m=1,2,\ldots,N. \tag{4.6.55}$$

It is obvious that equations (4.6.54) and (4.6.55) can be obtained from the buckling differential equation (4.6.29a) and its boundary conditions (4.6.29b) by using $\rho h\omega^2(\frac{b}{k\pi})^2$ and $\rho_1 A\omega^2(\frac{b}{k\pi})^2$ instead of σ and σ_1 respectively. The general solution for X_{mn} of equation (4.6.54) can also be expressed as equation (4.6.30) in which

$$\lambda_k^2=\frac{k^2\pi^2}{b^2}+\omega\sqrt{\frac{\rho h}{D}}, \qquad \mu_k^2=-\frac{k^2\pi^2}{b^2}+\omega\sqrt{\frac{\rho h}{D}}. \tag{4.6.56}$$

The frequency equation can be obtained from the buckling load equation (4.6.34) where λ_k, μ_k should be defined by equation (4.6.56) instead of equation (4.6.31) and

$$\eta_k=\frac{EI}{2D}-\frac{\rho_1 A}{2D}\omega^2\left(\frac{b}{k\pi}\right)^4 \tag{4.6.57}$$

instead of equation (4.6.33).

Noting the definitions of frequency parameters λ_k, μ_k and η_k shown in equations (4.6.56) and (4.6.57), the natural mode may also be expressed as equation (4.6.39). Some numerical results of the first three natural frequencies are given in Tables 4.6.7-4.6.10 where $(EI/aD)=20$, and $(\rho_1 A/\rho a h)=0.40$.

Table 4.6.7 Non-dimensional frequency parameter $(\rho h\, b^4 / D\pi^4)\,\omega^2$ of stiffened plates with one rib

β / order	1.0	1.2	1.4	1.6	1.8	2.0
1st	15.54 (17.14)*	17.8 (18.53)	20.19 (20.54)	23.17 (23.34)	27.05 (27.13)	32.11 (32.14)
2nd	25.00 (32.14)	45.70 (46.93)	47.98 (70.14)	49.05 (104.5)	50.33 (153.5)	51.88 (220.7)
3rd	45.87 (85.71)	47.01 (153.5)	78.15 (262.5)	126.3 (427.1)	194.9 (664.0)	289.0 (992.1)

* The results in the round brackets are quoted from reference [15]

Table 4.6.8 Non-dimensional frequency parameter $(\rho h\, b^4 / D\pi^4)\,\omega^2$ of stiffened plates with two ribs

β / order	1.0	1.2	1.4	1.6	1.8	2.0
1st	16.89 (17.14)*	18.43 (18.53)	20.49 (20.54)	23.31 (23.34)	27.11 (27.13)	32.14 (32.14)
2nd	31.99 (32.14)	46.92 (46.93)	48.03 (70.14)	49.06 (104.5)	50.33 (153.5)	51.89 (220.7)
3rd	46.51 (85.71)	47.02 (153.5)	61.45 (262.5)	69.37 (427.1)	80.13 (664.0)	94.42 (992.1)

* See the footnote of Table 4.6.7.

Table 4.6.9 Non-dimensional frequency parameter $(\rho h\, b^4 / D\, \pi^4)\, \omega^2$ of stiffened plates with three ribs

β order	1.0	1.2	1.4	1.6	1.8	2.0
1st	17.07 (17.14)*	18.51 (18.53)	20.53 (20.54)	23.33 (23.34)	27.12 (27.13)	32.14 (32.14)
2nd	32.11 (32.14)	46.93 (46.93)	48.04 (70.14)	49.06 (104.5)	50.33 (153.5)	51.89 (220.7)
3rd	46.59 (85.71)	48.04 (153.5)	62.01 (262.5)	70.34 (427.1)	81.65 (664.0)	96.63 (992.1)

* See the footnote of Table 4.6.7

Table 4.6.10 Non-dimensional frequency parameter $(\rho h\, b^4 / D\, \pi^4)\, \omega^2$ of the stiffened plates with four ribs

β order	1.0	1.2	1.4	1.6	1.8	2.0
1st	17.11 (17.14)*	18.53 (18.53)	20.54 (20.54)	23.34 (23.34)	27.13 (27.13)	32.14 (32.14)
2nd	32.13 (32.14)	46.93 (46.93)	70.10 (70.14)	104.5 (104.5)	153.5 (153.5)	220.6 (220.7)
3rd	86.65 (85.71)	153.2 (153.5)	261.8 (262.5)	425.9 (427.1)	662.0 (664.0)	989.1 (992.1)

* See the footnote of Table 4.6.7

Chapter 5

STRUCTURES WITH PERIODICITY IN TWO DIRECTIONS

Structures with cyclic periodicity in two directions seldom occur in practice. By way of exception, the infinitely long cylindrical shell with periodic stiffening ribs in axial and circular directions may be regarded as one. We shall consider the case when the displacements and stresses at the two infinite extreme ends are equal to zero. Therefore it may be imagined that the two extreme ends are connected in the same way as with other adjacent substructures. Moreover, for some structures with periodicity in two linear directions, their equivalent system with cyclic periodicity in two directions may be found [16–18]. Rectangular grids and space grids with simply supported ends may be taken as examples.

5.1 Governing Equation and Double U-Transformation

Figure 5.1.1 is a diagrammatic sketch of a structure with cyclic periodicity in the x- and y-directions, in which the right and left sides, and the upper and lower sides are imagined to be connected in the same way as with other adjacent substructures. For example, consider a rectangular plate with periodic stiffening ribs in two orthogonal directions and four edges simply supported. The plate and loading are extended twice by producing an anti-symmetrical image about the right side and lower side respectively. The new rectangular plate with four simply supported edges can be regarded as a cyclic periodic structure in two directions. The ribs on the right and lower edges of the new plate can be left out when the right and left sides and the upper and lower sides are imagined to be connected. Thus, the boundary condition of the original plate will be satisfied automatically and the supports may be removed.

Let M and N denote the numbers of substructures along the x-and y-directions shown in Figure 5.1.1. The cyclic periodic structure in two directions consists of $M \times N$ identical substructures. The total number of degrees of freedom in the whole structure is $M \times N \times J$, in which J denotes the number of degrees of freedom for the substructure. In Figure 5.1.1, a pair of integers (r,s) denotes a substructure in which r and s in the round brackets denote the ordinal numbers of the substructure along the x- and y-directions.

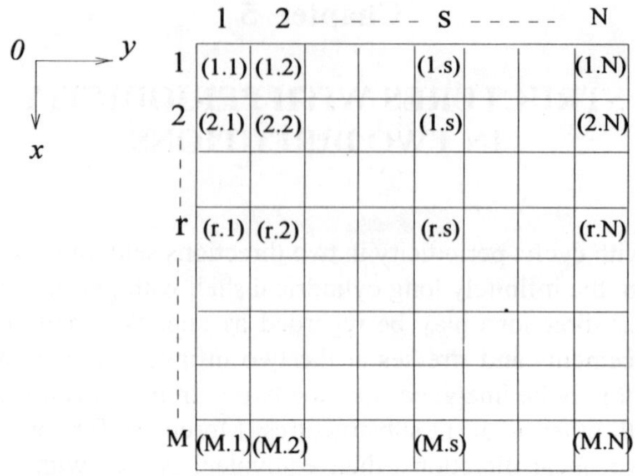

Figure 5.1.1 Diagrammatic sketch for a structure with periodicity in two directions

The degrees of freedom (DOF) for the whole structure are arranged in ascending order so that J degrees of freedom of substructure (1,1) are followed by J degrees of freedom of substructure(1,2), etc., until J DOF of substructure (1,N) are followed by J DOF of substructure(2,1) and so on. By the finite element method, the displacement and loading vectors of the whole structure, **W** and **F** may be expressed as

$$\mathbf{W} = \left[\mathbf{W}_1, \mathbf{W}_2, \cdots, \mathbf{W}_M\right]^T, \qquad \mathbf{F} = \left[\mathbf{F}_1, \mathbf{F}_2, \cdots, \mathbf{F}_M\right]^T \tag{5.1.1}$$

where

$$\mathbf{W}_r = \left[W_{r1}, W_{r2}, \cdots, W_{rN}\right]^T, \qquad \mathbf{F}_r = \left[F_{r1}, F_{r2}, \cdots, F_{rN}\right]^T$$

$$r=1,2,...,M \tag{5.1.2}$$

are vector components of **W** and **F**, while W_{rs} and F_{rs} (r=1,2,...M and s=1,2,....N) denote the displacement and loading vectors of substructure (r,s). The stiffness matrix **K** may be expressed by a symmetric cyclic matrix, *i.e.*,

$$\mathbf{K} = \begin{bmatrix} \mathbf{K}_{11}\mathbf{K}_{12}\cdots\mathbf{K}_{1M} \\ \mathbf{K}_{12}\mathbf{K}_{22}\cdots\mathbf{K}_{2M} \\ \vdots \quad \vdots \qquad \vdots \\ \mathbf{K}_{M1}\mathbf{K}_{M2}\cdots\mathbf{K}_{MM} \end{bmatrix} \tag{5.1.3}$$

with

$$\mathbf{K}_{ru} = \mathbf{K}_{ur}^{T}, \qquad r,u=1,2,...,M$$

$$\mathbf{K}_{11} = \mathbf{K}_{22} = \cdots = \mathbf{K}_{MM} \ ,$$

$$\mathbf{K}_{1r} = \mathbf{K}_{2,r+1} = \cdots = \mathbf{K}_{M-r+1,M} = \mathbf{K}_{M-r+2,1} = \cdots = \mathbf{K}_{M,r-1} \ , \quad r=2,3,...,M \tag{5.1.4}$$

and the submatrices of $\mathbf{K}$ are also cyclic matrices. In a general way, they can be expressed as

$$\mathbf{K}_{ru} = \begin{bmatrix} K_{ru}^{(1,1)} & K_{ru}^{(1,2)} & \cdots & K_{ru}^{(1,N)} \\ K_{ru}^{(2,1)} & K_{ru}^{(2,2)} & \cdots & K_{ru}^{(2,N)} \\ \vdots & \vdots & & \vdots \\ K_{ru}^{(N,1)} & K_{ru}^{(N,2)} & \cdots & K_{ru}^{(N,N)} \end{bmatrix}, \qquad r,u=1,2,...,M \tag{5.1.5}$$

together with cyclic periodicity conditions

$$K_{ru}^{(1,1)} = K_{ru}^{(2,2)} = \cdots = K_{ru}^{(N,N)},$$

$$K_{ru}^{(1,s)} = K_{ru}^{(2,s+1)} = \cdots = K_{ru}^{(N-s+1,N)} = K_{ru}^{(N-s+2,1)} = \cdots = K_{ru}^{(N,s-1)},$$

$$s=2,3,...,N \tag{5.1.6}$$

The mass matrix may be expressed in the same manner. The dynamic equation is

$$\mathbf{KW} + \mathbf{M}\ddot{\mathbf{W}} = \mathbf{F} \ . \tag{5.1.7}$$

The structure with cyclic periodicity in the x-and y-directions may be firstly regarded as one with rotational periodicity in the x-direction. For each rotating mode **W** (along the x-direction only), in a similar way to equation (1.2.4), can be expressed by using the eigenvector for the first substructure, *i.e.*,

$$\mathbf{W} = \mathbf{U}_r \mathbf{Z}_r \qquad r=1,2,...,M \tag{5.1.8}$$

where

$$\mathbf{U}_r = \frac{1}{\sqrt{M}} \left[\mathbf{I}_{N\bullet J}, e^{ir\psi_1} \mathbf{I}_{N\bullet J}, e^{i2r\psi_1} \mathbf{I}_{N\bullet J}, \cdots, e^{i(M-1)r\psi_1} \mathbf{I}_{N\bullet J} \right]^T \tag{5.1.9}$$

with $\psi_1 = 2\pi/M$ and $\mathbf{I}_{N\bullet J}$ denotes the unit matrix with N times J order, $\mathbf{Z}_r$ denotes an eigenvector of the substructures(1,1)-(1,N) with phase difference $r\psi_1$ of adjacent substructures in the x-direction. It has been proved[3] in chapter 1, that $\mathbf{Z}_r$ is the eigenvector of the following eigenvalue problem

$$\mathbf{k}_r \mathbf{Z}_r - \omega^2 \mathbf{m}_r \mathbf{Z}_r = \mathbf{0}, \qquad r=1,2,...,M \tag{5.1.10}$$

where

$$\mathbf{k}_r = \overline{\mathbf{U}}_r^T \mathbf{K} \mathbf{U}_r, \qquad \mathbf{m}_r = \overline{\mathbf{U}}_r^T \mathbf{M} \mathbf{U}_r,$$

i.e.,

$$\mathbf{k}_r = \sum_{u=1}^{M} \mathbf{K}_{1u} e^{i(u-1)r\psi_1},$$

$$\mathbf{m}_r = \sum_{u=1}^{M} \mathbf{M}_{1u} e^{i(u-1)r\psi_1}. \tag{5.1.11}$$

Noting that $\mathbf{k}_r$ and $\mathbf{m}_r$ are also cyclic matrices, therefore the rotating eigenvectors for $\mathbf{Z}_r$ of equation (5.1.10) are in existence, *i.e.*, $\mathbf{Z}_r$ may be expressed by a rotating vector as follows:

$$\mathbf{Z}_r = \mathbf{U}_s \mathbf{Z}_{rs} \qquad s=1,2,...,N \tag{5.1.12}$$

with

$$\mathbf{U}_s = \frac{1}{\sqrt{N}}\left[\mathbf{I}_J, e^{is\psi_2}\mathbf{I}_J, e^{i2s\psi_2}\mathbf{I}_J, \cdots, e^{i(N-1)s\psi_2}\mathbf{I}_J\right]^T \tag{5.1.13}$$

and $\psi_2 = 2\pi/N$. Substituting equation (5.1.12) into equation (5.1.10) and pre-multiplying by $\overline{\mathbf{U}}_s^T$ yields

$$\mathbf{k}_{rs}\mathbf{z}_{rs} - \omega^2 \mathbf{m}_{rs}\mathbf{z}_{rs} = 0 \qquad r=1,2,\ldots,M, \quad s=1,2,\ldots,N \tag{5.1.14}$$

where

$$\mathbf{k}_{rs} = \sum_{u=1}^{M}\sum_{v=1}^{N}\mathbf{K}_{1u}^{(1,v)} e^{i(u-1)r\psi_1} e^{i(v-1)s\psi_2}$$

$$\mathbf{m}_{rs} = \sum_{u=1}^{M}\sum_{v=1}^{N}\mathbf{M}_{1u}^{(1,v)} e^{i(u-1)r\psi_1} e^{i(v-1)s\psi_2} \tag{5.1.15}$$

and $\mathbf{k}_{rs}$ and $\mathbf{m}_{rs}$ are Hermitian matrices of order J. J eigenvectors of equation (5.1.14) will be denoted by $\mathbf{Z}_{rs}^j$ (j=1,2,...,J). Substituting equation (5.1.12) into equation (5.1.8), the rotating modes may be expressed as

$$\mathbf{W} = \mathbf{U}_{rs}\mathbf{Z}_{rs}^j \qquad r=1,2,\ldots,M \quad s=1,2,\ldots,N \quad j=1,2,\ldots,J \tag{5.1.16}$$

where

$$\mathbf{U}_{rs} = \mathbf{U}_r\mathbf{U}_s = \frac{1}{\sqrt{M}}\left[\mathbf{U}_s^T, e^{ir\psi_1}\mathbf{U}_s^T, e^{i2r\psi_1}\mathbf{U}_s^T, \cdots, e^{i(M-1)r\psi_1}\mathbf{U}_s^T\right]^T. \tag{5.1.17}$$

Each column vector of matrix $\mathbf{U}_{rs}$ is in agreement with the property of the rotating mode in two directions. For given values of r and s, J columns of $\mathbf{U}_{rs}$ represent a set of normal basis vectors in the mode subspace.

Expanding the possible displacement vector in a series of natural modes yields

$$W = \sum_{r=1}^{M}\sum_{s=1}^{N}\sum_{j=1}^{J} a_{rs}^j \mathbf{U}_{rs}\mathbf{Z}_{rs}^j \tag{5.1.18}$$

where a_{rs}^j denotes the mode coordinate. The possible displacement vector may be expanded in a series of components in the mode subspaces, *i.e.*,

$$W = \sum_{r=1}^{M}\sum_{s=1}^{N} \mathbf{U}_{rs}\mathbf{q}_{rs} \tag{5.1.19}$$

in which

$$\mathbf{q}_{rs} = \sum_{j=1}^{J} a_{rs}^{j}\,\mathbf{z}_{rs}^{j} \tag{5.1.20}$$

is referred to as the mode subspace co-ordinate. The displacement vector of the substructure (j,k) may be taken out of W shown in equation (5.1.19) as

$$W_{jk} = \frac{1}{\sqrt{MN}}\sum_{r=1}^{M}\sum_{s=1}^{N} e^{i(j-1)r\psi_1} e^{i(k-1)s\psi_2}\,\mathbf{q}_{rs} \quad j=1,2,...,M \quad k=1,2,...,N. \tag{5.1.21}$$

Equation (5.1.21) is referred to as the double U-transformation[16]. It may be verified that

$$\overline{\mathbf{U}}_{jk}^{T}\mathbf{U}_{rs} = \delta_{jr}\delta_{ks}\mathbf{I}_J \quad j,r=1,2,...,M \quad k,s=1,2,...,N \tag{5.1.22}$$

which represents the normality of the mode subspaces. The inverse transformation of double U-transformation can be obtained by pre-multiplying both sides of equation (5.1.19) by $\overline{\mathbf{U}}_{rs}^{T}$, *i.e.*,

$$\mathbf{q}_{rs} = \overline{\mathbf{U}}_{rs}^{T} W = \frac{1}{\sqrt{MN}}\sum_{j=1}^{M}\sum_{k=1}^{N} e^{-i(j-1)r\psi_1} e^{-i(k-1)s\psi_2}\, W_{jk}$$

$$r=1,2,..,M \quad s=1,2,...,N. \tag{5.1.23}$$

Substituting equation (5.1.19) into equation (5.1.7) and pre-multiplying by $\overline{\mathbf{U}}_{rs}^{T}$ yields

$$\mathbf{k}_{rs}\mathbf{q}_{rs} + \mathbf{m}_{rs}\ddot{\mathbf{q}}_{rs} = \mathbf{f}_{rs} \quad r=1,2,...,M \quad s=1,2,...,N \tag{5.1.24}$$

where matrices $\mathbf{k}_{rs}$, $\mathbf{m}_{rs}$ have been defined by equation (5.1.15) and

$$\mathbf{f}_{rs} = \overline{\mathbf{U}}_{rs}^{T}\mathbf{F} = \frac{1}{\sqrt{MN}}\sum_{j=1}^{M}\sum_{k=1}^{N} e^{-i(j-1)r\psi_1} e^{-i(k-1)s\psi_2}\mathbf{F}_{jk} \tag{5.1.25}$$

Consider now the initial conditions as

$$\mathbf{W}_{rs}\big|_{t=0} = \mathbf{W}_{rso}, \qquad \dot{\mathbf{W}}_{rs}\big|_{t=0} = \dot{\mathbf{W}}_{rso} \quad r=1,2,...,M \;\; s=1,2,...,N \tag{5.1.26}$$

where $\mathbf{W}_{rs0}$ and $\dot{\mathbf{W}}_{rs0}$ denote the initial displacement and velocity vectors of substructure (r,s). Substituting initial condition (5.1.26) into equation (5.1.23) yields

$$\mathbf{q}_{rs}\big|_{t=0} = \frac{1}{\sqrt{MN}}\sum_{j=1}^{M}\sum_{k=1}^{N} e^{-i(j-1)r\psi_1} e^{-i(k-1)s\psi_2}\mathbf{W}_{jk0}$$

$$\dot{\mathbf{q}}_{rs}\big|_{t=0} = \frac{1}{\sqrt{MN}}\sum_{j=1}^{M}\sum_{k=1}^{N} e^{-i(j-1)r\psi_1} e^{-i(k-1)s\psi_2}\dot{\mathbf{W}}_{jk0} \tag{5.1.27}$$

Equation (5.1.24) must be solved subject to the initial condition (5.1.27) for each mode subspace co-ordinate and then the dynamic response of the displacement for each substructure may be found by substituting the solutions for $\mathbf{q}_{rs}$ (r=1,2,...,M, s=1,2,...,N) into equation (5.1.21).

In this section, the double U-transformation has been derived from the mode method for discrete systems with cyclic periodicity in two directions. In the following sections the double U-transformation will be applied to both discrete and continuous systems.

5.2 Rectangular Orthogonal Networks - Lumped Mass System

The system to be considered is made up of both a rectangular cable network with $m \times n$ uniform mesh and $m \times n$ number of lumped masses M which are attached to the nodes of network as shown in Figure 5.2.1. The pretensioned cables will act as the coupling springs but their masses will be neglected. The nodes at the four boundary edges are fixed.

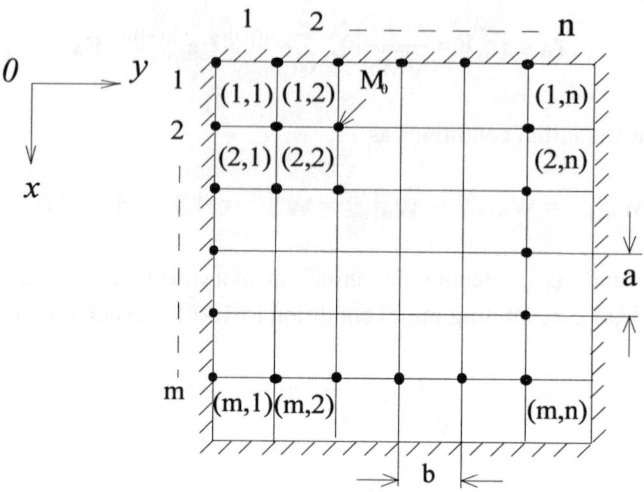

Figure 5.2.1 m × n network

At the outset, let us consider the extended network (see Figure 5.2.2) whose length and width are twice that of the actual ones. Moreover the anti-symmetrical loads about the right side and lower side of actual network must be applied on the corresponding extended part and then we regard the extended network as one having rotational periodicity in x- and y- directions. The boundary conditions of the original network can be satisfied automatically so that the fixed nodes can be replaced by the free nodes with mass M .

5.2.1 Dynamic Equation

Consider now the equivalent system with cyclic periodicity in x-and y- directions shown in Figure 5.2.2. The dynamic equations for all lumped mass may be expressed as

$$M\ddot{W}_{(j,k)} + (2K_1 + 2K_2)W_{(j,k)} - K_1(W_{(j+1,k)} + W_{(j-1,k)}) - K_2(W_{(j,k+1)} + W_{(j,k-1)}) = F_{(j,k)}$$

$$j=1,2,...,2m, \quad k=1,2,...,2n \tag{5.2.1}$$

with $W_{(2m+1,k)} \equiv W_{(1,k)}, W_{(0,k)} \equiv W_{(2m,k)}, W_{(j,2n+1)} \equiv W_{(j,1)}, W_{(j,0)} \equiv W_{(j,2n)}$ where $W_{(j,k)}$ and $F_{(j,k)}$ denote the transverse displacement and loading for the node (j,k) at left-upper corner of subsystem (j,k) and

$$K_1 = \frac{T_1}{a}, \qquad K_2 = \frac{T_2}{b} \tag{5.2.2}$$

in which T_1, T_2 denote the pretensions of the cables in x and y directions and a, b denote the spacing of y and x cables as shown in Figure 5.2.1. The nodal loads of extended network must satisfy the anti-symmetrical condition, *i.e.*,

$$F_{(j,k)} = F_{(2m-j+2,2n-k+2)} = -F_{(j,2n-k+2)} = -F_{(2m-j+2,k)} \quad j=2,3,\ldots,m \quad k=2,3,\ldots,n$$

$$F_{(j,k)} = 0, \qquad j=1,m+1 \text{ or } k=1, n+1. \tag{5.2.3}$$

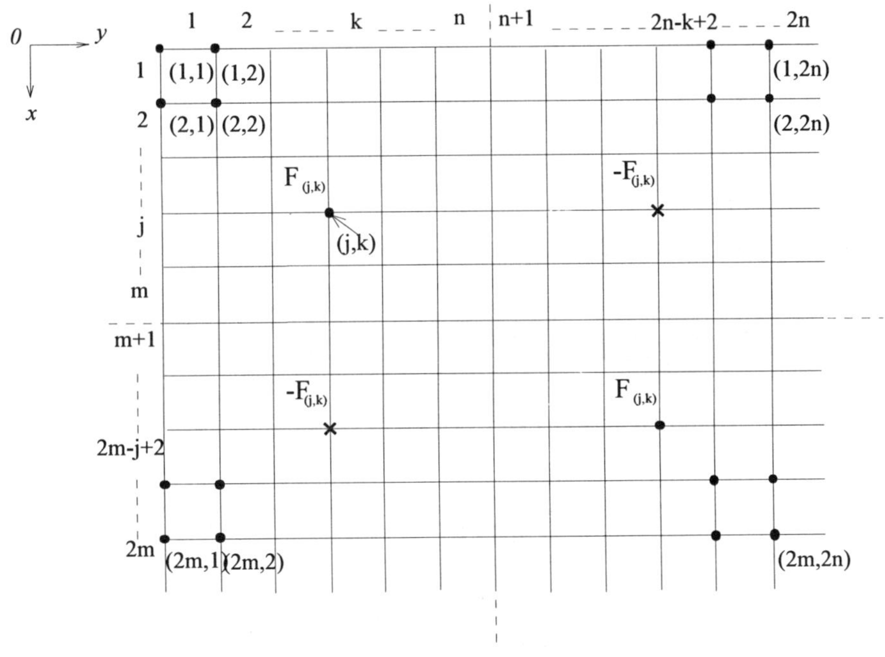

Figure 5.2.2 Extended network - equivalent system

In order to uncouple the simultaneous equation (5.2.1), the double U-transformation (5.1.21) need be used. Let

$$W_{(j,k)} = \frac{1}{\sqrt{2m}}\frac{1}{\sqrt{2n}}\sum_{r=1}^{2m}\sum_{s=1}^{2n} e^{i(j-1)r\psi_1} e^{i(k-1)s\psi_2} q_{(r,s)} \tag{5.2.4a}$$

or

$$q_{(r,s)} = \frac{1}{\sqrt{2m}}\frac{1}{\sqrt{2n}}\sum_{j=1}^{2m}\sum_{k=1}^{2n} e^{-i(j-1)r\psi_1} e^{-i(k-1)s\psi_2} W_{(j,k)} \tag{5.2.4b}$$

in which $\psi_1 = \pi/m, \psi_2 = \pi/n$.

Applying the double U transformation (5.2.4) to equation (5.2.1) results in

$$M\ddot{q}_{(r,s)}(t) + (2K_1 + 2K_2 - 2K_1\cos r\psi_1 - 2K_2\cos s\psi_2)q_{(r,s)}(t) = f_{(r,s)}(t) \tag{5.2.5}$$

where

$$f_{(r,s)} = \frac{1}{\sqrt{2m}}\frac{1}{\sqrt{2n}}\sum_{j=1}^{2m}\sum_{k=1}^{2n} e^{-i(j-1)r\psi_1} e^{-i(k-1)r\psi_2} F_{(j,k)} \tag{5.2.6}$$

Substituting the anti-symmetrical condition (5.2.3) into equation (5.2.6) yields

$$f_{(r,s)} = \frac{-4}{\sqrt{2m}\sqrt{2n}}\sum_{j=2}^{m}\sum_{k=2}^{n} \sin[(j-1)r\psi_1]\sin[(k-1)s\psi_2]F_{(j,k)} \tag{5.2.7}$$

The initial condition may be expressed as

$$q_{(r,s)}(0) = \frac{1}{\sqrt{2m}}\frac{1}{\sqrt{2n}}\sum_{j=1}^{2m}\sum_{k=1}^{2n} e^{-i(j-1)r\psi_1} e^{-i(k-1)s\psi_2} W_{(j,k)}(0)$$

$$\dot{q}_{(r,s)}(0) = \frac{1}{\sqrt{2m}}\frac{1}{\sqrt{2n}}\sum_{j=1}^{2m}\sum_{k=1}^{2n} e^{-i(j-1)r\psi_1} e^{-i(k-1)s\psi_2} \dot{W}_{(j,k)}(0) \tag{5.2.8}$$

in which $W_{(j,k)}(0), \dot{W}_{(j,k)}(0)$ denote the initial displacement and velocity of node (j,k).

5.2.2. Natural Vibration

Let

$$q_{(r,s)}(t) = q_{(r,s)} e^{i\omega t}, \qquad f_{(r,s)} = 0. \tag{5.2.9}$$

Substituting equation (5.2.9) into equation (5.2.5) yields the frequency equation

$$2K_1 + 2K_2 - 2K_1 \cos r\psi_1 - 2K_2 \cos s\psi_2 - M\omega^2 = 0,$$
$$r=1,2,\ldots,2m \quad s=1,2,\ldots,2n. \tag{5.2.10}$$

The natural frequencies can be obtained from equation (5.2.10) as

$$\omega^2_{(r,s)} = (2K_1 + 2K_2 - 2K_1 \cos r\psi_1 - 2K_2 \cos s\psi_2)/M \tag{5.2.11}$$

Consider the simple case of a square network under uniform tension in x- and y-directions, *i.e.*,

$$n = m,\ a = b = \frac{L}{n},\ T_1 = T_2 = T,\ \psi_1 = \psi_2 = \frac{\pi}{n}, K_1 = K_2 = \frac{nT}{L},\ M = 2a\rho \tag{5.2.12}$$

where L, ρ denote the length of the square network and mass per unit length of cable . Substituting equation (5.2.12) into equation (5.2.11) yields

$$\omega_{(r,s)} = \frac{n}{L}\sqrt{\frac{T}{\rho}}\sqrt{2 - \cos\frac{r\pi}{n} - \cos\frac{s\pi}{n}}\,. \tag{5.2.13}$$

The first four natural frequencies are given in Table 5.2.1.

Table 5.2.1 Natural frequencies ω for the orthogonal square cable networks - lumped mass model

Networks $(m\times n)$		3×3	4×4	5×5	6×6	7×7
Mode(r,s)*	(1,1)	3.0000	3.0615	3.0902	3.1058	3.1153
	(1,2)	4.2426	4.5482	4.6957	4.7774	4.8272
	(2,1)	4.2426	4.5482	4.6957	4.7774	4.8272
	(2,2)	5.1962	5.6569	5.8779	6.0000	6.0744
Multiplier		$(1/L)\sqrt{T/\rho}$				

* r and s $(r,s\le n)$ in the round brackets denote the numbers of half wave along the x- and y-directions for the mode.

It is obvious that for the rotating modes (r,s) (2m-r,s) (r,2n-s) and (2m-r,2n-s) the corresponding natural frequencies are the same. But there is only one mode which will satisfy the actual boundary conditions. Let

$$q_{(2m-r,s)} = q_{(r,2n-s)} = -q_{(2m-r,2n-s)} = -q_{(r,s)} = \text{real number} \tag{5.2.14}$$

with other mode subspace coordinates vanish. Substituting equation (5.2.14) into equation (5.2.4a) while neglecting a constant multiplier, results in

$$W_{(j,k)} = \sin\left[(j-1)r\,\psi_1\right]\sin\left[(k-1)s\,\psi_2\right] \tag{5.2.15}$$

This mode (r.s) satisfies the boundary condition for the original network, *i.e.*,

$$W_{(1,k)} = 0, \qquad W_{(j,1)} = 0, \qquad W_{(m+1,k)} = 0, \qquad W_{(j,n+1)} = 0. \tag{5.2.16}$$

5.2.3 Forced Vibration

Consider the harmonic vibration of the original system subjected to only one nodal load $F_{(j^*,k^*)} = F e^{i\omega t}$ with other nodal loads being equal to zero. For the equivalent system the anti-symmetrical loads can be expressed as

$$F_{(j^*,k^*)} = F_{(2m-j^*+2,2n-k^*+2)} = -F_{(j^*,2n-k^*+2)} = -F_{(2m-j^*+2,k^*)} = F e^{i\omega t} \tag{5.2.17}$$

with other loads being equal to zero. Substituting

$$q_{(r,s)}(t) = q_{(r,s)} e^{i\omega t}, \qquad f_{(r,s)}(t) = f_{(r,s)} e^{i\omega t}, \tag{5.2.18}$$

and equation (5.2.11) into equation (5.2.5) yields

$$q_{(r,s)} = f_{(r,s)}/M(\omega_{(rs)}^2 - \omega^2). \tag{5.2.19}$$

Substituting equation (5.2.17) into equation (5.2.6) yields

$$f_{(r,s)} = \frac{-4F}{\sqrt{2m}\sqrt{2n}} \sin[(j^* - 1)r\,\psi_1]\sin[(k^* - 1)s\,\psi_2]. \tag{5.2.20}$$

Substituting equations (5.2.19) with (5.2.20) into equation (5.2.4a) results in

$$W_{(j,k)} = \frac{4F}{mn}\sum_{r=1}^{m}\sum_{s=1}^{n} \frac{\sin(j-1)r\,\psi_1 \sin(j^* - 1)r\,\psi_1 \sin(k-1)s\,\psi_2 \sin(k^* - 1)s\,\psi_2}{M(\omega_{(rs)}^2 - \omega^2)}. \tag{5.2.21}$$

It is obvious that the displacements shown in equation (5.2.21) satisfy the boundary condition (5.2.16).

Consider now the dynamic response of the cable network subjected to arbitrary nodal loads. The solution for $q_{(r,s)}$ of equation (5.2.5) with initial condition (5.2.8) can be expressed as the Duhamel integral

$$q_{(r,s)} = \frac{1}{M\,\omega_{(r,s)}} \int_0^t f_{(r,s)}(\tau) \sin\omega_{(r,s)}(t-\tau)d\tau + q_{(r,s)}(0)\cos\omega_{(r,s)} t$$
$$+ \frac{\dot{q}_{(r,s)}(0)}{\omega_{(r,s)}} \sin\omega_{(r,s)} t, \tag{5.2.22}$$

in which $q_{(r,s)}$, $\dot{q}_{(r,s)}$ and $f_{(r,s)}$ are defined by equations (5.2.8) and (5.2.7). The dynamic response for nodal displacements can be found by inserting equation (5.2.22) in double U-transformation (5.2.4a).

5.3 Rectangular Orthogonal Networks - Continuous System

5.3.1 Governing Equation

The cable network considered and its equivalent system are shown in Figures 5.2.1 and 5.2.2 where distributed mass instead of lumped mass will be considered. A typical substructure (j,k) is shown in Figure 5.3.1.

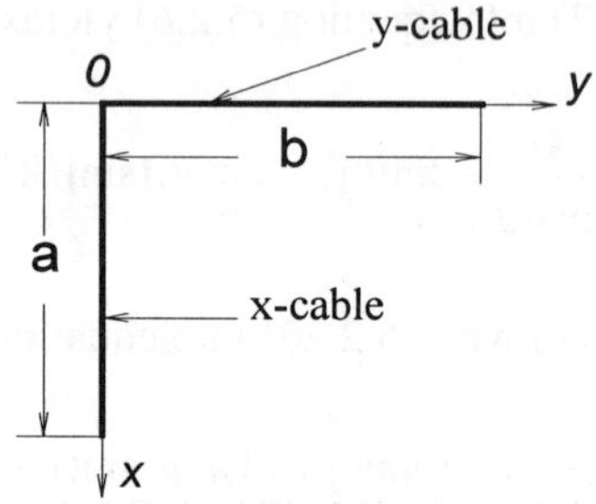

Figure 5.3.1 Substructure (j, k) with local coordinate system

In each substructure, a set of local coordinates oxy is established with the origin o at its nodal point and the x,y axes along two orthogonal cables . Let W_1, W_2 and ρ_1, ρ_2 denote the transverse displacements and mass per unit length of the x-and y-cables. The dynamic equations for all substructures are of the same form as shown below

$$\frac{1}{C_1^2}\frac{\partial^2 W_{1(j,k)}(x,t)}{\partial t^2} - \frac{\partial^2 W_{1(j,k)}(x,t)}{\partial x^2} = \frac{F_{1(j,k)}(x,t)}{T_1}, \qquad 0 < x < a$$

$$\frac{1}{C_2^2}\frac{\partial^2 W_{2(j,k)}(y,t)}{\partial t^2} - \frac{\partial^2 W_{2(j,k)}(y,t)}{\partial y^2} = \frac{F_{2(j,k)}(y,t)}{T_2}, \qquad 0 < y < b$$

$$j=1,2,..,2m \quad k=1,2,...,2n \tag{5.3.1}$$

where $F_{1(j,k)}, F_{2(j,k)}$ denote the distributed loads acting on the x-and y- cables of substructure (j,k) and

$$C_1 = \sqrt{T_1/\rho_1}, \qquad C_2 = \sqrt{T_2/\rho_2}\,. \tag{5.3.2}$$

Continuity across the joints requires the following conditions to be satisfied:

$$W_{1(j,k)}\Big|_{x=a} = W_{1(j+1,k)}\Big|_{x=0}, \qquad W_{2(j,k)}\Big|_{y=b} = W_{2(j,k+1)}\Big|_{y=0}$$

$$W_{1(j,k)}\Big|_{x=0} = W_{2(j,k)}\Big|_{y=0} \qquad j=1,2,\dots,2m \qquad k=1,2,\dots,2n \tag{5.3.3}$$

with $W_{1(2m+1,k)} \equiv W_{1(1,k)}, W_{2(j,2n+1)} \equiv W_{2(j,1)}$.

The equilibrium equations for all nodes can be expressed as

$$T_1\left(\frac{\partial}{\partial x} W_{1(j,k)}\Big|_{x=0} - \frac{\partial}{\partial x} W_{1(j-1,k)}\Big|_{x=a}\right) + T_2\left(\frac{\partial}{\partial y} W_{2(j,k)}\Big|_{y=0} - \frac{\partial}{\partial y} W_{2(j,k-1)}\Big|_{y=b}\right) = 0$$

$$j=1,2,\dots,2m \quad k=1,2,\dots,2n \tag{5.3.4}$$

with $W_{1(0,k)} \equiv W_{1(2m,k)}, W_{2(j,0)} \equiv W_{2(j,2n)}$.

In order to uncouple the continuity conditions (5.3.3) and the equilibrium equation (5.3.4), applying the double U-transformation to equations (5.3.1) (5.3.3) and (5.3.4), *i.e.*, let

$$W_{1(j,k)} = \frac{1}{\sqrt{2m}}\frac{1}{\sqrt{2n}}\sum_{r=1}^{2m}\sum_{s=1}^{2n} e^{i(j-1)r\psi_1} e^{i(k-1)s\psi_2} q_{1(r,s)}$$

$$W_{2(j,k)} = \frac{1}{\sqrt{2m}}\frac{1}{\sqrt{2n}}\sum_{r=1}^{2m}\sum_{s=1}^{2n} e^{i(j-1)r\psi_1} e^{i(k-1)s\psi_2} q_{2(r,s)}$$

$$j=1,2,\dots,2m \quad k=1,2,\dots,2n \tag{5.3.5}$$

with $\psi_1 = \pi/m, \psi_2 = \pi/n$ that will result in

$$\frac{1}{C_1^2}\frac{\partial^2 q_{1(r,s)}}{\partial t^2} - \frac{\partial^2 q_{1(r,s)}}{\partial x^2} = \frac{f_{1(r,s)}}{T_1} \qquad 0 < x < a$$

$$\frac{1}{C_2^2}\frac{\partial^2 q_{2(r,s)}}{\partial t^2}-\frac{\partial^2 q_{2(r,s)}}{\partial x^2}=\frac{f_{2(r,s)}}{T_2}$$

$$0< y < b \quad r=1,2,...,2m \quad s=1,2,...,2n \tag{5.3.6}$$

$$q_{1(r,s)}\Big|_{x=a}=e^{ir\psi_1}q_{1(r,s)}\Big|_{x=0}, \qquad q_{2(r,s)}\Big|_{y=b}=e^{is\psi_2}q_{2(r,s)}\Big|_{y=0},$$

$$q_{1(r,s)}\Big|_{x=0}=q_{2(r,s)}\Big|_{y=0}, \tag{5.3.7}$$

and

$$T_1\Big(\frac{\partial}{\partial x}q_{1(r,s)}\Big|_{x=0}-e^{-ir\psi_1}\frac{\partial}{\partial x}q_{1(r,s)}\Big|_{x=a}\Big)+T_2\Big(\frac{\partial}{\partial y}q_{2(r,s)}\Big|_{y=0}-e^{-is\psi_2}\frac{\partial}{\partial y}q_{2(r,s)}\Big|_{y=b}\Big)=0 \tag{5.3.8}$$

where $q_{1(r,s)} \equiv q_{1(r,s)}(x,t)$, $q_{2(r,s)} \equiv q_{2(r,s)}(y,t)$ and

$$f_{1(r,s)}=\frac{1}{\sqrt{2m}}\frac{1}{\sqrt{2n}}\sum_{j=1}^{2m}\sum_{k=1}^{2n}e^{-i(j-1)r\psi_1}e^{-i(k-1)s\psi_2}F_{1(j,k)},$$

$$f_{2(r,s)}=\frac{1}{\sqrt{2m}}\frac{1}{\sqrt{2n}}\sum_{j=1}^{2m}\sum_{k=1}^{2n}e^{-i(j-1)r\psi_1}e^{-i(k-1)s\psi_2}F_{2(j,k)}. \tag{5.3.9}$$

5.3.2 Natural Vibration

The natural vibration equation for generalized displacements $q_{1(r,s)}$ and $q_{2(r,s)}$ can be expressed as

$$\frac{d^2}{dx^2}q_{1(r,s)}(x)+\frac{\omega^2}{C_1^2}q_{1(r,s)}(x)=0, \quad 0< x < a$$

$$\frac{d^2}{dy^2}q_{2(r,s)}(y)+\frac{\omega^2}{C_2^2}q_{2(r,s)}(y)=0, \quad 0< y < b$$

$$r=1,2,...,2m \quad s=1,2,...,2n \tag{5.3.10}$$

where ω denotes the natural frequency and $q_{1(r,s)}(x), q_{2(r,s)}(y)$ denote the amplitude functions of the generalized displacements. The corresponding boundary conditions can be obtained from equations (5.3.7) and (5.3.8) as

$$q_{1(r,s)}(a) = e^{ir\psi_1} q_{1(r,s)}(0), \qquad q_{2(r,s)}(b) = e^{is\psi_2} q_{2(r,s)}(0),$$

$$q_{1(r,s)}(0) = q_{2(r,s)}(0),$$

$$T_1(q'_{1(r,s)}(0) - e^{-ir\psi_1} q'_{1(r,s)}(a)) + T_2(q'_{2(r,s)}(0) - e^{-is\psi_2} q'_{2(r,s)}(b)) = 0,$$

$$r=1,2,...,2m \quad s=1,2,...,2n. \tag{5.3.11}$$

The general solution for $q_{1(r,s)}$ and $q_{2(r,s)}$ of equation (5.3.10) may be expressed as

$$q_{1(r,s)}(x) = A_1 \sin\frac{\omega}{C_1}x + B_1 \cos\frac{\omega}{C_1}x,$$

$$q_{2(r,s)}(y) = A_2 \sin\frac{\omega}{C_2}y + B_2 \cos\frac{\omega}{C_2}y \tag{5.3.12}$$

where A_1, B_1, A_2 and B_2 denote four constants of integration. Substituting equation (5.3.12) into boundary condition (5.3.11) results in

$$A_1 \sin\frac{\omega}{C_1}a + B_1(\cos\frac{\omega}{C_1}a - e^{ir\psi_1}) = 0,$$

$$A_2 \sin\frac{\omega}{C_2}b + B_2(\cos\frac{\omega}{C_2}b - e^{is\psi_2}) = 0,$$

$$B_1 - B_2 = 0,$$

$$A_1\frac{T_1\omega}{C_1}(1 - e^{-ir\psi_1}\cos\frac{\omega}{C_1}a) + B_1\frac{T_1\omega}{C_1}e^{-ir\psi_1}\sin\frac{\omega}{C_1}a + A_2\frac{T_2\omega}{C_2}(1 - e^{-is\psi_2}\cos\frac{\omega}{C_2}b).$$

$$+ B_2\frac{T_2\omega}{C_2}e^{-is\psi_2}\sin\frac{\omega}{C_2}b = 0 \tag{5.3.13}$$

Let the coefficient determinant of the simultaneous equation (5.3.13) be equal to zero. Expanding this determinant yields the frequency equation[16]

$$\sin\frac{\omega}{C_1}a(\cos s\,\psi_2-\cos\frac{\omega}{C_2}b)\frac{T_2\omega}{C_2}+\sin\frac{\omega}{C_2}b(\cos r\,\psi_1-\cos\frac{\omega}{C_1}a)\frac{T_1\omega}{C_1}=0$$

$$r=1,2,\ldots,2m,\quad s=1,2,\ldots,2n. \tag{5. 3.14}$$

Consider now the case of a square network under uniform tension in x- and y-directions, *i.e.*,

$$n=m,\ a=b=\frac{L}{n},\ T_1=T_2=T,\ \psi_1=\psi_2=\frac{\pi}{n}, C_1=C_2=\sqrt{T/\rho}=C \tag{5.3.15}$$

Inserting equation (5.3.15) into equation (5.3.14), the frequency equation becomes

$$\cos\frac{\omega\pi}{\omega_0 n}=(\cos\frac{r\pi}{n}+\cos\frac{s\pi}{n})/2\quad r,s=1,2,\ldots,2n \tag{5.3.16}$$

where

$$\omega_0=\frac{\pi}{L}\sqrt{\frac{T}{\rho}} \tag{5.3.17}$$

ω_0 denotes the basic frequency of a single cable with length L and fixed ends. The first four natural frequencies are given in Table 5.3.1. The numerical results obtained by transfer matrix method[19] are also tabulated for comparison.

The corresponding mode shapes can be found by inserting the values of the natural frequency and the parameters shown in equation (5.3.15) into equation (5.3.13), after which the nontrivial solution for A_1, B_1, A_2 and B_2 of the homogeneous equation (5.3.13) can be expressed as

$$A_1=A_{rs}[e^{ir\psi}-\cos(\omega a/C)],\qquad A_2=A_{rs}[e^{is\psi}-\cos(\omega a/C)],$$

$$B_1=B_2=A_{rs}\sin(\omega a/C) \tag{5.3.18}$$

where A_{rs} is an undetermined multiplier.

Table 5.3.1 Natural frequencies ω for the orthogonal square cable networks - distributed mass model

Networks ($m\times n$)		3×3	4×4	5×5	6×6	7×7
Mode(r,s)*	(1,1)	1.0000	1.0000	1.0000	1.0000	1.0000
		(1.0000)*	(1.0000)	(1.0000)	(1.0000)	(1.0000)
	(1,2)	1.5000	1.5399	1.5559	1.5640	1.5687
		(1.5000)	(1.540)	(1.556)	(1.564)	(1.568)
	(2.1)	1.5000	1.5399	1.5559	1.5640	1.5687
		(1.5000)	(1.540)	(1.556)	(1.564)	(1.568)
	(2.2)	2.0000	2.0000	2.0000	2.0000	2.0000
		(2.0000)	(2.0000)	(2.0000)	(2.0000)	(2.0000)
Multiplier		$(\pi/L)\sqrt{T/\rho}$				

*The numerical results in the round brackets are obtained according to the the date in reference[19].

A set of eigenfunctions for $q_{1(r,s)}(x)$ and $q_{2(r,s)}(y)$ of equations (5.3.10), (5.3.11) can be found by substituting equation (5.3.18) into equation (5.3.12) with $C_1 = C_2 = C$ as

$$q_{1(r,s)}(x) = A_{rs}[e^{ir\psi}\sin(\omega x/C) + \sin(\omega(a-x)/C)],$$

$$q_{2(r,s)}(y) = A_{rs}[e^{is\psi}\sin(\omega y/C) + \sin((a-y)\omega/C)],$$

$$q_{1(j,k)}(x) = 0, \qquad q_{2(j,k)}(y) = 0, \quad j \neq r \text{ or } k \neq s \tag{5.3.19}$$

where ω is a positive root of the frequency equation (5.3.16) for a given set of r and s. Altogether there are four independent sets of eigenfunctions corresponding to an eigenvalue except the case of r,s=n,2n. They can be obtained by using (2n-r) and (2n-s) instead of r and s in equation (5.3.19), respectively, *i.e.*,

$$q_{1(2n-r,s)}(x) = A_{2n-r,s}[e^{-ir\psi}\sin\frac{\omega}{C}x + \sin\frac{\omega}{C}(a-x)],$$

$$q_{2(2n-r,s)}(y) = A_{2n-r,s}[e^{is\psi} \sin\frac{\omega}{C}y + \sin\frac{\omega}{C}(a-y)],$$

$$q_{1(j,k)}(x) = 0, \qquad q_{2(j,k)}(y) = 0, \quad j \neq 2n\text{-}r \text{ or } k \neq s \tag{5.3.20}$$

$$q_{1(r,2n-s)}(x) = A_{r,2n-s}[e^{ir\psi} \sin\frac{\omega}{C}x + \sin\frac{\omega}{C}(a-x)],$$

$$q_{2(r,2n-s)}(y) = A_{r,2n-s}[e^{-is\psi} \sin\frac{\omega}{C}y + \sin\frac{\omega}{C}(a-y)],$$

$$q_{1(j,k)}(x) = 0, \qquad q_{2(j,k)}(y) = 0, \quad j \neq r \text{ or } k \neq 2n\text{-}s \tag{5.3.21}$$

$$q_{1(2n-r,2n-s)}(x) = A_{2n-r,2n-s}[e^{-ir\psi} \sin\frac{\omega}{C}x + \sin\frac{\omega}{C}(a-x)],$$

$$q_{2(2n-r,2n-s)}(y) = A_{2n-r,2n-s}[e^{-is\psi} \sin\frac{\omega}{C}y + \sin\frac{\omega}{C}(a-y)]$$

$$q_{1(j,k)}(x) = 0, \qquad q_{2(j,k)}(y) = 0, \quad j \neq 2n\text{-}r \text{ or } k \neq 2n\text{-}s. \tag{5.3.22}$$

The linear combination of the above four sets of eigenfunctions is also a set of eigenfunctions. But there is only one set of eigenfunctions which will satisfy the actual boundary conditions. Let

$$A_{2n-r,s} = A_{r,2n-s} = -A_{2n-r,2n-s} = -A_{r,s} \tag{5.3.23}$$

By superposition of the four sets of eigenfunctions shown in equations (5.3.19)-(5.3.22), and substituting into equation (5.3.5) while neglecting a constant multiplier, one can obtain

$$W_{1(j,k)}(x) = \sin jr\psi\ \sin(k-1)s\psi\ \sin\frac{\omega}{C}x + \sin(j-1)r\psi\ \sin(k-1)s\psi\ \sin\frac{\omega}{C}(a-x),$$

$$W_{2(j,k)}(y) = \sin(j-1)r\psi\ \sin ks\psi\ \sin\frac{\omega}{C}y + \sin(j-1)r\psi\ \sin(k-1)s\psi\ \sin\frac{\omega}{C}(a-y),$$

$$r, s \neq n, 2n \quad j, k = 1, 2, \ldots, 2n \tag{5.3.24}$$

with $\psi = \pi/n$. This is the natural mode which satisfies the boundary conditions of the actual network, *i.e.*,

$$W_{2(j,1)}(0) = 0, \qquad W_{2(j,n)}(a) = 0, \qquad j=1,2,...,n$$

$$W_{1(1,k)}(0) = 0, \qquad W_{1(n,k)}(a) = 0, \qquad k=1,2,...,n. \tag{5.3.25}$$

Considering the cases of r=s=n and r=s=2n, the frequency equation (5.3.16) becomes

$$\cos\frac{\omega}{C}a = -1 \tag{5.3.26}$$

and

$$\cos\frac{\omega}{C}a = 1 \tag{5.3.27}$$

respectively. The solutions for ω of equations (5.3.26) and (5.3.27) can be expressed as

$$\omega = (2m+1)\pi\frac{c}{a}, \qquad m=0,1,2,... \tag{5.3.28}$$

and

$$\omega = 2m\pi\frac{c}{a}, \qquad m=1,2,... \tag{5.3.29}$$

respectively. Substituting equation (5.3.15) with equation (5.3.28) or (5.3.29), into equation (5.3.13) the similar eigenfunction may be found as

$$q_{1(n,n)}(x) = A_1\sin\frac{\omega}{c}x = A_1\sin\frac{(2m+1)\pi}{a}x,$$

$$q_{2(n,n)}(y) = A_2\sin\frac{\omega}{c}y = A_2\sin\frac{(2m+1)\pi}{a}y,$$

$$q_{1(j,k)}(x) = q_{2(j,k)}(y) = 0 . \quad j \neq n \text{ or } k \neq n, \quad m=0,1,2,... \tag{5.3.30}$$

or

$$q_{1(2n,2n)}(x) = A_1 \sin\frac{\omega}{c}x = A_1 \sin\frac{2m\pi}{a}x,$$

$$q_{2(2n,2n)}(y) = A_2 \sin\frac{\omega}{c}y = A_2 \sin\frac{2m\pi}{a}y,$$

$$q_{1(j,k)}(x) = q_{2(j,k)}(y) = 0 . \quad j \neq 2n \text{ or } k \neq 2n, \quad m=1,2,... \tag{5.3.31}$$

Corresponding natural modes can be obtained by substituting equations (5.3.30) and (5.3.31) into equation (5.3.5) , as

$$W_{1(j,k)}(x) = A_1 \cos(j-1)\pi \ \cos(k-1)\pi \ \sin\frac{(2m+1)\pi}{a}x,$$

$$W_{2(j,k)}(y) = A_2 \cos(j-1)\pi \ \cos(k-1)\pi \ \sin\frac{(2m+1)\pi}{a}y,$$

$$j,k=1,2,...,2n \tag{5.3.32}$$

and

$$W_{1(j,k)}(x) = A_1 \sin\frac{2m\pi}{a}x,$$

$$W_{2(j,k)}(y) = A_2 \sin\frac{2m\pi}{a}y \qquad j,k=1,2,...,2n \tag{5.3.33}$$

respectively, where A_1, A_2 denote two independent real constant multipliers. All nodes are the nodal points of the mode shapes shown in equations (5.3.32) and (5.3.33) and then these modes satisfy the boundary condition (5.3.25). Consequently, the deflection functions shown in equations (5.3.32) and (5.3.33) represent the natural modes of the actual network. These modes are the same as those of a single cable with length "a" and fixed ends.

For the cases of r=n, s=2n and r=2n, s=n, there is no mode satisfying the fixed-end condition of the actual network.

5.3.3 Dynamic Response

Now consider a rectangular network of uniform cables with rigidly held ends and $m \times n$ square meshes. All cables have pretension T and mass per unit length ρ. The cable spacing is "a". The dynamic equation (5.3.1) becomes[20]

$$-T\,W_{1(j,k),xx}(x,t) + \rho\,W_{1(j,k),tt}(x,t) = F_{1(j,k)}(x,t),$$

$$-T\,W_{2(j,k),yy}(y,t) + \rho\,W_{2(j,k),tt}(y,t) = F_{2(j,k)}(y,t),$$

$$j=1,2,...,2m \quad k=1,2,...,2n \tag{5.3.34}$$

in which the last subscripts x, y and t denote partial differentiation with respect to x, y and t, respectively. The continuity and equilibrium conditions (5.3.3), (5.3.4) may be written as

$$W_{1(j,k)}(a,t) = W_{1(j+1,k)}(0,t), \qquad W_{2(j,k)}(a,t) = W_{2(j,k+1)}(0,t)\,,$$

$$W_{1(j,k)}(0,t) = W_{2(j,k)}(0,t),$$

$$T\left[W_{1(j,k),x}(0,t) - W_{1(j-1,k),x}(a,t) + W_{2(j,k),y}(0,t) - W_{2(j,k-1),y}(a,t)\right] = 0\,,$$

$$j=1,2,...,2m \quad k=1,2,...,2n \tag{5.3.35}$$

It must be noted that the solution for $W_{1(j,k)}, W_{2(j,k)}$ of equations (5.3.34) and (5.3.35) includes an arbitrary constant which is independent of the loads. Such a constant representing the rigid body displacement should be neglected.

Applying the double U transformation (5.3.5) to equations (5.3.34) and (5.3.35), result in

$$-Tq_{1(r,s),xx}(x,t) + \rho\,q_{1(r,s),tt}(x,t) = f_{1(r,s)}(x,t),$$

$$-Tq_{2(r,s),yy}(y,t) + \rho\,q_{2(r,s),tt}(y,t) = f_{2(r,s)}(y,t),$$

$$r=1,2,...,2m \quad s=1,2,...,2n \tag{5.3.36}$$

and

$$q_{1(r,s)}(a,t) = e^{ir\psi_1} q_{1(r,s)}(0,t), \qquad q_{2(r,s)}(a,t) = e^{is\psi_2} q_{2(r,s)}(0,t)\,,$$

$$q_{1(r,s)}(0,t) = q_{2(r,s)}(0,t)\,,$$

$$q_{1(r,s),x}(0,t) - e^{-ir\psi_1} q_{1(r,s),x}(a,t) + q_{2(r,s),y}(0,t) - e^{-is\psi_2} q_{2(r,s),y}(a,t) = 0\,,$$

$$r=1,2,\ldots,2m \quad s=1,2,\ldots,2n \tag{5.3.37}$$

where $f_{1(r,s)}(x,t)$ and $f_{2(r,s)}(y,t)$ are defined by equation (5.3.9).

The initial condition for generalized displacements can be expressed as

$$q_{l(r,s)}\big|_{t=0} = \frac{1}{\sqrt{2m}}\frac{1}{\sqrt{2n}}\sum_{j=1}^{2m}\sum_{k=1}^{2n} e^{-i(j-1)r\psi_1} e^{-i(k-1)s\psi_2} W_{l0(j,k)}\,,$$

$$\dot{q}_{l(r,s)}\big|_{t=0} = \frac{1}{\sqrt{2m}}\frac{1}{\sqrt{2n}}\sum_{j=1}^{2m}\sum_{k=1}^{2n} e^{-i(j-1)r\psi_1} e^{-i(k-1)s\psi_2} \dot{W}_{l0(j,k)}\,,$$

$$l=1,2 \tag{5.3.38}$$

where $W_{10(j,k)}, \dot{W}_{10(j,k)}$ and $W_{20(j,k)}, \dot{W}_{20(j,k)}$ denote the initial displacement and velocity functions for the *x*- and *y*-cables of substructure (j,k). They must be anti-symmetric with respect to two symmetric planes of the equivalent system.

Equations (5.3.36) with (5.3.9) must be solved subject to the conditions (5.3.37) and (5.3.38), and then the dynamic response for deflections may be found by substituting the solutions for $q_{1(r,s)}(x,t)$ and $q_{2(r,s)}(y,t)$ of equation (5.3.36) into equation (5.3.5). As well known, the dynamic equations can be solved by using the mode method. The natural vibration of the cable network has been analyzed in the above section . For the present case, the frequency equation can be expressed as follows:

$$\cos\frac{\omega a}{C} = (\cos r\,\psi_1 + \cos s\,\psi_2)/2\,, \qquad r=1,2,\ldots,2m \quad s=1,2,\ldots,2n \tag{5.3.39}$$

with $\psi_1 = \pi/m, \psi_2 = \pi/n$ and C denotes the propagation velocity of the transverse waves along the cables, *i.e.*, $C = \sqrt{T/\rho}$. The characteristic roots may be found from frequency equation (5.3.39) as

$$\alpha_{(r,s)j} = (-1)^{j-1}\text{arc }\cos\left[\frac{1}{2}(\cos r\,\psi_1 + \cos s\,\psi_2)\right] + \left[j - \frac{1-(-1)^j}{2}\right]\pi\,,$$

$$r \neq m,2m \quad \text{or} \quad s \neq n,2n \quad j=1,2,\ldots,\infty \tag{5.3.40}$$

and

$$\alpha_{(m,n)j} = (2j-1)\pi\,, \quad \alpha_{(2m,2n)j} = 2j\pi\,, \quad j=1,2,\ldots,\infty \tag{5.3.41}$$

where $\alpha_{(r,s)j}$ is a non-dimensional frequency parameter, defined as

$$\alpha_{(r,s)j} = \omega_{(r,s)j}\, a/c \tag{5.3.42}$$

and $\omega_{(r,s)j}$ denotes the j-th positive root of frequency equation (5.3.39) for fixed values of r and s. The corresponding eigenfunctions may be expressed as

$$q^{(j)}_{1(r,s)}(x) = A_{(r,s)j}\left\{e^{ir\psi_1}\sin\left[\omega_{(r,s)j}\,x/c\right] + \sin\left[\omega_{(r,s)j}(a-x)/c\right]\right\}$$

$$q^{(j)}_{2(r,s)}(y) = A_{(r,s)j}\left\{e^{is\psi_2}\sin\left[\omega_{(r,s)j}\,y/c\right] + \sin\left[\omega_{(r,s)j}(a-y)/c\right]\right\}$$

$$r \neq m,2m \quad \text{or} \quad s \neq n,2n \quad j=1,2,\ldots,\infty \tag{5.3.43}$$

$$q^{(j)}_{1(m,n)}(x) = A_{(m,n)j}\sin\left[(2j-1)\pi\,x/a\right],$$

$$q^{(j)}_{2(m,n)}(y) = B_{(m,n)j}\sin\left[(2j-1)\pi\,y/a\right],$$

$$q^{(j)}_{1(2m,2n)}(x) = A_{(2m,2n)j}\sin(2j\pi\,x/a)\,,$$

$$q^{(j)}_{2(2m,2n)}(y) = B_{(2m,2n)j}\sin(2j\pi\,y/a)\,, \quad j=1,2,\ldots,\infty \tag{5.3.44}$$

where $A_{(r,s)j}$ and $B_{(r,s)j}$ are the undetermined multipliers which may be determined by using the following normalization conditions

$$\int_0^a \rho \bar{q}^{(j)}_{1(r,s)}(x)\, q^{(k)}_{1(r,s)}(x)dx + \int_0^a \rho \bar{q}^{(j)}_{2(r,s)}(y)\, q^{(k)}_{2(r,s)}(y)dy = \delta_{jk}$$

$$-\int_0^a T\bar{q}^{(j)}_{1(r,s)}(x)\, q^{(k)}_{1(r,s),xx}(x)dx - \int_0^a T\bar{q}^{(j)}_{2(r,s)}(y)\, q^{(k)}_{2(r,s),yy}(y)dy = \omega^2_{(r,s)k}\,\delta_{jk}$$

$$(s,r) \neq (m,n),\ (2m,2n) \qquad (5.3.45)$$

where δ_{jk} is the Kronecker symbol.

Noting that $q^{(j)}_{1(m,n)}(x), q^{(j)}_{2(m,n)}(y)$ and $q^{(j)}_{1(2m,2n)}(x)$, $q^{(j)}_{2(2m,2n)}(y)$ shown in equation (5.3.44) include two independent undetermined constants $A_{(m,n)j}, B_{(m,n)j}$ and $A_{(2m,2n)j}, B_{(2m,2n)j}$ respectively, $q^{(j)}_{1(m,n)}(x), q^{(j)}_{2(m,n)}(y)$ and $q^{(j)}_{1(2m,2n)}(x), q^{(j)}_{2(2m,2n)}(y)$ can be regarded as the linear combination of two sets of eigenfunctions. The normalization conditions can be rewritten as

$$\int_0^a \rho \bar{q}^{(j)}_{1(m,n)}(x)\, q^{(k)}_{1(m,n)}(x)dx = \delta_{jk},$$

$$\int_0^a \rho \bar{q}^{(j)}_{2(m,n)}(y)\, q^{(k)}_{2(m,n)}(y)dy = \delta_{jk} \qquad (5.3.46)$$

and

$$\int_0^a \rho \bar{q}^{(j)}_{1(2m,2n)}(x)\, q^{(k)}_{1(2m,2n)}(x)dx = \delta_{jk},$$

$$\int_0^a \rho \bar{q}^{(j)}_{2(2m,2n)}(y)\, q^{(k)}_{2(2m,2n)}(y)dy = \delta_{jk} \qquad (5.3.47)$$

instead of equation (5.3.45). The undetermined multipliers $A_{(m,n)j}, B_{(m,n)j}$ and $A_{(2m,2n)j}, B_{(2m,2n)j}$ may be determined by equation (5.3.46) and (5.3.47) respectively. By substituting equations (5.3.43) and (5.3.44) into the corresponding normalization conditions (5.3.45)-(5.3.47), the multipliers may be determined as

$$A_{(r,s)j} = \{\rho a[2 - \frac{1}{\alpha_{(r,s)j}}\sin 2\alpha_{(r,s)j} + (\frac{1}{\alpha_{(r,s)j}}\sin\alpha_{(r,s)j} - \cos\alpha_{(r,s)j})(\cos r\psi_1 + \cos s\psi_2)]\}^{-1/2}$$

$$(s,r) \neq (m,n),\ (2m,2n) \qquad (5.3.48)$$

$$A_{(m,n)j} = B_{(m,n)j} = A_{(2m,2n)j} = B_{(2m,2n)j} = \sqrt{2/\rho a}\,. \tag{5.3.49}$$

Consider now the mode method for solving the single substructure problem shown in equations (5.3.36) and (5.3.37). Let

$$\begin{Bmatrix} q_{1(r,s)}(x,t) \\ q_{2(r,s)}(y,t) \end{Bmatrix} = \sum_{j=1}^{\infty} a_{(r,s)j}(t) \begin{Bmatrix} q_{1(r,s)}^{(j)}(x) \\ q_{2(r,s)}^{(j)}(y) \end{Bmatrix} \tag{5.3.50}$$

where $a_{(r,s)j}$ denotes the mode coordinate and

$$a_{(m,n)j} = \begin{bmatrix} a_{1(m,n)j} & 0 \\ 0 & a_{2(m,n)j} \end{bmatrix}, \qquad a_{(2m,2n)j} = \begin{bmatrix} a_{1(2m,2n)j} & 0 \\ 0 & a_{2(2m,2n)j} \end{bmatrix} \tag{5.3.51}$$

but other mode coordinates represent the scalar functions of t. Noting that the equation (5.3.37) has been satisfied by the generalized displacements $q_{1(r,s)}(x,t)$ and $q_{2(r,s)}(y,t)$ shown in equation (5.3.50), substituting equation (5.3.50) into equation (5.3.36) yields

$$-T\sum_{j=1}^{\infty} a_{(r,s)j}(t)\, q_{1(r,s),xx}^{(j)}(x) + \rho \sum_{j=1}^{\infty} \ddot{a}_{(r,s)j}(t)\, q_{1(r,s)}^{(j)}(x) = f_{1(r,s)}(x,t)$$

$$\text{for } (r,s) \neq (m,n),\ (2m,2n) \tag{5.3.52}$$

$$-T\sum_{j=1}^{\infty} a_{(r,s)j}(t)\, q_{2(r,s),yy}^{(j)}(y) + \rho \sum_{j=1}^{\infty} \ddot{a}_{(r,s)j}(t)\, q_{2(r,s)}^{(j)}(y) = f_{2(r,s)}(y,t)\,,$$

$$\text{for } (r,s) \neq (m,n),\ (2m,2n) \tag{5.3.53}$$

$$-T\sum_{j=1}^{\infty} a_{1(r,s)j}(t)\, q_{1(r,s),xx}^{(j)}(x) + \rho \sum_{j=1}^{\infty} \ddot{a}_{1(r,s)j}(t)\, q_{1(r,s)}^{(j)}(x) = f_{1(r,s)}(x,t)$$

$$\text{for } (r,s)=(m,n)\,,\ (2m,2n) \tag{5.3.54}$$

$$-T\sum_{j=1}^{\infty} a_{2(r,s)j}(t)\, q_{2(r,s),yy}^{(j)}(y) + \rho \sum_{j=1}^{\infty} \ddot{a}_{2(r,s)j}(t)\, q_{2(r,s)}^{(j)}(y) = f_{2(r,s)}(y,t)\,,$$

$$\text{for } (r,s)=(m,n)\,,\ (2m,2n). \tag{5.3.55}$$

Multiplying equations (5.3.52), (5.3.54) and equations (5.3.53), (5.3.55) by $\overline{q}^{(j)}_{1(r,s)}$ and $\overline{q}^{(j)}_{2(r,s)}$, then integrating with respect to the space variables x and y respectively, between limits o and a, and finally adding both corresponding equations for the cases of (r,s) $\neq$ (m,n) and (2m,2n), lead to the single-degree-of-freedom equation

$$\omega^2_{(r,s)j} a_{(r,s)j}(t) + \ddot{a}_{(r,s)j}(t) = b_{(r,s)j}(t), \quad (r,s) \neq (m,n), (2m,2n) \qquad (5.3.56)$$

$$\omega^2_{(r,s)j} a_{1(r,s)j}(t) + \ddot{a}_{1(r,s)j}(t) = b_{1(r,s)j}(t)$$

$$\omega^2_{(r,s)j} a_{2(r,s)j}(t) + \ddot{a}_{2(r,s)j}(t) = b_{2(r,s)j}(t), \quad (r,s)=(m,n), (2m,2n) \qquad (5.3.57)$$

in which

$$b_{(r,s)j} = \int_0^a \overline{q}^{(j)}_{1(r,s)}(x) f_{1(r,s)}(x,t)dx + \int_0^a \overline{q}^{(j)}_{2(r,s)}(y) f_{2(r,s)}(y,t)dy$$

$$(r,s) \neq (m,n), (2m,2n) \qquad (5.3.58)$$

$$b_{1(r,s)j} = \int_0^a \overline{q}^{(j)}_{1(r,s)}(x) f_{1(r,s)}(x,t)dx,$$

$$b_{2(r,s)j} = \int_0^a \overline{q}^{(j)}_{2(r,s)}(y) f_{2(r,s)}(y,t)dy, \quad (r,s)=(m,n), (2m,2n) \qquad (5.3.59)$$

Upon applying the normalization conditions shown in equations (5.3.45)-(5.3.47) to equation (5.3.38), the initial conditions become

$$a_{(r,s)p}(0) = \frac{1}{\sqrt{2m}\sqrt{2n}} \sum_{j=1}^{2m} \sum_{k=1}^{2n} \Big\{ e^{-i(j-1)r\psi_1} e^{-i(k-1)s\psi_2}$$

$$\times \left[\int_0^a \rho \overline{q}^{(p)}_{1(r,s)} W_{10(j,k)} dx + \int_0^a \rho \overline{q}^{(p)}_{2(r,s)} W_{20(j,k)} dy \right] \Big\}$$

$$p=1,2,\ldots,\infty \qquad (r,s) \neq (m,n), (2m,2n)$$

$$a_{1(r,s)p}(0)=\frac{1}{\sqrt{2m}\sqrt{2n}}\sum_{j=1}^{2m}\sum_{k=1}^{2n}e^{-i(j-1)r\psi_1}e^{-i(k-1)s\psi_2}\int_0^a\rho\,\overline{q}_{1(r,s)}^{(p)}\,W_{10(j,k)}\,dx$$

$$a_{2(r,s)p}(0)=\frac{1}{\sqrt{2m}\sqrt{2n}}\sum_{j=1}^{2m}\sum_{k=1}^{2n}e^{-i(j-1)r\psi_1}e^{-i(j-1)s\psi_2}\int_0^a\rho\,\overline{q}_{2(r,s)}^{(p)}\,W_{20(j,k)}\,dy\,,$$

$$p=1,2,\ldots,\infty \qquad (r,s)=(m,n),\ (2m,2n) \tag{5.3.60}$$

and

$$\dot{a}_{(r,s)p}(0)=\frac{1}{\sqrt{2m}\sqrt{2n}}\sum_{j=1}^{2m}\sum_{k=1}^{2n}e^{-i(j-1)r\psi_1}e^{-i(k-1)s\psi_2}\left[\int_0^a\rho\,\overline{q}_{1(r,s)}^{(p)}\,\dot{W}_{10(j,k)}\,dx+\int_0^a\rho\,\overline{q}_{2(r,s)}^{(p)}\,\dot{W}_{20(j,k)}\,dy\right],$$

$$p=1,2,\ldots,\infty \qquad (r,s)\neq(m,n),\ (2m,2n)$$

$$\dot{a}_{1(r,s)p}(0)=\frac{1}{\sqrt{2m}\sqrt{2n}}\sum_{j=1}^{2m}\sum_{k=1}^{2n}e^{-i(j-1)r\psi_1}e^{-i(k-1)s\psi_2}\int_0^a\rho\,\overline{q}_{1(r,s)}^{(p)}\,\dot{W}_{10(j,k)}\,dx\,,$$

$$\dot{a}_{2(r,s)p}(0)=\frac{1}{\sqrt{2m}\sqrt{2n}}\sum_{j=1}^{2m}\sum_{k=1}^{2n}e^{-i(j-1)r\psi_1}e^{-i(k-1)s\psi_2}\int_0^a\rho\,\overline{q}_{2(r,s)}^{(p)}\,\dot{W}_{20(j,k)}\,dy\,,$$

$$p=1,2,\ldots,\infty \qquad (r,s)=(m,n),\ (2m,2n) \tag{5.3.61}$$

The solutions for $a_{(r,s)j}, a_{1(r,s)j}$ and $a_{2(r,s)j}$ of equations (5.3.56) and (5.3.57) with the initial conditions (5.3.60) and (5.3.61) may be expressed as a Duhamel integral

$$a_{(r,s)j}(t)=\frac{1}{\omega_{(r,s)j}}\int_0^t b_{(r,s)j}(\tau)\sin\omega_{(r,s)j}(t-\tau)d\tau+a_{(r,s)j}(0)\cos\omega_{(r,s)j}t$$

$$+\frac{\dot{a}_{(r,s)j}(0)}{\omega_{(r,s)j}}\sin\omega_{(r,s)j}t \qquad (r,s)\neq(m,n),\ (2m,2n)$$

$$a_{1(r,s)j}(t)=\frac{1}{\omega_{(r,s)j}}\int_0^t b_{1(r,s)j}(\tau)\sin\omega_{(r,s)j}(t-\tau)d\tau+a_{1(r,s)j}(0)\cos\omega_{(r,s)j}t+\frac{\dot{a}_{1(r,s)j}(0)}{\omega_{(r,s)j}}\sin\omega_{(r,s)j}t$$

$$a_{2(r,s)j}(t) = \frac{1}{\omega_{(r,s)j}} \int_0^t b_{2(r,s)j}(\tau) \sin \omega_{(r,s)j}(t-\tau) d\tau + a_{2(r,s)j}(0) \cos \omega_{(r,s)j} t + \frac{\dot{a}_{2(r,s)j}(0)}{\omega_{(r,s)j}} \sin \omega_{(r,s)j} t$$

$$(r,s)=(m,n),\ (2m,2n). \tag{5.3.62}$$

The generalized displacements $q_{1(r,s)}$ and $q_{2(r,s)}$ may be found by substituting equations (5.3.43) (5.3.44) (5.3.48) (5.3.49) and (5.3.62) into equation (5.3.50).

As an example, the response of a square cable network subjected to a concentrated force moving on the middle x-cable (m=n=even number) is to be analyzed. It is considered that the concentrated load of magnitude P acts at the end of the middle x-cable of the actual network at the initial time with zero initial conditions and thenceforth moves towards the other end at an uniform velocity v, as shown in Figure 5.3.2 (for m=n=4). With no loss of generality, let m=n=2l; *i.e.*,

$$\psi_1 = \psi_2 = \psi = \pi/2l \tag{5.3.63}$$

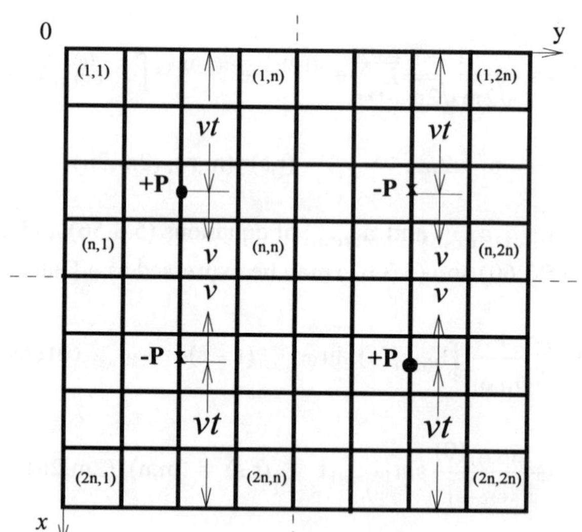

Figure 5.3.2 Square network subjected to a moving force

The (actual and fictitious) loading functions of the equivalent network may be expressed as

$$F_{1(k,l+1)}(x,t) = P\delta[x+(k-1)a-vt], \qquad F_{1(4l-k+1,l+1)}(x,t) = -P\delta[x-ka+vt],$$

$$F_{1(k,3l+1)}(x,t) = -P\delta[x+(k-1)a-vt], \qquad F_{1(4l-k+1,3l+1)}(x,t) = P\delta[x-ka+vt],$$

$$(k-1)\,t_0 < t \le k\,t_0\,, \quad k=1,2,\ldots,2l \tag{5.3.64}$$

with other loading functions vanishing, here $\delta(x)$ denotes the Dirac delta function and t_0 denotes the time to move through a substructure, *i.e.*,

$$t_0 = a/v\,. \tag{5.3.65}$$

Substituting equations (5.3.63) and (5.3.64) into equation (5.3.9) yields

$$f_{1(r,s)}(x,t) = \frac{Pi\sin(ls\psi)}{2l}\left\{e^{ikr\psi}\delta(x-ka+vt) - e^{-i(k-1)r\psi}\delta[x+(k-1)a-vt]\right\}$$

$$f_{2(r,s)}(y,t) \equiv 0, \qquad (k-1)\,t_0 < t \le k\,t_0\,, \quad k=1,2,\ldots,2l \tag{5.3.66}$$

Substituting equations (5.3.43) (5.3.66) and equations (5.3.44) (5.3.66) into equations (5.3.58) and (5.3.59), respectively, and then integrating with respect to the space variables x,y between the limits o and a, yield

$$b_{(r,s)j}(t) = -\{P\sin(ls\psi)/l\}\,A_{(r,s)j}\{\sin(kr\psi)\sin[\alpha_{(r,s)j}(1-k+t/t_0)]$$
$$+\sin[(k-1)r\psi]\sin[\alpha_{(r,s)j}(k-t/t_0)]\}$$

$$(r,s) \ne (2l,2l),\ (4l,4l), \quad (k-1)\,t_0 < t \le k\,t_0\,, \quad k=1,2,\ldots,2l \tag{5.3.67}$$

$$b_{1(r,s)j}(t) \equiv 0, \qquad b_{2(r,s)j}(t) \equiv 0\,, \qquad (r,s) = (2l,2l),\ (4l,4l) \tag{5.3.68}$$

Substituting equations (5.3.67) and (5.3.68) with zero initial conditions into equation (5.3.62), and then integrating with respect to τ , yields

$$a_{(r,s)j}(t) = -\{P\sin(ls\psi)/2l\omega^2_{(r,s)j}\} A_{(r,s)j} H_{(r,s)j}(t), \tag{5.3.69}$$

$$a_{1(r,s)j}(t) = a_{2(r,s)j}(t) = 0, \qquad (a_{(n,n)j} = a_{(2n,2n)j} = 0), \tag{5.3.70}$$

where $A_{(r,s)j}$ has been defined in equation (5.3.48) and

$$\begin{aligned}
H_{(r,s)j}(t) = \sum_{k=1}^{m} \Big\{ &\sin(kr\psi)\Big[\frac{\sin(\xi\omega_{(r,s)j}t_0)}{\xi}\cos[\omega_{(r,s)j}(k t_0 - t) - \eta\omega_{(r,s)j}t_0] \\
&- \frac{\sin(\eta\omega_{(r,s)j}t_0)}{\eta}\cos[\omega_{(r,s)j}(t - k t_0) + \xi\omega_{(r,s)j}t_0]\Big] \\
&+ \sin[(k-1)r\psi]\Big[\frac{\sin(\eta\omega_{(r,s)j}t_0)}{\eta}\cos[\omega_{(r,s)j}(k t_0 - t) - \eta\omega_{(r,s)j}t_0] \\
&- \frac{\sin(\xi\omega_{(r,s)j}t_0)}{\xi}\cos[\omega_{(r,s)j}(t - k t_0) + \xi\omega_{(r,s)j}t_0]\Big]\Big\} \\
&+ \sin[(m+1)r\psi]\Big[\frac{\sin[\xi\omega_{(r,s)j}(t - m t_0)]}{\xi}\cos[\eta\omega_{(r,s)j}(t - m t_0)] \\
&- \frac{\sin[\eta\omega_{(r,s)j}(t - m t_0)]}{\eta}\cos[\xi\omega_{(r,s)j}(t - m t_0)]\Big] \\
&+ \sin(mr\psi)\Big[\frac{\sin[\eta\omega_{(r,s)j}(t - m t_0)]}{\eta}\cos[\xi\omega_{(r,s)j}(mt_0 - t) + \alpha_{(r,s)j}] \\
&+ \frac{\sin[\xi\omega_{(r,s)j}(m t_0 - t)]}{\xi}\cos[\eta\omega_{(r,s)j}(t - m t_0) + \alpha_{(r,s)j}]\Big],
\end{aligned}$$

$$m t_0 < t \le (m+1) t_0, \quad m=0,1,2,\ldots,2l-1 \tag{5.3.71}$$

in which $\xi = (1 + v/c)/2$ and $\eta = (1 - v/c)/2$. It should be noted that the series on the right hand side of equation (5.3.71) vanishes when m equals zero.

Now the case of v=c will be considered in detail, *i.e.*, $\xi = 1$, $\eta \to 0$. For this case, equation (5.3.71) becomes

$$H_{(r,s)j}(t) = 2\,\alpha_{(r,s)j}\sin\alpha_{(r,s)j}\sum_{k=1}^{m-1}\left\{\sin(kr\psi)\sin\left[\alpha_{(r,s)j}(\frac{t}{t_0}-k)\right]\right\}$$

$$+\sin(mr\psi)\left\{\sin\left[\alpha_{(r,s)j}(m+1-\frac{t}{t_0})\right]+\alpha_{(r,s)j}(\frac{t}{t_0}-m)\cos\left[\alpha_{(r,s)j}(m+1-\frac{t}{t_0})\right]\right.$$

$$\left.-\alpha_{(r,s)j}\cos\left[\alpha_{(r,s)j}(\frac{t}{t_0}-m+1)\right]\right\}$$

$$+\sin[(m+1)r\psi]\left\{\sin\left[\alpha_{(r,s)j}(\frac{t}{t_0}-m)\right]-\alpha_{(r,s)j}(\frac{t}{t_0}-m)\cos\left[\alpha_{(r,s)j}(\frac{t}{t_0}-m)\right]\right\}$$

$$m\,t_0 < t \le (m+1)\,t_0, \quad m=0,1,\ldots,2l\text{-}1 \tag{5.3.72}$$

Substituting equations (5.3.69) (5.3.70) and (5.3.43) into equation (5.3.50) yields

$$\begin{Bmatrix} q_{1(r,s)}(x,t) \\ q_{2(r,s)}(y,t) \end{Bmatrix} = -\frac{P\sin(ls\psi)}{2l}\sum_{j=1}^{\infty}\frac{A^2_{(r,s)j}}{\omega^2_{(r,s)j}}H_{(r,s)j}(t)\begin{Bmatrix} e^{ir\psi}\sin(\alpha_{(r,s)j}x/a)+\sin[\alpha_{(r,s)j}(1-x/a)] \\ e^{is\psi}\sin(\alpha_{(r,s)j}y/a)+\sin[\alpha_{(r,s)j}(1-y/a)] \end{Bmatrix}$$

$$\tag{5.3.73}$$

where $A_{(r,s)j}$ and $\alpha_{(r,s)j}$ are defined by equations (5.3.48) and (5.3.40), respectively. The dynamic deflection response for every substructure may be found by inserting equation (5.3.73) into equaion (5.3.5). The deflection response at the center node of the actual network can be expressed as

$$W_{1(l+1,l+1)}(0,t) = \frac{1}{4l}\sum_{r=1}^{4l}\sum_{s=1}^{4l}e^{ilr\psi}\,e^{ils\psi}\,q_{1(r,s)}(0,t)\,. \tag{5.3.74}$$

Substituting equation (5.3.73) into this expression yields

$$W_{1(l+1,l+1)}(0,t) = \frac{Pa}{2T}W(l,t) \tag{5.3.75}$$

where T denotes the pretension of the cables and

$$W(l,t)=\frac{1}{l^2}\sum_{r=1,3}^{2l-1}\sum_{s=1,3}^{2l-1}(-1)^{(r-1)/2}\sum_{j=1}^{\infty}\frac{\sin\alpha_{(r,s)j}}{\alpha_{(r,s)j}\Delta_{(r,s)j}}H_{(r,s)j}(t) \tag{5.3.76}$$

with

$$\Delta_{(r,s)j}=2\alpha_{(r,s)j}-\sin 2\alpha_{(r,s)j}+(\sin\alpha_{(r,s)j}-\alpha_{(r,s)j}\cos\alpha_{(r,s)j})(\cos r\psi+\cos s\psi)\,. \tag{5.3.77}$$

As a numerical example, $W_{l(l+1,l+1)}(0,t)$ with l=5 was calculated over a period in which the concentrated load moves from one end of the middle x-cable to the other end of the same cable (*i.e.*, $0<t\le 10\,t_0$). The dynamic deflection curve of the center node vs. time is shown in Figure 5.3.3.

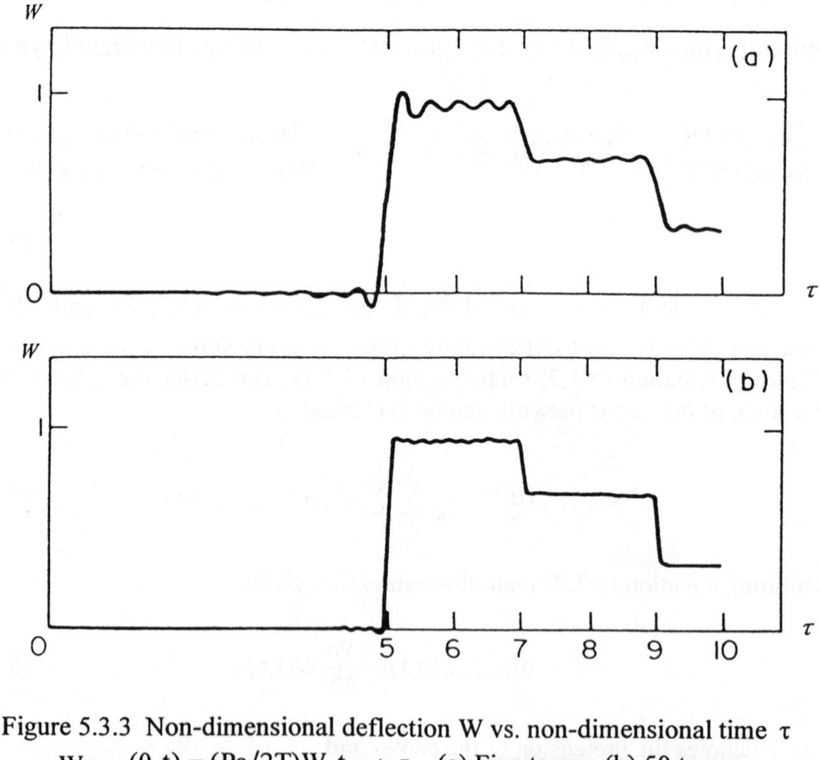

Figure 5.3.3 Non-dimensional deflection W vs. non-dimensional time τ $W_{l(6,6)}(0,t)=(Pa/2T)W, t=t_0\tau$, (a) Five terms; (b) 50 terms.

In Figure 5.3.3, cases (a) and (b) are those in which the first five and 50 terms of the infinite series on the right hand side of equation (5.3.76) were taken into account, respectively, and the remaining terms were neglected. The curves of the deflection response were calculated step by step, with each time step being equal to $0.05\ t_0$.

5.4 Rectangular Diagonal Networks - Lumped Mass System

The system to be considered is made up of two sets of pretensioned straight cables intersecting each other and meeting at spot-welded nodes so that no slippage occurs at the joints. Consider now a $m \times n$ diagonal network with fixed ends at four edges as shown in Figure 5.4.1. The pretension and mass per unit length of the cables are denoted by T and ρ respectively. Its equivalent system with cyclic periodicity in longitudinal and transverse directions can be created in the same way as that for orthogonal network. The equivalent system is a $2m \times 2n$ diagonal network with cyclic periodicity in two directions, as shown in Figure 5.4.2. A typical substructure (j, k) is shown in Figure 5.4.3 which includes two cables with length "a" and two nodes with lumped mass M ($M = \rho\, a$). In each substructure, a set of local coordinates oxy is established with the origin o at its center node and the x,y axes along two cables respectively.

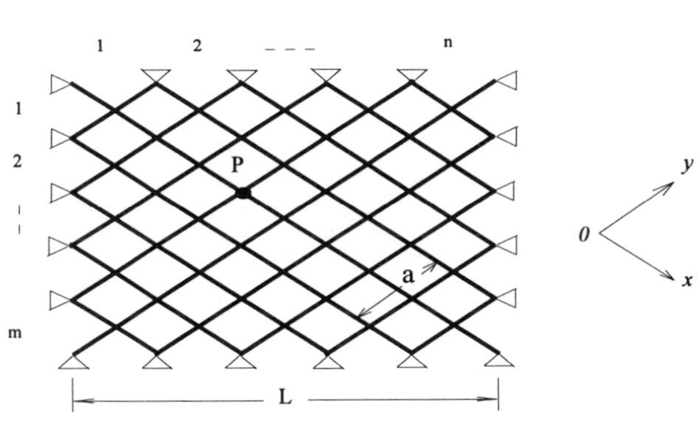

Figure 5.4.1 $m \times n$ Diagonal network

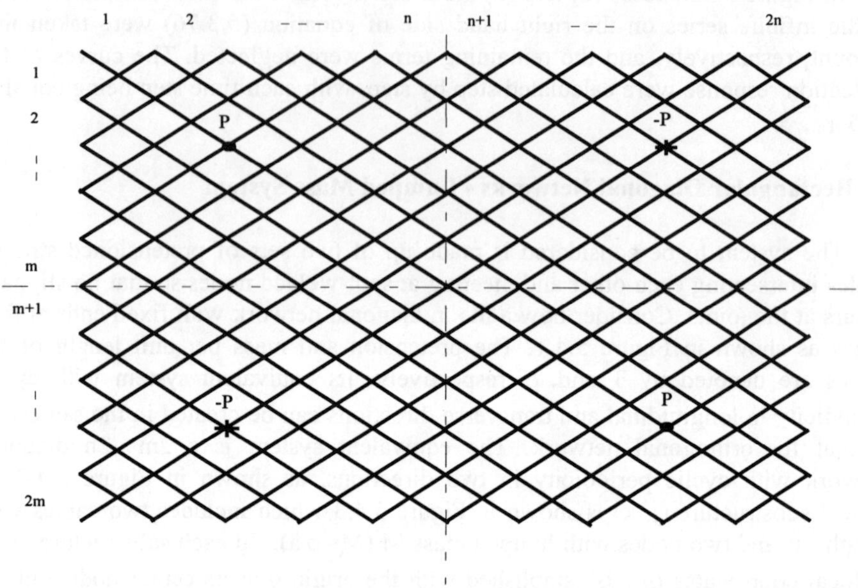

Figure 5.4.2 Equivalent network

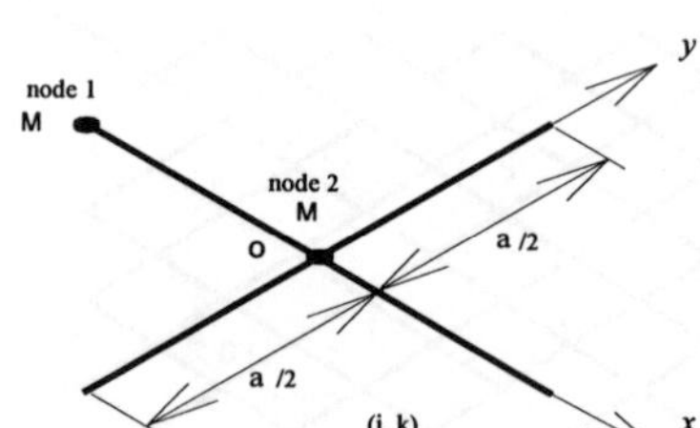

Figure 5.4.3 Substructure (j,k)

The dynamic equations for all substructures are of the same form as shown below:

$$\frac{2T}{a}\left[W_{2(j,k)}+W_{2(j-1,k-1)}+W_{2(j-1,k)}+W_{2(j,k-1)}-4\,W_{1(j,k)}\right]+F_{1(j,k)}=M\,\ddot{W}_{1(j,k)}\,,$$

$$\frac{2T}{a}\left[W_{1(j,k)}+W_{1(j+1,k+1)}+W_{1(j+1,k)}+W_{1(j,k+1)}-4\,W_{2(j,k)}\right]+F_{2(j,k)}=M\,\ddot{W}_{2(j,k)}\,,$$

$$j=1,2,...,2m,\quad k=1,2,...,2n \tag{5.4.1}$$

where $W_{1(j,k)}, F_{1(j,k)}$ and $W_{2(j,k)}, F_{2(j,k)}$ denote the transverse displacement, loading functions of the node 1 and node 2 for substructure (j,k).

Now applying the double U transformation (5.3.5) to equation (5.4.1), results in

$$M\ddot{q}_{1(r,s)}+\frac{2T}{a}\left[4\,q_{1(r,s)}-(1+e^{-ir\psi_1})(1+e^{-is\psi_2})\,q_{2(r,s)}\right]=f_{1(r,s)}\,,$$

$$M\ddot{q}_{2(r,s)}+\frac{2T}{a}\left[4\,q_{2(r,s)}-(1+e^{ir\psi_1})(1+e^{is\psi_2})\,q_{1(r,s)}\right]=f_{2(r,s)}\,,$$

$$r=1,2,...,2m\quad s=1,2,...,2n \tag{5.4.2}$$

where $\psi_1=\pi/m, \psi_2=\pi/n$ and $f_{1(r,s)}, f_{2(r,s)}$ can be expressed as equation (5.3.9). Consider now the natural vibration. The corresponding governing equation becomes

$$\begin{bmatrix} \frac{Ma}{2T}\omega^2-4 & (1+e^{-ir\psi_1})(1+e^{-is\psi_2}) \\ (1+e^{ir\psi_1})(1+e^{is\psi_2}) & \frac{Ma}{2T}\omega^2-4 \end{bmatrix}\begin{Bmatrix} q_{1(r,s)} \\ q_{2(r,s)} \end{Bmatrix}=\begin{Bmatrix} 0 \\ 0 \end{Bmatrix},$$

$$r=1,2...,2m,\ s=1,2...,2n \tag{5.4.3}$$

where ω denotes the natural frequency and $q_{1(r,s)}, q_{2(r,s)}$ amplitude of generalized displacements. Let the coefficient determinant of the simultaneous equation (5.4.3) be equal to zero and then expanding this determinant yields the frequency equation

$$(\frac{Ma}{2T}\omega^2)^2-8\frac{Ma}{2T}\omega^2+16-4(1+\cos r\,\psi_1)(1+\cos s\psi_2)=0$$

$$r=1,2,...,2m,\quad s=1,2,...,2n. \tag{5.4.4}$$

The roots of the frequency equation (5.4.4) can be obtained as

$$\omega^2 = \frac{4T}{Ma}\left[2 \pm \sqrt{(1+\cos r\,\psi_1)(1+\cos s\,\psi_2)}\right],$$

$$r=1,2,..,2m, \quad s=1,2,...,2n. \tag{5.4.5}$$

Consider the simple case of a square network with $n \times n$ substructures, *i.e.*, $\psi_1 = \psi_2 = \pi/n, a = \sqrt{2}L/n$, where L denotes the length of the actual network. For this case equation (5.4.5) may be rewritten as

$$\omega = \frac{\sqrt{2}n}{\pi}\omega_0\sqrt{2 \pm \sqrt{(1+\cos(r\pi/n))(1+\cos(s\pi/n))}}$$

$$r=1,2,...,2n \quad s=1,2,...,2n \tag{5.4.6}$$

where

$$\omega_0 = \frac{\pi}{L}\sqrt{\frac{T}{\rho}} \tag{5.4.7}$$

which represents the basic frequency of a single cable with length L and fixed ends. The first four natural frequencies are given in Table 5.4.1.

Table 5.4.1 Natural frequencies ω for the diagonal square cable networks - Lumped mass system

Networks	$(m \times n)$	(3×3)	(4×4)	(5×5)	(6×6)	(7×7)
Mode(r,s)	(1,1)	0.9549	0.9745	0.9836	0.9886	0.9916
	(1,2)	1.438	1.499	1.528	1.544	1.554
	(2,1)	1.438	1.499	1.528	1.544	1.554
	(2,2)	1.654	1.801	1.871	1.910	1.934
Multiplier		$(\pi / L)\sqrt{T/\rho}$				

5.5 Rectangular Diagonal Networks - Continuous System

The diagonal network considered and its equivalent system can also be shown as Figures 5.4.1 and 5.4.2 but the distributed mass model will be adopted instead of the lumped mass model in this section.

For the present case, the natural vibration equations for substructure (j,k) shown in Figure 5.4.3 where the lumped masses should be replaced by distributed mass, may be expressed as [21]

$$W_{1(j,k),xx}(x) + \frac{\omega^2}{C^2} W_{1(j,k)}(x) = \frac{P_{(j,k)}}{T}\delta(x), \qquad -\frac{a}{2} < x < \frac{a}{2}$$

$$W_{2(j,k),yy}(y) + \frac{\omega^2}{C^2} W_{2(j,k)}(y) = -\frac{P_{(j,k)}}{T}\delta(y), \qquad -\frac{a}{2} < y < \frac{a}{2}$$

$$j=1,2,..,2m, \quad k=1,2,...,2n \tag{5.5.1}$$

where $W_{1(j,k)}(x)$, $W_{2(j,k)}(y)$ denote the amplitude functions of the x- and y- cables for substructure (j,k). oxy is the local co-ordinate system as shown in Figure 5.4.3. ω is the circular frequency. $P_{(j,k)}$ is the action between x-and y- cables for substructure (j,k) and

$$C^2 = T/\rho \tag{5.5.2}$$

where T and ρ denote the pretension and mass per unit length of the x-and y- cables. Continuity across the joints requires the following conditions to be satisfied:

$$W_{1(j,k)}(\frac{a}{2}) = W_{2(j+1,k)}(\frac{a}{2}), \quad W_{1(j,k)}(\frac{a}{2}) = W_{2(j,k+1)}(-\frac{a}{2}),$$

$$W_{1(j,k)}(\frac{a}{2}) = W_{1(j+1,k+1)}(-\frac{a}{2}), \quad W_{1(j,k)}(0) = W_{2(j,k)}(0),$$

$$j=1,2,..,2m, \quad k=1,2,..,2n \tag{5.5.3}$$

in which " a " denotes the length of x-and y- cables in substructure. The equilibrium equations for all joints between the adjacent substructure can be expressed as:

$$T\left[W_{1(j+1,k+1),x}(-\frac{a}{2}) - W_{1(j,k),x}(\frac{a}{2})\right] + T\left[W_{2(j,k+1),y}(-\frac{a}{2}) - W_{2(j+1,k),y}(\frac{a}{2})\right] = 0$$

$$j=1,2,...,2m, \quad k=1,2,...,2n \tag{5.5.4}$$

where

$$W_{2(2m+1,k)}(y) \equiv W_{2(1,k)}(y), \quad W_{2(j,2n+1)}(y) \equiv W_{2(j,1)}(y),$$

$$W_{1(2m+1,k)}(x) \equiv W_{1(1,k)}(x), \quad W_{1(j,2n+1)}(x) \equiv W_{1(j,1)}(x). \tag{5.5.5}$$

Obviously, the anti-symmetric mode with respect to the two symmetric planes of the equivalent network must satisfy the boundary conditions of the actual network. According to the property of a rotationally periodic structure, there is an integer number of waves along the periodic directions of the equivalent network for each natural mode except the rigid body mode $(W_{1(j,k)} = W_{2(j,k)} = \text{constant})$. The anti-symmetric mode about two symmetry planes of the equivalent network is in existence corresponding to each natural frequency except for $\omega = 0$. The anti-symmetric modes automatically satisfy the boundary conditions of the actual network.

In order to uncouple the continuity conditions (5.5.3) and the equilibrium equation (5.5.4), applying the double U-transformation (5.3.5) to equations (5.5.1),(5.5.3) and (5.5.4), result in

$$q_{1(r,s),xx}(x) + \frac{\omega^2}{C^2} q_{1(r,s)}(x) = \frac{P_{(r,s)}}{T} \delta(x),$$

$$q_{2(r,s),yy}(y) + \frac{\omega^2}{C^2} q_{1(r,s)}(y) = -\frac{P_{(r,s)}}{T} \delta(y),$$

$$r=1,2,..,2m, \quad s=1,2,...,2n \tag{5.5.6}$$

$$q_{1(r,s)}(\frac{a}{2}) = e^{ir\psi_1} q_{2(r,s)}(\frac{a}{2}), \quad q_{1(r,s)}(\frac{a}{2}) = e^{is\psi_2} q_{2(r,s)}(-\frac{a}{2}),$$

$$q_{1(r,s)}(\frac{a}{2}) = e^{i(r\psi_1+s\psi_2)} q_{1(r,s)}(-\frac{a}{2}), \quad q_{1(r,s)}(0) = q_{2(r,s)}(0). \tag{5.5.7}$$

and

$$T\left[e^{i(r\psi_1+s\psi_2)}q_{1(r,s),x}(-\frac{a}{2})-q_{1(r,s),x}(\frac{a}{2})\right]+T\left[e^{is\psi_2}q_{2(r,s),y}(-\frac{a}{2})-e^{ir\psi_1}q_{2(r,s),y}(\frac{a}{2})\right]=0$$

(5.5.8)

where $\psi_1=\pi/m, \psi_2=\pi/n$ and

$$p_{(r,s)}=\frac{1}{\sqrt{2m}\sqrt{2n}}\sum_{j=1}^{2m}\sum_{k=1}^{2n}e^{-i(j-1)r\psi_1}e^{-i(k-1)s\psi_2}P_{(j,k)} \qquad (5.5.9)$$

The general solution for $q_{1(r,s)}$ and $q_{2(r,s)}$ of equation (5.5.6) may be expressed as

$$q_{1(r,s)}(x)=\begin{cases}A_L\sin\frac{\omega}{C}x+B_L\cos\frac{\omega}{C}x, & -\frac{a}{2}\le x\le 0\\ A_R\sin\frac{\omega}{C}x+B_R\cos\frac{\omega}{C}x, & 0\le x\le\frac{a}{2}\end{cases}$$

$$q_{2(r,s)}(y)=\begin{cases}C_L\sin\frac{\omega}{C}y+D_L\cos\frac{\omega}{C}y, & -\frac{a}{2}\le y\le 0\\ C_R\sin\frac{\omega}{C}y+D_R\cos\frac{\omega}{C}y, & 0\le y\le\frac{a}{2}\end{cases}$$

r=1,2,...2m, s=1,2,...,2n (5.5.10)

where $A_L, B_L, A_R, B_R, C_L, D_L, C_R$ and D_R are eight constants of integration. In addition to the properties of $\delta(x)$ and $\delta(y)$ in equation (5.5.6) the following equation are obtained

$$q_{1(r,s),x}(x)\Big|_{0^-}^{0^+}=p_{(r,s)}/T,\quad q_{1(r,s)}(0^+)=q_{1(r,s)}(0^-),$$

$$q_{2(r,s),y}(y)\Big|_{0^-}^{0^+}=-p_{(r,s)}/T,\quad q_{2(r,s)}(0^+)=q_{2(r,s)}(0^-).$$

r=1,2,..,2m s=1,2,...,2n (5.5.11)

Substituting equation (5.5.10) into equation (5.5.7) (5.5.8) (5.5.11) and eliminating p(r,s) from equation (5.5.11) then letting the coefficient determinant of the simultaneous equations be equal to zero, the frequency equation as shown below will result:

$$\cos\frac{\omega a}{C}=\frac{1}{2}\left(\cos\frac{r\pi}{m}+\cos\frac{s\pi}{n}+\cos\frac{r\pi}{m}\cos\frac{s\pi}{n}-1\right) \tag{5.5.12}$$

The corresponding eigenfunctions can be obtained from the nontrivial solution of the homogeneous equation. For $(r,s)\neq(m,n),(2m,n),(m,2n)$ and $(2m,2n)$, the eigenfunctions may be expressed as

$$q_{1(r,s)}(x)=\begin{cases} A_{r,s}\left\{\left[\sin\frac{\omega}{C}x+\sin\frac{\omega}{C}(x+a)\right]e^{i(r\psi_1+s\psi_2)/2} -\frac{1}{2}\left[\sum\limits_{u=0}^{1}\sum\limits_{v=0}^{1}e^{i(\frac{1}{2}-u)r\psi_1}e^{i(\frac{1}{2}-v)s\psi_2}\right]\sin\frac{\omega}{C}x\right\}, & -\frac{a}{2}\le x\le 0 \\ A_{r,s}\left\{\left[-\sin\frac{\omega}{C}x+\sin\frac{\omega}{C}(a-x)\right]e^{i(r\psi_1+s\psi_2)/2} +\frac{1}{2}\left[\sum\limits_{u=0}^{1}\sum\limits_{v=0}^{1}e^{i(\frac{3}{2}-u)r\psi_1}e^{i(\frac{3}{2}-v)s\psi_2}\right]\sin\frac{\omega}{C}x\right\}, & 0\le x\le\frac{a}{2} \end{cases}$$

$$q_{2(r,s)}(y)=\begin{cases} A_{r,s}\left\{\left[\sin\frac{\omega}{C}y+\sin\frac{\omega}{C}(y+a)\right]e^{i(r\psi_1+s\psi_2)/2} -\frac{1}{2}\left[\sum\limits_{u=0}^{1}\sum\limits_{v=0}^{1}e^{i(\frac{3}{2}-u)r\psi_1}e^{i(\frac{1}{2}-v)s\psi_2}\right]\sin\frac{\omega}{C}y\right\}, & -\frac{a}{2}\le y\le 0 \\ A_{r,s}\left\{\left[-\sin\frac{\omega}{C}y+\sin\frac{\omega}{C}(a-y)\right]e^{i(r\psi_1+s\psi_2)/2} +\frac{1}{2}\left[\sum\limits_{u=0}^{1}\sum\limits_{v=0}^{1}e^{i(\frac{1}{2}-u)r\psi_1}e^{i(\frac{3}{2}-v)s\psi_2}\right]\sin\frac{\omega}{C}y\right\}, & 0\le y\le\frac{a}{2} \end{cases}$$

$$q_{1(j,k)}(x)=0,\quad q_{2(j,k)}(y)=0,\quad j\neq r \text{ or } k\neq s \tag{5.5.13}$$

where ω is a positive root of the frequency equation (5.5.12) for a given set of *r* and *s*, and $A_{r,s}$ is an arbitrary constant multiplier.

Altogether there are four independent sets of eigenfunctions corresponding to a same eigenvalue. They can be obtained by substituting (2m-r) and (2n-s) instead of r and s in equation (5.5.13), respectively, *i.e.*,

$$q_{1(2m-r,s)}(x), \qquad q_{2(2m-r,s)}(y),$$

$$q_{1(r,2n-s)}(x), \qquad q_{2(r,2n-s)}(y),$$

$$q_{1(2m-r,2n-s)}(x), \qquad q_{2(2m-r,2n-s)}(y). \tag{5.5.14}$$

besides equation (5.5.13).

The linear combination of the above four sets of eigenfunctions is also a set of eigenfunction. But there is only one set of eigenfunction satisfying the actual boundary conditions. Let

$$A_{2m-r,s} = A_{r,2n-s} = -A_{2m-r,2n-s} = -A_{r,s}\ . \tag{5.5.15}$$

By superposition of the four sets of eigenfunction and substituting into equation (5.3.5) while neglecting a constant multiplier, one can obtain

$$W_{1(j,k)}(x) = \begin{cases} \left[\sin\dfrac{\omega}{C}x + \sin\dfrac{\omega}{C}(x+a)\right]\sin(j-\dfrac{1}{2})r\,\psi_1\sin(k-\dfrac{1}{2})s\,\psi_2 \\ \quad -\dfrac{1}{2}\sin\dfrac{\omega}{C}x\displaystyle\sum_{u=0}^{1}\sum_{v=0}^{1}\sin(j-\dfrac{1}{2}-u)r\,\psi_1\sin(k-\dfrac{1}{2}-v)s\,\psi_2, \\ \hfill -\dfrac{a}{2}\le x\le 0 \\ \left[-\sin\dfrac{\omega}{C}x + \sin\dfrac{\omega}{C}(a-x)\right]\sin(j-\dfrac{1}{2})r\,\psi_1\sin(k-\dfrac{1}{2})s\,\psi_2 \\ \quad +\dfrac{1}{2}\sin\dfrac{\omega}{C}x\displaystyle\sum_{u=0}^{1}\sum_{v=0}^{1}\sin(j+\dfrac{1}{2}-u)r\,\psi_1\sin(k+\dfrac{1}{2}-v)s\,\psi_2, \\ \hfill 0\le x\le\dfrac{a}{2} \end{cases}$$

$$
W_{2(j,k)}(y)=\begin{cases}
\left[\sin\dfrac{\omega}{C}y+\sin\dfrac{\omega}{C}(y+a)\right]\sin(j-\dfrac{1}{2})r\,\psi_1\sin(k-\dfrac{1}{2})s\,\psi_2 \\
\quad -\dfrac{1}{2}\sin\dfrac{\omega}{C}y\sum_{u=0}^{1}\sum_{v=0}^{1}\sin(j+\dfrac{1}{2}-u)r\,\psi_1\sin(k-\dfrac{1}{2}-v)s\,\psi_2, \\
\hfill -\dfrac{a}{2}\le y\le 0 \\
\left[-\sin\dfrac{\omega}{C}y+\sin\dfrac{\omega}{C}(a-y)\right]\sin(j-\dfrac{1}{2})r\,\psi_1\sin(k-\dfrac{1}{2})s\,\psi_2 \\
\quad +\dfrac{1}{2}\sin\dfrac{\omega}{C}y\sum_{u=0}^{1}\sum_{v=0}^{1}\sin(j-\dfrac{1}{2}-u)r\,\psi_1\sin(k+\dfrac{1}{2}-v)s\,\psi_2, \\
\hfill 0\le y\le \dfrac{a}{2}
\end{cases}
\tag{5.5.16}
$$

It can be verified that the amplitude functions of deflection shown in equation (5.5.16) satisfy the boundary conditions of the original network, *i.e.*,

$$
W_{1(1,k)}(-\frac{a}{2})=0,\quad W_{1(j,1)}(-\frac{a}{2})=0,\quad W_{1(m,k)}(\frac{a}{2})=0,\quad W_{1(j,n)}(\frac{a}{2})=0,
$$

$$
W_{2(1,k)}(\frac{a}{2})=0,\quad W_{2(j,1)}(-\frac{a}{2})=0,\quad W_{2(m,k)}(-\frac{a}{2})=0,\quad W_{2(j,n)}(\frac{a}{2})=0,
$$

$$
j=1,2,..,m,\qquad k=1,2,...,n. \tag{5.5.17}
$$

Consequently, the deflection functions shown in equation (5.5.16) represent the natural modes of the actual network except (r,s)=(m,n),(2m,n),(m,2n),(2m,2n).

When (r,s)=(m,n),(2m,n),(m,2n) , the frequency equation becomes

$$
\cos\frac{\omega a}{C}=-1 \tag{5.5.18}
$$

and then the natural frequencies can be expressed as

$$\omega = \frac{(2t+1)\pi C}{a}, \qquad t=0,1,2,... \tag{5.5.19}$$

The corresponding eigenfunctions are

$$q_{1(r,s)}(x) = B\cos\frac{(2t+1)\pi}{a}x$$

$$q_{2(r,s)}(y) = B\cos\frac{(2t+1)\pi}{a}y \tag{5.5.20}$$

Substituting equation (5.5.20) into equation (5.3.5) and neglecting an arbitrary constant multiplier, yields the natural modes

$$W_{1(j,k)}(x) = \cos(j-1)r\,\psi_1\cos(k-1)s\,\psi_2\cos\frac{(2t+1)\pi}{a}x$$

$$W_{2(j,k)}(y) = \cos(j-1)r\,\psi_1\cos(k-1)s\,\psi_2\cos\frac{(2t+1)\pi}{a}y \tag{5.5.21}$$

When (r,s)=(2m,2n), the natural frequencies and modes can be obtained by the same way, as

$$\omega = \frac{2t\pi C}{a}, \qquad t=1,2,... \tag{5.5.22}$$

and

$$W_{1(j,k)}(x) = A\sin\frac{2t\pi}{a}x,$$

$$W_{2(j,k)}(y) = B\sin\frac{2t\pi}{a}y, \qquad j=1,2,..,m \quad k=1,2,...,n \tag{5.5.23}$$

where A and B are two independent constants.

Consider the case of a square network with length L and $n \times n$ mesh, *i.e.*,

$$m = n, \qquad \psi_1 = \psi_2 = \pi/n, \qquad a = \sqrt{2}\,L/n. \tag{5.5.24}$$

The numerical results of the first four natural frequencies are given in Table 5.5.1.

Table 5.5.1 Natural frequencies ω for the diagonal square cable networks - distributed mass system

Networks $(m \times n)$		3×3	4×4	5×5	6×6	7×7
Mode(r,s)	(1,1)	0.9758	0.9871	0.9914	0.9942	0.9956
	(1,2)	1.517	1.546	1.559	1.566	1.570
	(2,1)	1.517	1.546	1.559	1.566	1.570
	(2,2)	1.780	1.886	1.930	1.952	1.965
Multiplier			$(\pi / L)\sqrt{T/\rho}$			

5.6 Rectangular Grids

Consider now a rectangular grid with all edges simply supported. The grid is made up of two orthogonal sets of beams connected at the nodes so that no moments are transmitted from one set to the other. Let the flexural rigidity and mass per unit length of the beams in the x- and y- directions be denoted by EI_1, ρ_1 and EI_2, ρ_2, respectively. The geometric parameters a, b, n, m, and local coordinate system oxy have the same definitions as those in Figures 5.2.1 and 5.3.1. A typical substructure (j,k) also is shown in the same Figure 5.3.1 where the members are now beams instead of cables.

Now, in a manner similar to the previous section 5.3 for analyzing rectangular network , we consider the equivalent grid, which is a rotational periodic structure in two directions. The dynamic equations for all substructures are of the same form, *i.e.*,

$$EI_1 \frac{\partial^4}{\partial x^4} W_{1(j,k)}(x,t) + \rho_1 \frac{\partial^2}{\partial t^2} W_{1(j,k)}(x,t) = F_{1(j,k)}(x,t), \qquad 0 < x < a$$

$$EI_2 \frac{\partial^4}{\partial y^4} W_{2(j,k)}(y,t) + \rho_2 \frac{\partial^2}{\partial t^2} W_{2(j,k)}(y,t) = F_{2(j,k)}(y,t), \qquad 0 < y < b$$

$$j=1,2,...,2m, \qquad k=1,2,...,2n \tag{5.6.1}$$

where $W_{1(j,k)}, W_{2(j,k)}$ and $F_{1(j,k)}, F_{2(j,k)}$ denote the deflections and loading functions of the x-and y- beams for substructure (j,k).

Continuity across the nodes requires the following conditions to be satisfied.

$$W_{1(j,k)}\Big|_{x=a} = W_{1(j+1,k)}\Big|_{x=0},\quad W_{1(j,k),x}\Big|_{x=a} = W_{1(j+1,k),x}\Big|_{x=0},$$

$$W_{1(j,k),xx}\Big|_{x=a} = W_{1(j+1,k),xx}\Big|_{x=0},\quad W_{2(j,k)}\Big|_{y=b} = W_{2(j,k+1)}\Big|_{y=0},$$

$$W_{2(j,k),y}\Big|_{y=b} = W_{2(j,k+1),y}\Big|_{y=0},\quad W_{2(j,k),yy}\Big|_{y=b} = W_{2(j,k+1),yy}\Big|_{y=0},$$

$$W_{1(j,k)}\Big|_{x=0} = W_{2(j,k)}\Big|_{y=0},\qquad j=1,2,\ldots,2m,\quad k=1,2,\ldots,2n \tag{5.6.2}$$

with $W_{1(2m+1,k)} \equiv W_{1(1,k)}$ and $W_{2(j,2n+1)} \equiv W_{2(j,1)}$. The equilibrium equation for the shear forces at the node can be written as

$$EI_1\left[W_{1(j,k),xxx}\Big|_{x=0} - W_{1(j-1,k),xxx}\Big|_{x=a}\right] + EI_2\left[W_{2(j,k),yyy}\Big|_{y=0} - W_{2(j,k-1),yyy}\Big|_{y=b}\right] = P_{(j,k)},$$

$$j=1,2,\ldots,2m,\quad k=1,2,\ldots,2n \tag{5.6.3}$$

where $P_{(j,k)}$ denotes concentrated load applied at the node of substructure (j,k) and $W_{1(0,k)} \equiv W_{1(2m,,k)}$, $W_{2(j,0)} \equiv W_{2(j,2n)}$.

Applying the double U-transformation shown in equation (5.3.5) to equations (5.6.1)-(5.6.3) results in

$$EI_1\frac{\partial^4}{\partial x^4}q_{1(r,s)}(x,t) + \rho_1\frac{\partial^2}{\partial t^2}q_{1(r,s)}(x,t) = f_{1(r,s)}(x,t),\qquad 0<x<a$$

$$EI_2\frac{\partial^4}{\partial y^4}q_{2(r,s)}(y,t) + \rho_2\frac{\partial^2}{\partial t^2}q_{2(r,s)}(y,t) = f_{2(r,s)}(y,t),\qquad 0<y<b$$

$$j=1,2,\ldots,2m,\quad k=1,2,\ldots,2n \tag{5.6.4}$$

and

$$q_{1(r,s)}\Big|_{x=a} = e^{ir\psi_1}q_{1(r,s)}\Big|_{x=0},\quad q_{1(r,s),x}\Big|_{x=a} = e^{ir\psi_1}q_{1(r,s),x}\Big|_{x=0},$$

$$q_{1(r,s),xx}\big|_{x=a} = e^{ir\psi_1} q_{1(r,s),xx}\big|_{x=0}, \quad q_{2(r,s)}\big|_{y=b} = e^{is\psi_2} q_{2(r,s)}\big|_{y=0},$$

$$q_{2(r,s),y}\big|_{y=b} = e^{is\psi_2} q_{2(r,s),y}\big|_{y=0}, \quad q_{2(r,s),yy}\big|_{y=b} = e^{is\psi_2} q_{2(r,s),yy}\big|_{y=0},$$

$$q_{1(r,s)}\big|_{x=0} = q_{2(r,s)}\big|_{y=0}. \tag{5.6.5}$$

$$EI_1\left[q_{1(r,s),xxx}\big|_{x=0} - e^{-ir\psi_1} q_{1(r,s),xxx}\big|_{x=a}\right] + EI_2\left[q_{2(r,s),yyy}\big|_{y=0} - e^{-is\psi_2} q_{2(r,s),yyy}\big|_{y=b}\right] = p_{(r,s)}$$

$$r=1,2,\ldots,2m, \quad s=1,2,\ldots,2n \tag{5.6.6}$$

where $f_{1(r,s)}, f_{2(r,s)}$ have been defined by equation (5.3.9) and

$$p_{(r,s)} = \frac{1}{\sqrt{2m}\sqrt{2n}} \sum_{j=1}^{2m} \sum_{k=1}^{2n} e^{-i(j-1)r\psi_1} e^{-i(k-1)s\psi_2} P_{(j,k)} \tag{5.6.7}$$

in which $P_{(j,k)}, F_{1(j,k)}, F_{2(j,k)}$ must satisfy the anti-symmetrical conditions respectively.

5.6.1 Static Problems

Consider now the rectangular grillwork with square mesh and same rigidity of x-and y-beams, *i.e.*,

$$a = b, \qquad EI_1 = EI_2 \equiv EI. \tag{5.6.8}$$

It is assumed that the static loads apply at the nodes. For present case the equation (5.6.4) becomes

$$EI\frac{d^4}{dx^4} q_{1(r,s)}(x) = 0, \qquad 0 < x < a$$

$$EI\frac{d^4}{dy^4} q_{2(r,s)}(y) = 0, \qquad 0 < y < b$$

$$r=1,2,\ldots,2m, \quad s=1,2,\ldots,2n \tag{5.6.9}$$

The general solution for $q_{1(r,s)}(x)$ and $q_{2(r,s)}(y)$ of equation (5.6.9) can be written as

$$q_{1(r,s)}(x) = A_{1(r,s)} + B_{1(r,s)}x + C_{1(r,s)}x^2 + D_{1(r,s)}x^3,$$

$$q_{2(r,s)}(y) = A_{2(r,s)} + B_{2(r,s)}y + C_{2(r,s)}y^2 + D_{2(r,s)}y^3. \qquad (5.6.10)$$

Substituting equation (5.6.10) into equations (5.6.5) and (5.6.6) yield

$$(1-e^{ir\psi_1})A_{1(r,s)} + a\,B_{1(r,s)} + a^2 C_{1(r,s)} + a^3 D_{1(r,s)} = 0,$$

$$(1-e^{ir\psi_1})B_{1(r,s)} + 2a\,C_{1(r,s)} + 3a^2 D_{1(r,s)} = 0,$$

$$2(1-e^{ir\psi_1})C_{1(r,s)} + 6a\,D_{1(r,s)} = 0,$$

$$(1-e^{is\psi_2})A_{2(r,s)} + a\,B_{2(r,s)} + a^2 C_{2(r,s)} + a^3 D_{2(r,s)} = 0,$$

$$(1-e^{is\psi_2})B_{2(r,s)} + 2a\,C_{2(r,s)} + 3a^2 D_{2(r,s)} = 0$$

$$2(1-e^{is\psi_2})C_{2(r,s)} + 6a\,D_{2(r,s)} = 0,$$

$$A_{1(r,s)} = A_{2(r,s)}. \qquad (5.6.11)$$

$$EI\left[(1-e^{-ir\psi_1})6\,D_{1(r,s)} + (1-e^{-is\psi_2})6\,D_{2(r,s)}\right] = p_{(r,s)} \qquad (5.6.12)$$

where $p_{(r,s)}$ has been defined by equation (5.6.7). Let us discuss now a $m \times n$ rectangular grillwork subjected to a concentrated load of magnitude P acting at the nodal point (j^*, k^*). The anti-symmetrical loads applied at four nodes of the equivalent grillwork shown in Figure 5.2.2 may be expressed as

$$F_{(j^*,k^*)} = P, \qquad F_{(j^*,2n-k^*+2)} = -P,$$

$$F_{(2m-j^*+2,k^*)} = -P, \qquad F_{(2m-j^*+2,2n-k^*+2)} = P \qquad (5.6.13)$$

and other nodal loads are equal to zero. Substituting equation (5.6.13) into equation (5.6.7) results in

$$P_{(r,s)} = \frac{-4P}{\sqrt{2m}\sqrt{2n}} \sin(j^* - 1)r\,\psi_1 \sin(k^* - 1)s\,\psi_2 \,. \qquad (5.6.14)$$

The solution for $A_{1(r,s)}\,A_{2(r,s)}$ to $D_{1(r,s)}\,D_{2(r,s)}$ of equations (5.6.11) and (5.6.12) with equation (5.6.14) can be found as

$$A_{1(r,s)} = -\frac{P\,a^3\theta_{rs}}{3EI\sqrt{2m}\sqrt{2n}}(2+\cos r\,\psi_1)(2+\cos s\,\psi_2)\,,$$

$$B_{1(r,s)} = -\frac{P\,a^2\theta_{rs}}{EI\sqrt{2m}\sqrt{2n}}\, i\sin r\,\psi_1 (2+\cos s\,\psi_2)\,,$$

$$C_{1(r,s)} = \frac{P\,a\,\theta_{rs}}{EI\sqrt{2m}\sqrt{2n}}(1-\cos r\,\psi_1)(2+\cos s\,\psi_2)\,,$$

$$D_{1(r,s)} = -\frac{P\,\theta_{rs}}{3EI\sqrt{2m}\sqrt{2n}}(1-\cos r\,\psi_1)(2+\cos s\,\psi_2)(1-e^{ir\psi_1})\,,$$

$$A_{2(r,s)} = A_{1(r,s)}\,,$$

$$B_{2(r,s)} = -\frac{P\,a^2\theta_{rs}}{EI\sqrt{2m}\sqrt{2n}}\, i\sin s\,\psi_2 (2+\cos r\,\psi_1)\,,$$

$$C_{2(r,s)} = \frac{P\,a\,\theta_{rs}}{EI\sqrt{2m}\sqrt{2n}}(1-\cos s\,\psi_2)(2+\cos r\,\psi_1)\,,$$

$$D_{2(r,s)} = -\frac{P\,\theta_{rs}}{3EI\sqrt{2m}\sqrt{2n}}(1-\cos s\,\psi_2)(2+\cos r\,\psi_1)(1-e^{is\psi_2}) \qquad (5.6.15)$$

where

$$\theta_{rs} = \frac{\sin(j^*-1)r\,\psi_1 \sin(k^*-1)s\,\psi_2}{(1-\cos r\,\psi_1)^2(2+\cos s\,\psi_2)+(1-\cos s\,\psi_2)^2(2+\cos r\,\psi_1)} \qquad (5.6.16)$$

The deflections of the nodes can be expressed as

$$W_{(j,k)} = P\,\delta_{(j,k)(j^*,k^*)} = \frac{1}{\sqrt{2m}\sqrt{2n}}\sum_{r=1}^{2m}\sum_{s=1}^{2n} e^{i(j-1)r\psi_1} e^{i(k-1)s\psi_2} q_{1(r,s)}(0)\,, \qquad (5.6.17)$$

with

$$q_{1(r,s)}(0) = A_{1(r,s)} = -\frac{P a^3 \theta_{rs}}{3EI\sqrt{2m}\sqrt{2n}}(2+\cos r\,\psi_1)(2+\cos s\,\psi_2)\,. \qquad (5.6.18)$$

Substituting equation (5.6.18) into equation (5.6.17) results in

$$\delta_{(j,k)(j^*,k^*)} = \frac{a^3}{3EI}\frac{1}{mn}\sum_{r=1}^{m-1}\sum_{s=1}^{n-1}[\sin(j-1)r\,\psi_1 \sin(k-1)s\,\psi_2 \sin(j^*-1)r\,\psi_1 \sin(k^*-1)s\,\psi_2$$

$$\times\frac{(2+\cos r\,\psi_1)(2+\cos s\,\psi_2)}{(1-\cos r\,\psi_1)^2(2+\cos s\,\psi_2)+(1-\cos s\,\psi_2)^2(2+\cos r\,\psi_1)}]$$

$$(5.6.19)$$

with $\psi_1 = \pi/m$ and $\psi_2 = \pi/n$. Symbol $\delta_{(j,k)(j^*,k^*)}$ is referred to as the influence coefficient which means the deflection at node (j,k) caused by unit force acting at the node (j^*,k^*). It is obvious that the reciprocal relation

$$\delta_{(j,k)(j^*,k^*)} = \delta_{(j^*,k^*)(j,k)} \qquad (5.6.20)$$

is satisfied in this case.

The bending moments at the nodes of the x- and y- beams can be obtained as

$$M_{x(j,k)} = -EI\,W_{1(j,k),xx}(0)$$

$$= \frac{2Pa}{mn}\sum_{r=1}^{m-1}\sum_{s=1}^{n-1}[\sin(j-1)r\,\psi_1 \sin(k-1)s\,\psi_2 \sin(j^*-1)r\,\psi_1 \sin(k^*-1)s\,\psi_2$$

$$\times\frac{(1-\cos r\,\psi_1)(2+\cos s\,\psi_2)}{(1-\cos r\,\psi_1)^2(2+\cos s\,\psi_2)+(1-\cos s\,\psi_2)^2(2+\cos r\,\psi_1)}],$$

$$M_{y(j,k)} = -EI\,W_{2(j,k),yy}(0)$$

$$= \frac{2Pa}{mn} \sum_{r=1}^{m-1} \sum_{s=1}^{n-1} [\sin(j-1)r\,\psi_1 \sin(k-1)s\,\psi_2 \sin(j^*-1)r\,\psi_1 \sin(k^*-1)s\,\psi_2$$

$$\times \frac{(1-\cos s\,\psi_2)(2+\cos r\,\psi_1)}{(1-\cos r\,\psi_1)^2(2+\cos s\,\psi_2)+(1-\cos s\,\psi_2)^2(2+\cos r\,\psi_1)}]. \tag{5.6.21}$$

Consider now a 2×3 grid subjected to a concentrated load of magnitude P acting at the node (2,3) as shown in Figure 5.6.1, *i.e.*,

$$m = 2, \qquad n = 3, \qquad j^* = 2, \qquad k^* = 3. \tag{5.6.22}$$

The nodal deflections can be found by using the formula (5.6.19), as

$$W_{(2,2)} = \frac{7Pa^3}{144EI}, \qquad W_{(2,3)} = \frac{13Pa^3}{144EI}.$$

Substituting equation (5.6.22) into equation (5.6.21) gives

$$M_{x(2,2)} = \frac{7}{48}Pa, \quad M_{y(2,2)} = -\frac{1}{24}Pa, \quad M_{x(2,3)} = \frac{13}{48}Pa, \quad M_{y(2,3)} = \frac{5}{24}Pa.$$

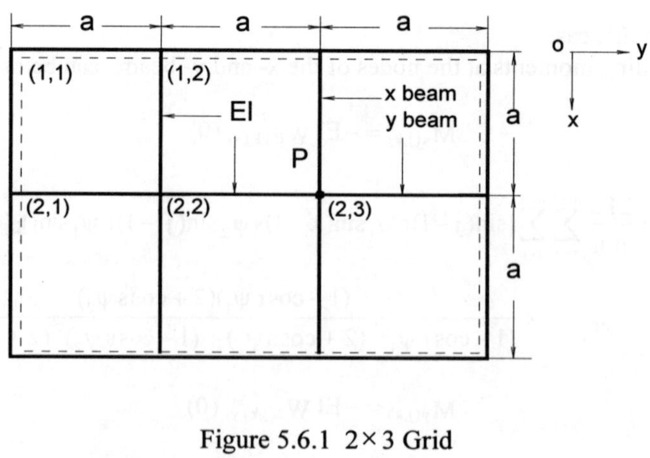

Figure 5.6.1 2× 3 Grid

The same results are obtained by analyzing the grid with stiffness method, demonstrating the exactness of the present solution. The proposed method has been extended to the case taking into account the torsional rigidity of members[17].

5.6.2 Natural Vibration - Lumped Mass System

The governing equation can be expressed as

$$EI_1 \frac{d^4}{dx^4} W_{1(j,k)}(x) = 0, \qquad 0 < x < a$$

$$EI_2 \frac{d^4}{dy^4} W_{2(j,k)}(y) = 0, \qquad 0 < y < b$$

$$j=1,2,\ldots,2m \qquad k=1,2,\ldots,2n \tag{5.6.23}$$

and the equilibrium equation of node is

$$EI_1\left[W_{1(j,k),xxx}(0) - W_{1(j-1,k),xxx}(a)\right] + EI_2\left[W_{2(j,k),yyy}(0) - W_{2(j,k-1),yyy}(b)\right]$$
$$= \omega^2 M\, W_{1(j,k)}(0) \tag{5.6.24}$$

where $W_{1(j,k)}(x), W_{2(j,k)}(y)$ denote the natural mode for x- and y-beams of substructure (j,k), ω denotes the natural frequency and M the lumped mass of each node, *i.e.*,

$$M = \rho_1 a + \rho_2 b\,. \tag{5.6.25}$$

The continuity conditions can also be expressed as equation (5.6.2). Applying the double U-transformation (5.3.5) to equations (5.6.23) (5.6.24) and (5.6.2) yield

$$EI_1 \frac{d^4}{dx^4} q_{1(r,s)}(x) = 0, \qquad 0 < x < a$$

$$EI_2 \frac{d^4}{dy^4} q_{2(r,s)}(y) = 0, \qquad 0 < y < b \tag{5.6.26}$$

$$EI_1\left[q_{1(r,s),xxx}(0)-e^{-ir\psi_1}q_{1(r,s),xxx}(a)\right]+EI_2\left[q_{2(r,s),yyy}(0)-e^{-is\psi_2}q_{2(r,s),yyy}(b)\right]$$
$$=\omega^2 M q_{1(r,s)}(0) \tag{5.6.27}$$

with equation (5.6.5).

The general solution for $q_{1(r,s)}(x)$ and $q_{2(r,s)}(y)$ of equation (5.6.26) can also be written as equation (5.6.10). Now consider the case of $a=b$ and $EI_1=EI_2=EI$. Substituting equations (5.6.10) and (5.6.8) into equations (5.6.5) and (5.6.27), give equation (5.6.11) and

$$EI\left[(1-e^{-ir\psi_1})6D_{1(r,s)}+(1-e^{-is\psi_2})6D_{2(r,s)}\right]=\omega^2 M A_{1(r,s)} \tag{5.6.28}$$

In a manner similar to solve the simultaneous equations (5.6.11) and (5.6.12), following relation may be obtained by solving equations (5.6.11) and (5.6.28) as

$$A_{1(r,s)}=\frac{\omega^2 M a^3}{12EI}\frac{(2+\cos r\psi_1)(2+\cos s\psi_2)}{(1-\cos r\psi_1)^2(2+\cos s\psi_2)+(1-\cos s\psi_2)^2(2+\cos r\psi_1)}A_{1(r,s)} \tag{5.6.29}$$

and then the natural frequencies may be expressed as

$$\omega_{r,s}^2=\frac{12EI[(1-\cos r\psi_1)^2(2+\cos s\psi_2)+(1-\cos s\psi_2)^2(2+\cos r\psi_1)]}{Ma^3(2+\cos r\psi_1)(2+\cos s\psi_2)}$$
$$r=1,2,...,m-1, \quad s=1,2,...,n-1 \tag{5.6.30}$$

Consider an uniform square grillage with square mesh, *i.e.*,

$$n=m, a=b=\frac{L}{n}, EI_1=EI_2=EI, \psi_1=\psi_2=\frac{\pi}{n}, \rho_1=\rho_2=\rho, M=2\rho a. \tag{5.6.31}$$

The equations (5.6.30) can be rewritten as

$$\omega_{r,s} = \frac{\alpha_{r,s}^2}{L^2}\sqrt{\frac{EI}{\rho}} \tag{5.6.32}$$

where L denotes the length of square grillage and

$$\alpha_{r,s}^2 = \sqrt{6}n^2\left[\frac{(1-\cos r\psi)^2(2+\cos s\psi)+(1-\cos s\psi)^2(2+\cos r\psi)}{(2+\cos r\psi)(2+\cos s\psi)}\right]^{\frac{1}{2}}$$

$$r,s=1,2,\ldots,n-1 \tag{5.6.33}$$

with $\psi = \pi/n$. $\alpha_{r,s}$ is a non-dimensional parameter of natural frequency. The first four natural frequencies are given in Table 5.6.1.

Table 5.6.1 Frequency parameter $\alpha_{r,s}$ for the uniform square grids - Lumped mass system

Grids $m \times n$	2×2	3×3	4×4	5×5	6×6	7×7
Mode(r,s) (1,1)	3.1302	3.1399	3.1411	3.1414	3.1415	3.1415
(1,2)		5.2807	5.3458	5.3579	5.3615	5.3628
(2,1)		5.2807	5.3458	5.3579	5.3615	5.3628
(2,2)		6.1793	6.2603	6.2754	6.2795	6.2815

Consider now the natural mode corresponding to natural frequency $\omega_{r,s}$. By substituting

$$q_{1(r,s)}(0) = A_{1(r,s)} = 1, \quad q_{1(2n-r,s)}(0) = q_{1(r,2n-s)}(0) = -q_{1(2n-r,2n-s)}(0) = -1 \tag{5.6.34}$$

with other generalized displacements vanishing, into double U-transformation (5.3.5) yields the natural mode for the nodes of the actual grid with all ends simply supported

$$W_{(j,k)} = W_{1(j,k)}(0) = \sin(j-1)r\psi \sin(k-1)s\psi\,,$$

$$j=1,2,\ldots,m \quad k=1,2,\ldots,n \tag{5.6.35}$$

in which an arbitrary multiplier has been neglected and $\psi = \pi/n$.

5.6.3 Natural Vibration - Distributed Mass System

Let

$$W_{1(j,k)}(x,t) = W_{1(j,k)}(x)e^{i\omega t}, \quad W_{2(j,k)}(y,t) = W_{2(j,k)}(y)e^{i\omega t},$$

$$F_{1(j,k)}(x,t) = F_{2(j,k)}(y,t) = 0. \tag{5.6.36}$$

Substituting equation (5.6.36) into equation (5.6.1) yields

$$\frac{d^4}{dx^4}W_{1(j,k)}(x) - \beta_1^4 W_{1(j,k)}(x) = 0, \quad 0 < x < a$$

$$\frac{d^4}{dy^4}W_{2(j,k)}(y) - \beta_2^4 W_{2(j,k)}(y) = 0, \quad 0 < y < a$$

$$j=1,2,\ldots,2m \quad k=1,2,\ldots,2n \tag{5.6.37}$$

where $W_{1(j,k)}(x)$ and $W_{2(j,k)}(y)$ denote the amplitude functions of the x- and y-beams for substructure (j,k) and

$$\beta_1^4 = \rho_1\omega^2/EI_1, \qquad \beta_2^4 = \rho_2\omega^2/EI_2. \tag{5.6.38}$$

Continuity conditions can also be expressed as equation (5.6.2) and the equilibrium condition for the nodes may be obtained by substituting $P_{(j,k)} = 0$ into equation (5.6.3).

Applying the double U-transformation shown in equation (5.3.5) to equations (5.6.37), (5.6.2) and (5.6.3) where $P_{(j,k)}$ should be equal to zero, yield

$$\frac{d^4}{dx^4}q_{1(r,s)}(x) - \beta_1^4 q_{1(r,s)}(x) = 0, \quad 0 < x < a$$

$$\frac{d^4}{dy^4}q_{2(r,s)}(y) - \beta_2^4 q_{2(r,s)}(y) = 0, \quad 0 < y < a$$

$$r=1,2,\ldots,2m \qquad s=1,2,\ldots,2n \tag{5.6.39}$$

with equations (5.6.5) and (5.6.6) in which $p_{(r,s)} = 0$. The general solution for $q_{1(r,s)}(x)$ and $q_{2(r,s)}(y)$ of equation (5.6.39) can be written as

$$q_{1(r,s)}(x) = A_1 \sin\beta_1 x + B_1 \cos\beta_1 x + C_1 \operatorname{sh}\beta_1 x + D_1 \operatorname{ch}\beta_1 x$$

$$q_{2(r,s)}(y) = A_2 \sin\beta_2 y + B_2 \cos\beta_2 y + C_2 \operatorname{sh}\beta_2 y + D_2 \operatorname{ch}\beta_2 y \tag{5.6.40}$$

Substituting equation (5.6.40) into equations (5.6.5) and (5.6.6), give

$$\begin{gathered}
A_1 \sin\beta_1 a + B_1 \cos\beta_1 a + C_1 \operatorname{sh}\beta_1 a + D_1 \operatorname{ch}\beta_1 a = e^{ir\psi_1}(B_1 + D_1)\,, \\
A_1 \cos\beta_1 a - B_1 \sin\beta_1 a + C_1 \operatorname{ch}\beta_1 a + D_1 \operatorname{sh}\beta_1 a = e^{ir\psi_1}(A_1 + C_1)\,, \\
-A_1 \sin\beta_1 a - B_1 \cos\beta_1 a + C_1 \operatorname{sh}\beta_1 a + D_1 \operatorname{ch}\beta_1 a = e^{ir\psi_1}(-B_1 + D_1)\,, \\
A_2 \sin\beta_2 b + B_2 \cos\beta_2 b + C_2 \operatorname{sh}\beta_2 b + D_2 \operatorname{ch}\beta_2 b = e^{is\psi_2}(B_2 + D_2)\,, \\
A_2 \cos\beta_2 b - B_2 \sin\beta_2 b + C_2 \operatorname{ch}\beta_2 b + D_2 \operatorname{sh}\beta_2 b = e^{is\psi_2}(A_2 + C_2)\,, \\
-A_2 \sin\beta_2 b - B_2 \cos\beta_2 b + C_2 \operatorname{sh}\beta_2 b + D_2 \operatorname{ch}\beta_2 b = e^{is\psi_2}(-B_2 + D_2)\,, \\
B_1 + D_1 = B_2 + D_2\,, \\
EI_1\beta_1^3\left[-A_1 + C_1 - e^{-ir\psi_1}(-A_1 \cos\beta_1 a + B_1 \sin\beta_1 a + C_1 \operatorname{ch}\beta_1 a + D_1 \operatorname{sh}\beta_1 a)\right] \\
+ EI_2\beta_2^3\left[-A_2 + C_2 - e^{-is\psi_2}(-A_2 \cos\beta_2 b + B_2 \sin\beta_2 b + C_2 \operatorname{ch}\beta_2 b + D_2 \operatorname{sh}\beta_2 b)\right] = 0
\end{gathered} \tag{5.6.41}$$

From the first six equation in equation (5.6.41), the unknowns B_1, C_1, D_1 and B_2, C_2, D_2 can be expressed in terms of A_1 and A_2, respectively, as

$$B_1 = \frac{\sin\beta_1 a}{e^{ir\psi_1} - \cos\beta_1 a} A_1\,, \qquad C_1 = \frac{\operatorname{ch}\beta_1 a - e^{ir\psi_1}}{e^{ir\psi_1} - \cos\beta_1 a} \bullet \frac{\cos\beta_1 a - \cos r\,\psi_1}{\operatorname{ch}\beta_1 a - \cos r\,\psi_1} A_1\,,$$

$$D_1 = \frac{-\operatorname{sh}\beta_1 a}{e^{ir\psi_1} - \cos\beta_1 a} \bullet \frac{\cos\beta_1 a - \cos r\,\psi_1}{\operatorname{ch}\beta_1 a - \cos r\,\psi_1} A_1\,. \tag{5.6.42}$$

and

$$B_2 = \frac{\sin\beta_2 b}{e^{is\psi_2} - \cos\beta_2 b} A_2 , \quad C_2 = \frac{\mathrm{ch}\,\beta_2 b - e^{is\psi_2}}{e^{is\psi_2} - \cos\beta_2 b} \cdot \frac{\cos\beta_2 b - \cos s\psi_2}{\mathrm{ch}\,\beta_2 b - \cos s\psi_2} A_2 ,$$

$$D_2 = \frac{-\mathrm{sh}\,\beta_2 b}{e^{is\psi_2} - \cos\beta_2 b} \cdot \frac{\cos\beta_2 b - \cos s\psi_2}{\mathrm{ch}\,\beta_2 b - \cos s\psi_2} A_2 . \tag{5.6.43}$$

Substituting equations (5.6.42) and (5.6.43) into the last two equations in equation (5.6.41), results in

$$a_{11} A_1 + a_{12} A_2 = 0 ,$$

$$a_{21} A_1 + a_{22} A_2 = 0 . \tag{5.6.44}$$

where

$$a_{11} = \frac{1}{e^{ir\psi_1} - \cos\beta_1 a} \left[\sin\beta_1 a - \frac{\mathrm{sh}\,\beta_1 a(\cos\beta_1 a - \cos r\psi_1)}{\mathrm{ch}\,\beta_1 a - \cos r\psi_1} \right] ,$$

$$a_{12} = \frac{-1}{e^{is\psi_2} - \cos\beta_2 b} \left[\sin\beta_2 b - \frac{\mathrm{sh}\,\beta_2 b(\cos\beta_2 b - \cos s\psi_2)}{\mathrm{ch}\,\beta_2 b - \cos s\psi_2} \right] ,$$

$$a_{21} = EI_1\beta_1^3 \frac{\cos\beta_1 a - \cos r\psi_1}{e^{ir\psi_1} - \cos\beta_1 a} , \qquad a_{22} = EI_2\beta_2^3 \frac{\cos\beta_2 b - \cos s\psi_2}{e^{is\psi_2} - \cos\beta_2 b} \tag{5.6.45}$$

Let the coefficient determinant of the homogeneous equation (5.6.44) be equal to zero, *i.e.*,

$$\begin{vmatrix} a_{11} & a_{12} \\ a_{21} & a_{22} \end{vmatrix} = 0 \tag{5.6.46}$$

Inserting equation (5.6.45) into equation (5.6.46) and expanding this determinant yields the frequency equation

$$EI_2\beta_2^3(\cos\beta_2 b-\cos s\,\psi_2)\left[\sin\beta_1 a-\frac{\mathrm{sh}\,\beta_1 a(\cos\beta_1 a-\cos r\,\psi_1)}{\mathrm{ch}\,\beta_1 a-\cos r\,\psi_1}\right]$$

$$+EI_1\beta_1^3(\cos\beta_1 a-\cos r\,\psi_1)\left[\sin\beta_2 b-\frac{\mathrm{sh}\,\beta_2 b(\cos\beta_2 b-\cos s\,\psi_2)}{\mathrm{ch}\,\beta_2 b-\cos s\,\psi_2}\right]=0$$

(5.6.47)

Consider the case of an uniform square grillage, *i.e.*,

$$n=m,\quad a=b=\frac{L}{n},\quad EI_1=EI_2,\quad \beta_1=\beta_2=\beta,\quad \psi_1=\psi_2=\frac{\pi}{n}. \tag{5.6.48}$$

Substituting equation (5.6.48) into equation (5.6.47) , the frequency equation becomes

$$(\cos\frac{\alpha}{n}-\cos\frac{s\pi}{n})\left[\sin\frac{\alpha}{n}-\mathrm{sh}\frac{\alpha}{n}(\cos\frac{\alpha}{n}-\cos\frac{r\pi}{n})\Big/(\mathrm{ch}\frac{\alpha}{n}-\cos\frac{r\pi}{n})\right]$$

$$+(\cos\frac{\alpha}{n}-\cos\frac{r\pi}{n})\left[\sin\frac{\alpha}{n}-\mathrm{sh}\frac{\alpha}{n}(\cos\frac{\alpha}{n}-\cos\frac{s\pi}{n})\Big/(\mathrm{ch}\frac{\alpha}{n}-\cos\frac{s\pi}{n})\right]=0$$

(5.6.49)

with $\alpha\equiv\beta L$.

The relation between the natural frequency and non-dimensional parameter α can be expressed as

$$\omega=\frac{\alpha^2}{L^2}\sqrt{\frac{EI}{\rho}}\,. \tag{5.6.50}$$

The first four natural frequencies are given in Table 5.6.2. The mode shapes for the grillage can be obtained in a similar manner to the procedure outlined for the cable network case.

Table 5.6.2 Frequency parameter α for the uniform square grids - distributed mass system

Grids	$m \times n$	2×2	3×3	4×4	5×5	6×6	7×7
Mode(r,s)	(1,1)	π	π	π	π	π	π
	(1,2)	4.8579	5.3346	5.3574	5.3618	5.3631	5.3636
	(2,1)	4.8579	5.3346	5.3574	5.3618	5.3631	5.3636
	(2,2)	2π	2π	2π	2π	2π	2π

Substituting the characteristic root ω of equation (5.6.47) into equation (5.6.41), a nontrivial solution may be expressed as

$$A_1 = (e^{ir\psi_1} - \cos\beta_1 a)\Big/\left[(\cos\beta_1 a - \cos r\,\psi_1)EI_1\beta_1^3\right],$$

$$B_1 = \sin\beta_1 a\Big/\left[(\cos\beta_1 a - \cos r\,\psi_1)EI_1\beta_1^3\right],$$

$$C_1 = (\mathrm{ch}\,\beta_1 a - e^{ir\psi_1})\Big/\left[(\mathrm{ch}\,\beta_1 a - \cos r\,\psi_1)EI_1\beta_1^3\right],$$

$$D_1 = -\,\mathrm{sh}\,\beta_1 a\Big/\left[(\mathrm{ch}\,\beta_1 a - \cos r\,\psi_1)EI_1\beta_1^3\right],$$

$$A_2 = -\,(e^{is\psi_2} - \cos\beta_2 b)\Big/\left[(\cos\beta_2 b - \cos s\,\psi_2)EI_2\beta_2^3\right]$$

$$B_2 = -\sin\beta_2 b\Big/\left[(\cos\beta_2 b - \cos s\,\psi_2)EI_2\beta_2^3\right],$$

$$C_2 = (e^{is\psi_2} - \mathrm{ch}\,\beta_2 b)\Big/\left[(\mathrm{ch}\,\beta_2 b - \cos s\,\psi_2)EI_2\beta_2^3\right],$$

$$D_2 = \mathrm{sh}\,\beta_2 b\Big/\left[(\mathrm{ch}\,\beta_2 b - \cos s\,\psi_2)EI_2\beta_2^3\right]. \qquad (5.6.51)$$

Substituting equation (5.6.51) into equation (5.6.40), results in

$$q_{1(r,s)}(x) = A_{r,s}\left\{\left[e^{ir\psi_1}\sin\beta_1 x + \sin\beta_1(a-x)\right]\Big/\left[(\cos\beta_1 a - \cos r\,\psi_1)EI_1\beta_1^3\right]\right.$$

$$-\left[e^{ir\psi_1}\operatorname{sh}\beta_1 x+\operatorname{sh}\beta_1(a-x)\right]\Big/\left[(\operatorname{ch}\beta_1 a-\cos r\,\psi_1)EI_1\beta_1^3\right]\Big\}\,,$$

$$q_{2(r,s)}(y)=-A_{r,s}\Big\{\left[e^{is\psi_2}\sin\beta_2 y+\sin\beta_2(b-y)\right]\Big/\left[(\cos\beta_2 b-\cos s\,\psi_2)EI_2\beta_2^3\right]$$

$$-\left[e^{is\psi_2}\operatorname{sh}\beta_2 y+\operatorname{sh}\beta_2(b-y)\right]\Big/\left[(\operatorname{ch}\beta_2 b-\cos s\,\psi_2)EI_2\beta_2^3\right]\Big\}\,. \tag{5.6.52}$$

with other generalized displacements vanishing.

Altogether there are four independent sets of eigenfunctions corresponding to the same eigenvalue except r=m, 2m and s=n, 2n. They can be obtained from equation (5.6.52) by using 2m-r and 2n-s instead of r and s respectively.

Let

$$A_{2m-r,s}=A_{r,2n-s}=-A_{2m-r,2n-s}=-A_{r,s} \tag{5.6.53}$$

By superposition of the four sets of eigenfunctions and substituting into equation (5.3.5) as well as neglecting the constant multiplier, we can arrive at

$$W_{1(j,k)}(x)=\frac{\sin(k-1)s\,\psi_2}{EI_1\beta_1^3}\Big[\frac{\sin jr\,\psi_1\sin\beta_1 x+\sin(j-1)r\,\psi_1\sin\beta_1(a-x)}{\cos\beta_1 a-\cos r\,\psi_1}$$

$$-\frac{\sin jr\,\psi_1\operatorname{sh}\beta_1 x+\sin(j-1)r\,\psi_1\operatorname{sh}\beta_1(a-x)}{\operatorname{ch}\beta_1 a-\cos r\,\psi_1}\Big],$$

$$W_{2(j,k)}(y)=\frac{-\sin(j-1)r\,\psi_1}{EI_2\beta_2^3}\Big[\frac{\sin ks\,\psi_2\sin\beta_2 y+\sin(k-1)s\,\psi_2\sin\beta_2(b-y)}{\cos\beta_2 b-\cos s\,\psi_2}$$

$$-\frac{\sin ks\,\psi_2\operatorname{sh}\beta_2 y+\sin(k-1)s\,\psi_2\operatorname{sh}\beta_2(b-y)}{\operatorname{ch}\beta_2 b-\cos s\,\psi_2}\Big]$$

$$j=1,2,\ldots,m \qquad k=1,2,\ldots,n \tag{5.6.54}$$

This set of eigenfunctions satisfies the boundary conditions of the actual grid with all edges simply supported, *i.e.*,

$$W_{1(j,1)}(x)=0\,, \qquad j=1,2,\ldots,m$$

$$W_{1(1,k)}(0)=0, \qquad W''_{1(1,k)}(0)=0,$$

$$W_{1(m,k)}(a)=0, \qquad W''_{1(m,k)}(a)=0, \qquad k=1,2,...,n$$

$$W_{2(1,k)}(x)=0, \qquad k=1,2,...,n$$

$$W_{2(j,1)}(0)=0, \qquad W''_{2(j,i)}(0)=0,$$

$$W_{2(j,n)}(b)=0, \qquad W''_{2(j,n)}(b)=0, \qquad j=1,2,...,m \qquad (5.6.55)$$

Therefore the set of functions shown in equation (5.6.54) represents a natural mode of the rectangular grid with rectangular mesh and four edges simply supported.

5.7 Double Layer Grids

Consider the double layer grid with m by n panels as shown in Figure 5.7.1 which is made up of eight sets of bars hinge-jointed at the nodes so that there are no bending moment and shearing force in the cross-sections of the bars. Here only the case of simply supported boundary condition is considered, *i.e.,* the grid is supported at all nodes located at the boundary of the top layer with the transverse and tangential displacements being restrained at the supports. Four sets of bars in the top and bottom layers are of the same modulus of elasticity E_1, area of cross-section A_1 and length a; and the other four sets of bars forming the diagonals are of modulus of elasticity E_2, area of cross-section A_2 and length b.

A global coordinate system oxyz with the oxy plane parallel to the top and bottom layers and the x and y axes along the length and width directions, respectively, is established as shown in Figure 5.7.1. A typical substructure (j,k) is shown in Figure 5.7.2 in which j and k in the round brackets denote the ordinal numbers of the substructure along the *x*-and *y*-directions . Each substructure is made up of two nodes and eight bars whose serial numbers are as shown in Figure 5.7.2(b).

At the outset, let us consider the equivalent system as shown in Figure 5.7.3 whose length and width are twice the actual size of the original grid, *i.e.,* it is a 2m by 2n double layer grid. Moreover, in order to make the loading pattern anti-symmetric with respect to two symmetric planes of the equivalent grid, fictitious loads must be applied as shown in Figure 5.7.3. Consequently the deformation is also anti-symmetric. Each pair of symmetric nodes, (A_j, A_j') with j=1,2,...,2n-1; (B_k, B_k') with k=1,2,...,2m-1, and $(C,C_a),(C_b,C')$, $(C,C_b),(C_a,C')$ as shown in Figure 5.7.3 on the boundary of the equivalent system will have the same normal displacement (along x or y axis) and other displacement components identically vanishing (boundary restriction). So that every pair of boundary nodes may be imagined to be jointed by a fictitious hinge.

The simply supported conditions for the actual system can be satisfied automatically in its equivalent system. In order to make it a perfect cyclic periodic system, the attached bars shown by dotted lines in Figure 5.7.3 must be added. The anti-symmetric deformation is not affected by such extra attached bars which are necessary to build up the cyclic periodic system.

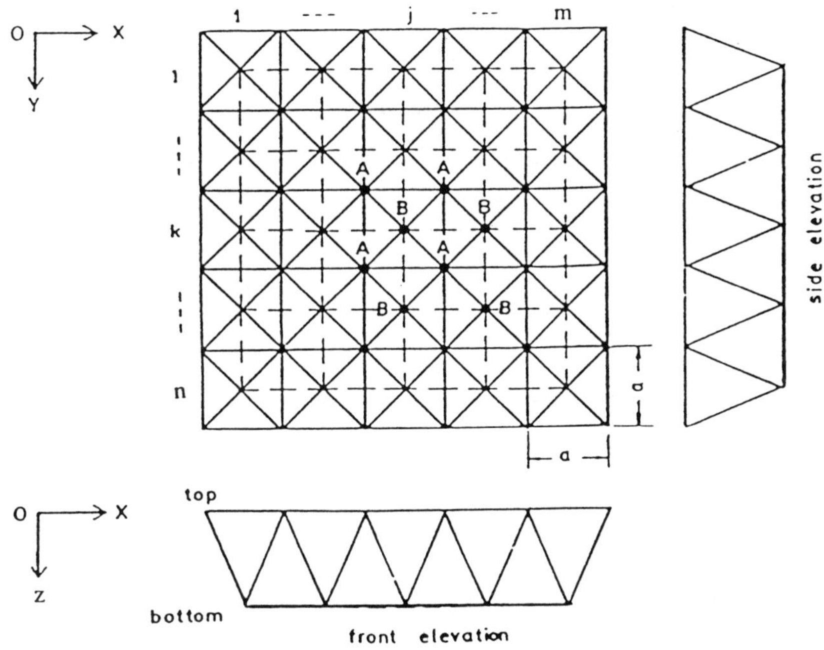

Figure 5.7.1 Double layer grid with $m \times n$ panels

(a)

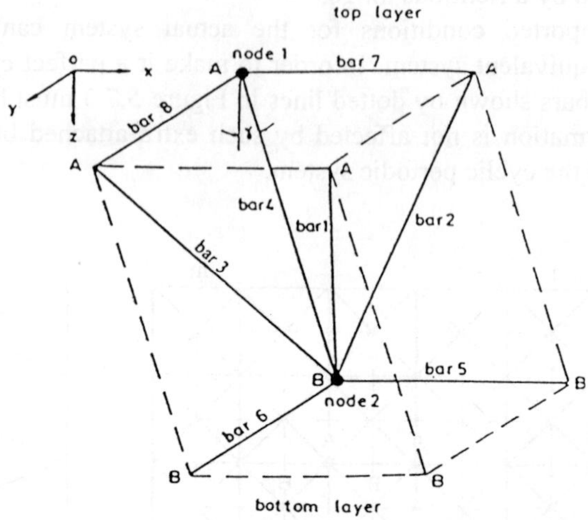

(b)

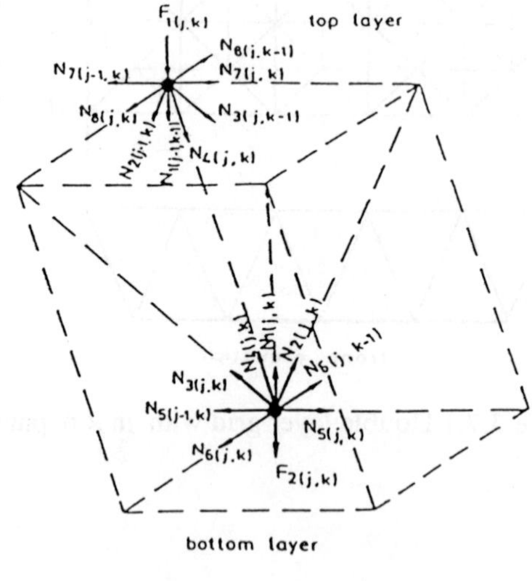

Figure 5.7.2 Substructure (j, k)
(a) Members of substructure (b) Forces acting at the nodes

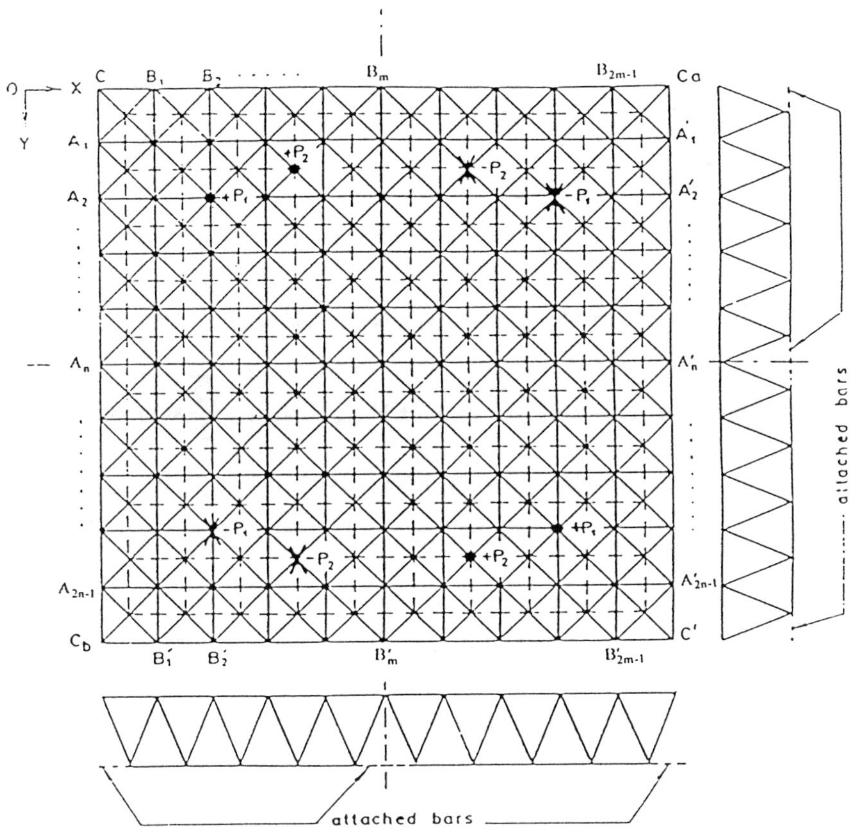

Figure 5.7.3 Equivalent system with $2m \times 2n$ panels

5.7.1 Static Problems

Consider now the equivalent system of $2m \times 2n$ double layer grid with cyclic periodicity in two directions instead of the actual original $m \times n$ system. The forces acting at the nodes of substructure (j,k) are shown in Figure 5.7.2(b). The equilibrium equations for two nodes take the form [18] :

$$N_{7(j,k)} - N_{7(j-1,k)} - (N_{1(j-1,k-1)} + N_{2(j-1,k)} - N_{3(j,k-1)} - N_{4(j,k)})\frac{\sqrt{2}}{2}\sin\gamma = 0,$$

$$N_{8(j,k)} - N_{8(j,k-1)} - (N_{1(j-1,k-1)} - N_{2(j-1,k)} + N_{3(j,k-1)} - N_{4(j,k)})\frac{\sqrt{2}}{2}\sin\gamma = 0,$$

$$(N_{1(j-1,k-1)} + N_{2(j-1,k)} + N_{3(j,k-1)} + N_{4(j,k)})\cos\gamma + F_{1(j,k)} = 0,$$

$$N_{5(j,k)} - N_{5(j-1,k)} + (N_{1(j,k)} + N_{2(j,k)} - N_{3(j,k)} - N_{4(j,k)})\frac{\sqrt{2}}{2}\sin\gamma = 0,$$

$$N_{6(j,k)} - N_{6(j,k-1)} + (N_{1(j,k)} - N_{2(j,k)} + N_{3(j,k)} - N_{4(j,k)})\frac{\sqrt{2}}{2}\sin\gamma = 0,$$

$$-(N_{1(j,k)} + N_{2(j,k)} + N_{3(j,k)} + N_{4(j,k)})\cos\gamma + F_{2(j,k)} = 0,$$

$$j=1,2,\ldots,2m, \quad k=1,2,\ldots,2n \tag{5.7.1}$$

where $N_{i(j,k)}$ (i=1,2,...,8) denotes the axial force in the cross-section of the i-th bar for substructure (j,k); $F_{i(j,k)}$ (i=1,2) denotes the nodal load acting at the i-th node for substructure (j,k); substructures (o,k), (j,0) and (0,0) should be replaced by (2m,k), (j,2n) and (2m,2n) respectively; γ $(0 < \gamma < \pi/2)$ denotes the angle between the diagonal and z axis as shown in Figure 5.7.2(a), *i.e.*,

$$\gamma = \arcsin\frac{\sqrt{2}a}{2b}. \tag{5.7.2}$$

The stress-displacement relation may be written as

$$N_{1(j,k)} = k_2[(u_{1(j+1,k+1)} - u_{2(j,k)})\frac{\sqrt{2}}{2}\sin\gamma + (v_{1(j+1,k+1)} - v_{2(j,k)})\frac{\sqrt{2}}{2}\sin\gamma$$
$$-(w_{1(j+1,k+1)} - w_{2(j,k)})\cos\gamma]$$

$$N_{2(j,k)} = k_2[(u_{1(j+1,k)} - u_{2(j,k)})\frac{\sqrt{2}}{2}\sin\gamma - (v_{1(j+1,k)} - v_{2(j,k)})\frac{\sqrt{2}}{2}\sin\gamma$$
$$- (w_{1(j+1,k)} - w_{2(j,k)})\cos\gamma]$$

$$N_{3(j,k)} = k_2[-(u_{1(j,k+1)} - u_{2(j,k)})\frac{\sqrt{2}}{2}\sin\gamma + (v_{1(j,k+1)} - v_{2(j,k)})\frac{\sqrt{2}}{2}\sin\gamma$$
$$- (w_{1(j,k+1)} - w_{2(j,k)})\cos\gamma]$$

$$N_{4(j,k)} = k_2[-(u_{1(j,k)} - u_{2(j,k)})\frac{\sqrt{2}}{2}\sin\gamma - (v_{1(j,k)} - v_{2(j,k)})\frac{\sqrt{2}}{2}\sin\gamma$$
$$- (w_{1(j,k)} - w_{2(j,k)})\cos\gamma]$$

$$N_{5(j,k)} = k_1(u_{2(j+1,k)} - u_{2(j,k)}),$$
$$N_{6(j,k)} = k_1(v_{2(j,k+1)} - v_{2(j,k)}),$$
$$N_{7(j,k)} = k_1(u_{1(j+1,k)} - u_{1(j,k)}),$$
$$N_{8(j,k)} = k_1(v_{1(j,k+1)} - v_{1(j,k)})$$
$$j=1,2,...,2m, \quad k=1,2,...,2n \tag{5.7.3}$$

where $k_1 = E_1A_1 / a$, $k_2 = E_2A_2 / b$, $u_{i(j,k)}$, $v_{i(j,k)}$, $w_{i(j,k)}$ (i=1,2) denote the displacement components in the coordinate system oxyz for the i-th node of substructure (j,k).

The system of $2m \times 2n$ equations shown in equations (5.7.1) and (5.7.3) may be uncoupled by using the double U-transformation technique, *i.e.*, let

$$X_{(j,k)} = \frac{1}{\sqrt{2m}\sqrt{2n}}\sum_{r=1}^{2m}\sum_{s=1}^{2n} e^{i(j-1)r\psi_1} e^{i(k-1)s\psi_2}\tilde{X}_{(r,s)}$$
$$j=1,2,...,2n, \quad k=1,2,...,2n \tag{5.7.4}$$

or

$$\widetilde{X}_{(r,s)} = \frac{1}{\sqrt{2m}\sqrt{2n}} \sum_{j=1}^{2m} \sum_{k=1}^{2n} e^{-i(j-1)r\psi_1} e^{-i(k-1)s\psi_2} X_{(j,k)}$$

$$r=1,2,...,2m, \quad s=1,2,...,2n \tag{5.7.5}$$

with $i=\sqrt{-1}, \psi_1 = \pi/m$ and $\psi_2 = \pi/n$. The equations (5.7.1) and (5.7.3) becomes

$$(1-e^{-ir\psi_1})\widetilde{N}_{7(r,s)} - [e^{-ir\psi_1}e^{-is\psi_2}\,\widetilde{N}_{1(r,s)} + e^{-ir\psi_1}\widetilde{N}_{2(r,s)} - e^{-is\psi_2}\widetilde{N}_{3(r,s)} - \widetilde{N}_{4(r,s)}]\frac{\sqrt{2}}{2}\sin\gamma = 0$$

$$(1-e^{-is\psi_2})\widetilde{N}_{8(r,s)} - [e^{-ir\psi_1}e^{-is\psi_2}\,\widetilde{N}_{1(r,s)} - e^{-ir\psi_1}\widetilde{N}_{2(r,s)} + e^{-is\psi_2}\widetilde{N}_{3(r,s)} - \widetilde{N}_{4(r,s)}]\frac{\sqrt{2}}{2}\sin\gamma = 0$$

$$[e^{-ir\psi_1}e^{-is\psi_2}\,\widetilde{N}_{1(r,s)} + e^{-ir\psi_1}\widetilde{N}_{2(r,s)} + e^{-is\psi_2}\widetilde{N}_{3(r,s)} + \widetilde{N}_{4(r,s)}]\cos\gamma + \widetilde{F}_{1(r,s)} = 0\,,$$

$$(1-e^{-ir\psi_1})\widetilde{N}_{5(r,s)} + [\widetilde{N}_{1(r,s)} + \widetilde{N}_{2(r,s)} - \widetilde{N}_{3(r,s)} - \widetilde{N}_{4(r,s)}]\frac{\sqrt{2}}{2}\sin\gamma = 0\,,$$

$$(1-e^{-is\psi_2})\widetilde{N}_{6(r,s)} + [\widetilde{N}_{1(r,s)} - \widetilde{N}_{2(r,s)} + \widetilde{N}_{3(r,s)} - \widetilde{N}_{4(r,s)}]\frac{\sqrt{2}}{2}\sin\gamma = 0\,,$$

$$-[\widetilde{N}_{1(r,s)} + \widetilde{N}_{2(r,s)} + \widetilde{N}_{3(r,s)} + \widetilde{N}_{4(r,s)}]\cos\gamma + \widetilde{F}_{2(r,s)} = 0$$

$$r=1,2,...,2m \qquad s=1,2,...,2n \tag{5.7.6}$$

and

$$\widetilde{N}_{1(r,s)} = k_2 e^{ir\psi_1} e^{is\psi_2}[\widetilde{u}_{1(r,s)}\frac{\sqrt{2}}{2}\sin\gamma + \widetilde{v}_{1(r,s)}\frac{\sqrt{2}}{2}\sin\gamma - \widetilde{w}_{1(r,s)}\cos\gamma]$$

$$-k_2[\widetilde{u}_{2(r,s)}\frac{\sqrt{2}}{2}\sin\gamma + \widetilde{v}_{2(r,s)}\frac{\sqrt{2}}{2}\sin\gamma - \widetilde{w}_{2(r,s)}\cos\gamma]\,,$$

$$\widetilde{N}_{2(r,s)} = k_2 e^{ir\psi_1}[\widetilde{u}_{1(r,s)}\frac{\sqrt{2}}{2}\sin\gamma - \widetilde{v}_{1(r,s)}\frac{\sqrt{2}}{2}\sin\gamma - \widetilde{w}_{1(r,s)}\cos\gamma]$$

$$-k_2[\widetilde{u}_{2(r,s)}\frac{\sqrt{2}}{2}\sin\gamma - \widetilde{v}_{2(r,s)}\frac{\sqrt{2}}{2}\sin\gamma - \widetilde{w}_{2(r,s)}\cos\gamma]\,,$$

$$\widetilde{N}_{3(r,s)} = k_2 e^{is\psi_2}[-\widetilde{u}_{1(r,s)} \frac{\sqrt{2}}{2}\sin\gamma + \widetilde{v}_{1(r,s)} \frac{\sqrt{2}}{2}\sin\gamma - \widetilde{w}_{1(r,s)} \cos\gamma]$$

$$+ k_2[\widetilde{u}_{2(r,s)} \frac{\sqrt{2}}{2}\sin\gamma - \widetilde{v}_{2(r,s)} \frac{\sqrt{2}}{2}\sin\gamma + \widetilde{w}_{2(r,s)} \cos\gamma],$$

$$\widetilde{N}_{4(r,s)} = -k_2[(\widetilde{u}_{1(r,s)} + \widetilde{v}_{1(r,s)} - \widetilde{u}_{2(r,s)} - \widetilde{v}_{2(r,s)})\frac{\sqrt{2}}{2}\sin\gamma + (\widetilde{w}_{1(r,s)} - \widetilde{w}_{2(r,s)})\cos\gamma],$$

$$\widetilde{N}_{5(r,s)} = k_1(e^{ir\psi_1} - 1)\widetilde{u}_{2(r,s)},$$

$$\widetilde{N}_{6(r,s)} = k_1(e^{is\psi_2} - 1)\widetilde{v}_{2(r,s)},$$

$$\widetilde{N}_{7(r,s)} = k_1(e^{ir\psi_1} - 1)\widetilde{u}_{1(r,s)}$$

$$\widetilde{N}_{8(r,s)} = k_1(e^{is\psi_2} - 1)\widetilde{v}_{1(r,s)} \qquad r=1,2,...,2m \qquad s=1,2,...,2n \tag{5.7.7}$$

where $\widetilde{F}_{P(r,s)}$ are defined by equation (5.7.5), *i.e.*,

$$\widetilde{F}_{P(r,s)} = \frac{1}{\sqrt{2m}\sqrt{2n}}\sum_{j=1}^{2m}\sum_{k=1}^{2n} e^{-i(j-1)r\psi_1} e^{-i(k-1)s\psi_2} F_{P(r,s)}$$

$$p=1,2 \quad r=1,2,...,2m \qquad s=1,2,...,2n \tag{5.7.8}$$

Noting that the loading pattern must be anti-symmetric for the equivalent system, *i.e.*,

$$F_{1(2m-j+2,2n-k+2)} = F_{1(j,k)},$$

$$F_{1(2m-j+2,k)} = F_{1(j,2n-k+2)} = -F_{1(j,k)}, \qquad j=2,3,...,m \quad k=2,3,...,n$$

$$F_{1(1,k)} = F_{1(j,1)} = 0, \qquad j=1,2,...,m+1 \quad k=1,2,...,n+1$$

$$F_{2(2m-j+1,2n-k+1)} = F_{2(j,k)},$$

$$F_{2(2m-j+1,k)} = F_{2(j,2n-k+1)} = -F_{2(j,k)} \quad j=1,2,...,m \quad k=1,2,...,n \tag{5.7.9}$$

equation (5.7.8) becomes

$$\widetilde{F}_{1(r,s)} = \frac{-4}{\sqrt{2m}\sqrt{2n}} \sum_{j=1}^{m}\sum_{k=1}^{n} \sin[(j-1)r\,\psi_1]\sin[(k-1)s\,\psi_2]F_{1(j,k)},$$

$$\widetilde{F}_{2(r,s)} = \frac{-4}{\sqrt{2m}\sqrt{2n}} e^{ir\psi_1/2}e^{is\psi_2/2} \sum_{j=1}^{m}\sum_{k=1}^{n} \sin[(j-1/2)r\,\psi_1]\sin[(k-1/2)s\,\psi_2]F_{2(j,k)},$$

$$r=1,2,\ldots,2m \qquad s=1,2,\ldots,2n \tag{5.7.10}$$

Equations (5.7.6) and (5.7.7) together with (5.7.10) represent the governing equation of the static problems for the double layer grids. Noting that $\widetilde{F}_{1(r,s)} = \widetilde{F}_{2(r,s)} = 0$ when r=2m or s=2n; the nontrivial solutions of equations (5.7.6) and (5.7.7) for r=2m or s=2n should be neglected.

Substituting equation (5.7.7) into equation (5.7.6) yields

$$\begin{bmatrix} K_{11(r,s)} & K_{12(r,s)} \\ K_{21(r,s)} & K_{22(r,s)} \end{bmatrix} \begin{Bmatrix} X_{1(r,s)} \\ X_{2(r,s)} \end{Bmatrix} = \begin{Bmatrix} b_{1(r,s)} \\ b_{2(r,s)} \end{Bmatrix}$$

$$r=1,2,\ldots,2m-1, \quad s=1,2,\ldots,2n-1 \tag{5.7.11}$$

in which

$$\{X_{1(r,s)}\} = \left[\widetilde{u}_{1(r,s)}, \widetilde{v}_{1(r,s)}, \beta\widetilde{w}_{1(r,s)}\right]^T,$$

$$\{X_{2(r,s)}\} = \left[\widetilde{u}_{2(r,s)}, \widetilde{v}_{2(r,s)}, \beta\widetilde{w}_{2(r,s)}\right]^T,$$

$$\{b_{1(r,s)}\} = \left[0,0,-\sqrt{2}\widetilde{F}_{1(r,s)} / (k_2 \sin\gamma\cos\gamma)\right]^T,$$

$$\{b_{2(r,s)}\} = \left[0,0,-\sqrt{2}\widetilde{F}_{2(r,s)} / (k_2 \sin\gamma\cos\gamma)\right]^T,$$

$$\left[K_{11(r,s)}\right] = \left[K_{22(r,s)}\right] = \begin{bmatrix} 2\alpha(\cos r\psi_1 - 1) - 4 & & 0 \\ & 2\alpha(\cos s\psi_2 - 1) - 4 & \\ 0 & & -4 \end{bmatrix},$$

$$\left[K_{21(r,s)}\right] = \left[\overline{K}_{12(r,s)}\right]$$

$$
= \begin{bmatrix} (1+e^{ir\psi_1})(1+e^{is\psi_2}) & & \text{symmetry} \\ (1-e^{ir\psi_1})(1-e^{is\psi_2}) & (1+e^{ir\psi_1})(1+e^{is\psi_2}) & \\ (1-e^{ir\psi_1})(1+e^{is\psi_2}) & (1+e^{ir\psi_1})(1-e^{is\psi_2}) & (1+e^{ir\psi_1})(1+e^{is\psi_2}) \end{bmatrix}
\tag{5.7.12}
$$

with two non-dimensional parameters as given below

$$
\alpha = 2\,K_1/(K_2 \sin^2\gamma), \qquad \beta = \sqrt{2}\mathrm{ctg}\gamma\,. \tag{5.7.13}
$$

Eliminating $\{X_{1(r,s)}\}$ form equation (5.7.11), gives

$$
[A_{(r,s)}]\{X_{2(r,s)}\} = [d_{4(r,s)}, d_{5(r,s)}, d_{6(r,s)}]^T, \tag{5.7.14}
$$

where

$$
[A_{(r,s)}] = [K_{22(r,s)}] - [K_{21(r,s)}][K_{11(r,s)}]^{-1}[K_{12(r,s)}] \tag{5.7.15}
$$

$$
[d_{4(r,s)}, d_{5(r,s)}, d_{6(r,s)}]^T = \{b_{2(r,s)}\} - [K_{21(r,s)}][K_{11(r,s)}]^{-1}\{b_{1(r,s)}\} \tag{5.7.16}
$$

Substituting equation (5.7.12) into equations (5.7.16) and (5.7.15) yields

$$
d_{4(r,s)} = \frac{-\sqrt{2}\widetilde{F}_{1(r,s)}}{4\,k_2 \sin\gamma\cos\gamma}(1-e^{ir\psi_1})(1+e^{is\psi_2}),
$$

$$
d_{5(r,s)} = \frac{-\sqrt{2}\widetilde{F}_{1(r,s)}}{4\,k_2 \sin\gamma\cos\gamma}(1+e^{ir\psi_1})(1-e^{is\psi_2}),
$$

$$
d_{6(r,s)} = \frac{-\sqrt{2}}{k_2 \sin\gamma\cos\gamma}[\widetilde{F}_{2(r,s)} + \frac{1}{4}\widetilde{F}_{1(r,s)}(1+e^{ir\psi_1})(1+e^{is\psi_2})] \tag{5.7.17}
$$

and

$$\left[A_{(r,s)}\right]=\begin{bmatrix} a_{11} & a_{12} & a_{13} \\ a_{21} & a_{22} & a_{23} \\ a_{31} & a_{32} & a_{33} \end{bmatrix} \tag{5.7.18}$$

with the elements of matrix $[A_{(r,s)}]$ as given by:

$$\begin{aligned}
a_{11} &= 4(1-\cos r\,\psi_1)(1-\cos s\,\psi_2)/B_s + 4(1+\cos r\,\psi_1)(1+\cos s\,\psi_2)/A_r \\
&\quad + (1-\cos r\,\psi_1)(1-\cos s\,\psi_2) - A_r, \\
a_{22} &= 4(1+\cos r\,\psi_1)(1+\cos s\,\psi_2)/B_s + 4(1-\cos r\,\psi_1)(1-\cos s\,\psi_2)/A_r \\
&\quad + (1+\cos r\,\psi_1)(1-\cos s\,\psi_2) - B_s, \\
a_{33} &= 4(1+\cos r\,\psi_1)(1-\cos s\,\psi_2)/B_s + 4(1-\cos r\,\psi_1)(1+\cos s\,\psi_2)/A_r \\
&\quad + (1+\cos r\,\psi_1)(1+\cos s\,\psi_2) - 4, \\
a_{12} &= a_{21} = (1-4/B_s-4/A_r)\sin r\psi_1 \sin s\psi_2, \\
\bar{a}_{31} &= a_{13} = i\sin r\psi_1[4(1+\cos s\psi_2)/A_r - 4(1-\cos s\psi_2)/B_s - (1+\cos s\psi_2)], \\
\bar{a}_{32} &= a_{23} = i\sin s\psi_2[4(1+\cos r\psi_1)/B_s - 4(1-\cos r\psi_1)/A_r - (1+\cos r\psi_1)].
\end{aligned} \tag{5.7.19}$$

and

$$A_r = 4+2\alpha(1-\cos r\,\psi_1), \qquad B_s = 4+2\alpha(1-\cos s\,\psi_2). \tag{5.7.20}$$

The solution for $\{X_{2(r,s)}\}$ of equation (5.7.14) may be written as

$$\begin{Bmatrix} \tilde{u}_{2(r,s)} \\ \tilde{v}_{2(r,s)} \\ \beta\tilde{w}_{2(r,s)} \end{Bmatrix} = \left[A_{(r,s)}\right]^{-1} \begin{Bmatrix} d_{4(r,s)} \\ d_{5(r,s)} \\ d_{6(r,s)} \end{Bmatrix}$$

$$r=1,2,\ldots,2m-1, \qquad s=1,2,\ldots,2n-1 \tag{5.7.21}$$

In a similar way to the above, the solution for $\{X_{1(r,s)}\}$ of equation (5.7.11) may be expressed as

$$\begin{Bmatrix} \tilde{u}_{1(r,s)} \\ \tilde{v}_{1(r,s)} \\ \beta\tilde{w}_{1(r,s)} \end{Bmatrix} = \left[\overline{A}_{(r,s)}\right]^{-1} \begin{Bmatrix} d_{1(r,s)} \\ d_{2(r,s)} \\ d_{3(r,s)} \end{Bmatrix}$$

$$r=1,2,...,2m-1, \qquad s=1,2,...,2n-1 \tag{5.7.22}$$

with

$$d_{1(r,s)} = \frac{-\sqrt{2}\tilde{F}_{2(r,s)}}{4\,k_2 \sin\gamma\, \cos\gamma}(1-e^{-ir\psi_1})(1+e^{-is\psi_2}),$$

$$d_{2(r,s)} = \frac{-\sqrt{2}\tilde{F}_{2(r,s)}}{4\,k_2 \sin\gamma\, \cos\gamma}(1+e^{-ir\psi_1})(1-e^{-is\psi_2}),$$

$$d_{3(r,s)} = \frac{-\sqrt{2}}{k_2 \sin\gamma\, \cos\gamma}[\tilde{F}_{1(r,s)} + \frac{1}{4}\tilde{F}_{2(r,s)}(1+e^{-ir\psi_1})(1+e^{-is\psi_2})] \tag{5.7.23}$$

The nodal displacements in each substructure may now be found by substituting equations (5.7.21) and (5.7.22) into equation (5.7.4) and then the axial force for every bar can be obtained easily from the stress-displacement relation (5.7.3).

To take an example, conside now a 2×2 double layer grid whose members are of the same mechanical and geometric properties and all interior nodes are subjected to equal concentrated loads of magnitude P as shown in Figure 5.7.4.

For present case, the related parameters can be expressed specifically as

$$m = n = 2,\ \psi_1 = \psi_2 = \pi/2,\ \gamma = \pi/4$$

$$k_1 = k_2 = k,\ \alpha = 4,\ \beta = \sqrt{2},$$

$$F_{1(2,2)} = P,\ F_{1(j,k)} = 0, \qquad j \neq 2 \ \text{ or } \ k \neq 2,$$

$$F_{2(j,k)} = P, \qquad j,k=1,2. \tag{5.7.24}$$

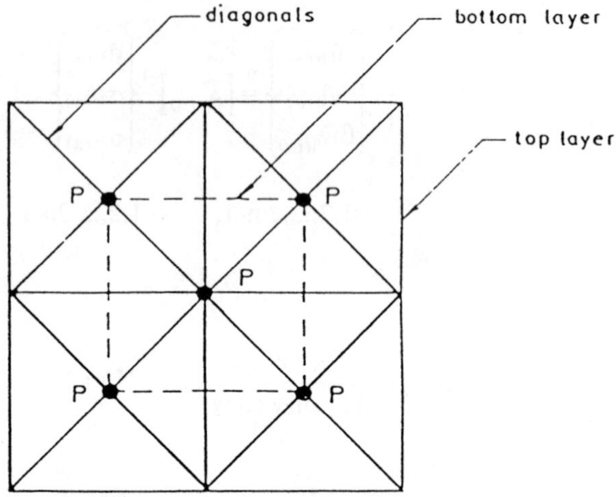

Figure 5.7.4 2×2 double layer grid

Substituting equation (5.7.24) into equation (5.7.10) yields

$$\widetilde{F}_{1(r,s)} = -P\sin\frac{r\pi}{2}\sin\frac{s\pi}{2},$$

$$\widetilde{F}_{2(r,s)} = -Pe^{ir\pi/4}e^{is\pi/4}\sum_{j=1}^{2}\sum_{k=1}^{2}\sin[(j-\frac{1}{2})\frac{r\pi}{2}]\sin[(k-\frac{1}{2})\frac{s\pi}{2}]$$

$$r,s=1,2,3,4. \tag{5.7.25}$$

Noting that $\widetilde{F}_{1(r,s)}=0$ and $\widetilde{F}_{2(r,s)}=0$ when r=2,4 or s=2,4, the cases of r,s=1,3 need be considered only.

Substituting relative parameters shown in equation (5.7.24) into equation (5.7.19) with equation (5.7.20) gives

$$[A_{(r,s)}] = -\frac{1}{3}\begin{bmatrix} 31 & -\sin\frac{r\pi}{2}\sin\frac{s\pi}{2} & 3i\sin\frac{r\pi}{2} \\ -\sin\frac{r\pi}{2}\sin\frac{s\pi}{2} & 31 & 3i\sin\frac{s\pi}{2} \\ -3i\sin\frac{r\pi}{2} & -3i\sin\frac{s\pi}{2} & 7 \end{bmatrix}$$

$$r,s=1,3. \qquad (5.7.26)$$

The inverse matrix of $[A_{(r,s)}]$ is

$$[A_{(r,s)}]^{-1} = -\frac{1}{128}\begin{bmatrix} 13 & \sin\frac{r\pi}{2}\sin\frac{s\pi}{2} & -6i\sin\frac{r\pi}{2} \\ \sin\frac{r\pi}{2}\sin\frac{s\pi}{2} & 13 & -6i\sin\frac{s\pi}{2} \\ 6i\sin\frac{r\pi}{2} & 6i\sin\frac{s\pi}{2} & 60 \end{bmatrix}$$

$$r,s=1,3. \qquad (5.7.27)$$

Inserting relative parameters shown in equation (5.7.24) into equations (5.7.23) and (5.7.17) yields

$$d_{1(r,s)} = -\frac{\sqrt{2}}{2k}\tilde{F}_{2(r,s)}(1+i\sin\frac{r\pi}{2})(1-i\sin\frac{s\pi}{2}),$$

$$d_{2(r,s)} = -\frac{\sqrt{2}}{2k}\tilde{F}_{2(r,s)}(1-i\sin\frac{r\pi}{2})(1+i\sin\frac{s\pi}{2}),$$

$$d_{3(r,s)} = -\frac{2\sqrt{2}}{k}\tilde{F}_{1(r,s)} - \frac{\sqrt{2}}{2k}\tilde{F}_{2(r,s)}(1-i\sin\frac{r\pi}{2})(1-i\sin\frac{s\pi}{2}),$$

$$d_{4(r,s)} = -\frac{\sqrt{2}}{2k}\tilde{F}_{1(r,s)}(1-i\sin\frac{r\pi}{2})(1+i\sin\frac{s\pi}{2})$$

$$d_{5(r,s)} = -\frac{\sqrt{2}}{2k}\tilde{F}_{1(r,s)}(1+i\sin\frac{r\pi}{2})(1-i\sin\frac{s\pi}{2}),$$

$$d_{6(r,s)} = -\frac{2\sqrt{2}}{k}\widetilde{F}_{2(r,s)} - \frac{\sqrt{2}}{2k}\widetilde{F}_{1(r,s)}(1 + i\sin\frac{r\pi}{2})(1 + i\sin\frac{s\pi}{2})$$

$$r,s=1,3. \tag{5.7.28}$$

Substituting equations (5.7.27) and (5.7.28) into equations (5.7.22) and (5.7.21) yields

$$\begin{Bmatrix} \widetilde{u}_{1(r,s)} \\ \widetilde{v}_{1(r,s)} \\ \sqrt{2}\widetilde{w}_{1(r,s)} \end{Bmatrix} = \frac{3\sqrt{2}}{32k}\widetilde{F}_{1(r,s)}\begin{Bmatrix} i\sin(r\pi/2) \\ i\sin(s\pi/2) \\ 10 \end{Bmatrix}$$

$$+\frac{\sqrt{2}}{64k}\widetilde{F}_{2(r,s)}\begin{Bmatrix} 5[1+i\sin(r\pi/2)][1-i\sin(s\pi/2)] \\ 5[1-i\sin(r\pi/2)][1+i\sin(s\pi/2)] \\ 18[1-i\sin(r\pi/2)][1-i\sin(s\pi/2)] \end{Bmatrix}$$

$$\begin{Bmatrix} \widetilde{u}_{2(r,s)} \\ \widetilde{v}_{2(r,s)} \\ \sqrt{2}\widetilde{w}_{2(r,s)} \end{Bmatrix} = \frac{3\sqrt{2}}{32k}\widetilde{F}_{2(r,s)}\begin{Bmatrix} -i\sin(r\pi/2) \\ -i\sin(s\pi/2) \\ 10 \end{Bmatrix}$$

$$+\frac{\sqrt{2}}{64k}\widetilde{F}_{1(r,s)}\begin{Bmatrix} 5[1-i\sin(r\pi/2)][1+i\sin(s\pi/2)] \\ 5[1+i\sin(r\pi/2)][1-i\sin(s\pi/2)] \\ 18[1+i\sin(r\pi/2)][1+i\sin(s\pi/2)] \end{Bmatrix}$$

$$r=1,3. \tag{5.7.29}$$

By substituting equation (5.7.29) into the double U-transformation (5.7.4), the nodal displacements can be found and the exact results are summarized in Table 5.7.1. Then the axial forces for each bar can be obtained from the stress-displacement relation and the results are shown in Table 5.7.2.

Table 5.7.1 Nodal displacements

	Node 1			Node 2		
Substructures	u_1	v_1	w_1	u_2	v_2	w_2
(1,1)	0	0	0	$-11\sqrt{2}$	$-11\sqrt{2}$	78
(1,2)	$26\sqrt{2}$	0	0	$-11\sqrt{2}$	$11\sqrt{2}$	78
(2,1)	0	$26\sqrt{2}$	0	$11\sqrt{2}$	$-11\sqrt{2}$	78
(2,2)	0	0	132	$11\sqrt{2}$	$11\sqrt{2}$	78
(1,3)	0	0	0	$(11\sqrt{2})^*$	$(11\sqrt{2})$	(-78)
(2,3)	0	$-26\sqrt{2}$	0	$(-11\sqrt{2})$	$(11\sqrt{2})$	(-78)
(3,1)	0	0	0	$(11\sqrt{2})$	$(11\sqrt{2})$	(-78)
(3,2)	$-26\sqrt{2}$	0	0	$(11\sqrt{2})$	$(-11\sqrt{2})$	(-78)
(3,3)	0	0	0	$(-11\sqrt{2})$	$(-11\sqrt{2})$	(78)
Multiplier		$P/(64k)$				

* The nodes corresponding to the values in the round brackets do not belong to the actual structure.

Table 5.7.2 Axial forces for bars

Substructures		(1,1)	(1,2)	(2,1)	(2,2)
	N_1	-8	13	13	14
	N_2	13	-8	14	13
	N_3	13	14	-8	13
	N_4	14	13	13	-8
Axial forces	N_5	11	11	$(0)^*$	(0)
	N_6	11	(0)	11	(0)
	N_7	0	-13	0	-13
	N_8	0	0	-13	-13
Multiplier			$\sqrt{2}P/32$		

* The bars corresponding to the values (identically vanishing) in the round brackets are attached bars which do not belong to the actual structure.

5.7.2 Natural Vibration

Consider now the natural vibration of the double layer grids with lumped mass at the nodes. The lumped mass denoted by M is attached to every node. The equation of natural vibration may be obtained directly from equilibrium equation (5.7.1) by using the inertia forces instead of the nodal loads, *i.e.*,

$$F_{1(j,k)} = M\omega^2 w_{1(j,k)}, \; F_{2(j,k)} = M\omega^2 w_{2(j,k)} \tag{5.7.30}$$

where ω denotes the natural frequency.

The inertia forces in the x- and y-directions have been neglected in this section. The uncoupled equation in terms of the generalized displacements can be obtained by substituting

$$\widetilde{F}_{1(j,k)} = M\omega^2 \widetilde{w}_{1(j,k)}, \; \widetilde{F}_{2(j,k)} = M\omega^2 \widetilde{w}_{2(j,k)} \tag{5.7.31}$$

into equation (5.7.11), *i.e.*,

$$\begin{bmatrix} K_{11(r,s)} & K_{12(r,s)} \\ K_{21(r,s)} & K_{22(r,s)} \end{bmatrix} \begin{Bmatrix} X_{1(r,s)} \\ X_{2(r,s)} \end{Bmatrix} = \omega^2 \begin{bmatrix} M_1 & 0 \\ 0 & M_1 \end{bmatrix} \begin{Bmatrix} X_{1(r,s)} \\ X_{2(r,s)} \end{Bmatrix} \tag{5.7.32}$$

where

$$[M_1] = \begin{bmatrix} 0 & 0 & 0 \\ 0 & 0 & 0 \\ 0 & 0 & -M/(k_2\cos^2\gamma) \end{bmatrix}. \tag{5.7.33}$$

The equation (5.7.32) represents the eigenvalue problem of the Hermitian matrices. Consequently the eigenvalues of equation (5.7.32) are real numbers. The frequency equation may be found by letting the characteristic determinant vanish, *i.e.*,

$$\begin{vmatrix} K_{11(r,s)} - \omega^2 M_1 & K_{12(r,s)} \\ K_{21(r,s)} & K_{22(r,s)} - \omega^2 M_1 \end{vmatrix} = 0 \quad r=1,2,..,m, \quad s=1,2,...,n. \tag{5.7.34}$$

Consider now the specific case as shown in the example for static problem, *i.e.*,

$$m = n = 2,\ \psi_1 = \psi_2 = \pi/2,\ \gamma = \pi/4$$

$$k_1 = k_2 = k,\ \alpha = 4,\ \beta = \sqrt{2}, \tag{5.7.35}$$

The solutions for ω of equation (5.7.34) can be expressed as shown in Table 5.7.3.

Table 5.7.3 Natural frequencies ω

ω (r,s)	1st	2nd	3rd	4th	5th
(1,1)	$\sqrt{2/3}$				$\sqrt{8/3}$
(1,2)		$\sqrt{52/29}$			
(2,1)			$\sqrt{52/29}$		
(2,2)				$\sqrt{2}$	
Multiplier			$\sqrt{K/M}$		

Chapter 6

EXACT SOLUTION OF THE FINITE ELEMENT EQUATION

When structures with regular shape and uniform property are divided into uniform finite element meshes such structures can be regarded as periodic structures. The governing equation may be divided into many number of independent equations with the same degrees of freedom as one node by using the U-transformation. Then the exact solutions for displacement, natural frequency and buckling load of the corresponding equations may be found. The convergence rates of the finite element solution can be determined easily for particular problems.

6.1 Beam Element

In order to explain the method adopted in this chapter more clearly, the one dimensional problem of a beam will be studied firstly. A simply supported beam is divided into n equal elements and then the beam may be regarded as a periodic structure with n substructures as shown in Figure 6.1.1(a).

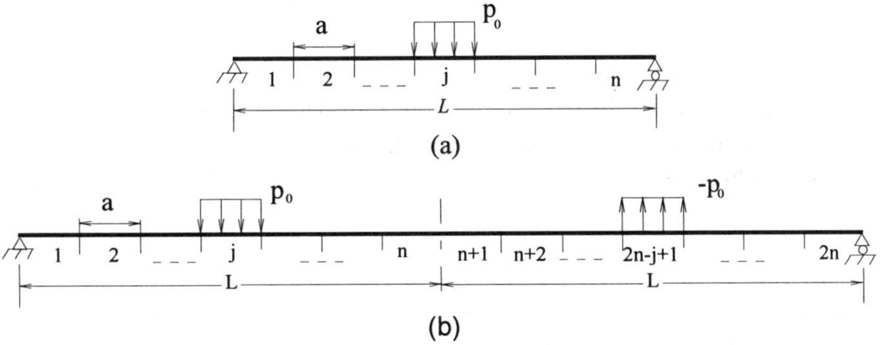

Figure 6.1.1 Simply supported beam
(a) Actual system with n elements;
(b) Equivalent system with 2n elements

Similar to that described in section 4.1.1, an equivalent system with cyclic periodicity may be produced as shown in Figure 6.1.1(b). The total potential energy of the cyclic periodic structure with 2n substructures may be expressed as[22]

$$\Pi = \sum_{j=1}^{2n} \pi_j \tag{6.1.1}$$

in which π_j denotes the potential energy of the j-th element. In general , it can be expressed as

$$\pi_j = \frac{1}{2}\{\bar{\delta}\}_j^T [K]^e \{\delta\}_j - \frac{1}{2}(\{\bar{\delta}\}_j^T \{F\}_j + \{\delta\}_j^T \{\bar{F}\}_j) \tag{6.1.2}$$

where $[K]^e$, $\{F\}_j$ and $\{\delta\}_j$ denote the element stiffness matrix , loading and displacement vectors respectively and the subscript j denotes the serial number of the element . The superior bar denotes complex conjugation. In equation (6.1.2), it may be unnecessary to have the superior bar ; but it will be much convenient for deriving the variational equation with complex variables.

The element displacement vectors $\{\delta\}_j$ is made up of the left and right nodal displacement vectors, *i.e.*,

$$\{\delta\}_j = \begin{Bmatrix} \delta_1 \\ \delta_2 \end{Bmatrix}_j = [w_{j1}, \theta_{j1}, w_{j2}, \theta_{j2}]^T \tag{6.1.3}$$

where $\{\delta_1\}, \{\delta_2\}$ denote the left and right nodal displacement vectors, respectively, and they are made up of the deflection and slope at the nodes.

The continuity across the node requires the following condition to be satisfied:

$$\{\delta_2\}_j = \{\delta_1\}_{j+1} \qquad j=1,2,...,2n \tag{6.1.4}$$

where the subscript 2n+1 should be replaced by 1 due to cyclic periodicity.

Applying the U-transformation to equations (6.1.1), (6.1.2) and (6.1.4), *i.e.*, substituting the following equation

$$\{\delta\}_j = \frac{1}{\sqrt{2n}} \sum_{m=1}^{2n} e^{i(j-1)m\psi} \{q\}_m \tag{6.1.5}$$

with $\psi = \pi/n$, $i = \sqrt{-1}$ and

$$\{q\}_m = \begin{Bmatrix} q_1 \\ q_2 \end{Bmatrix}_m \tag{6.1.6}$$

into equations (6.1.1), (6.1.2) and (6.1.4) results in

$$\Pi = \sum_{m=1}^{2n} \frac{1}{2} \{\bar{q}\}_m^T [K]^e \{q\}_m - \sum_{m=1}^{2n} \frac{1}{2} (\{\bar{q}\}_m^T \{f\}_m + \{q\}_m^T \{\bar{f}\}_m) \tag{6.1.7}$$

and

$$\{q_2\}_m = e^{im\psi} \{q_1\}_m \qquad m=1,2,\ldots,2n \tag{6.1.8}$$

where

$$\{f\}_m = \frac{1}{\sqrt{2n}} \sum_{j=1}^{2n} e^{-i(j-1)m\psi} \{F\}_j . \tag{6.1.9}$$

It is convenient that the continuity condition (6.1.8) may be rewritten as

$$\{q\}_m = [T]_m \{q_1\}_m \tag{6.1.10}$$

where

$$[T]_m = \begin{bmatrix} 1 & 0 & e^{im\psi} & 0 \\ 0 & 1 & 0 & e^{im\psi} \end{bmatrix}^T . \tag{6.1.11}$$

Substituting equation (6.1.10) into equation (6.1.7), yields

$$\Pi = \sum_{m=1}^{2n} \frac{1}{2} \{\bar{q}_1\}_m^T [K]_m^* \{q_1\}_m - \sum_{m=1}^{2n} \frac{1}{2} (\{\bar{q}_1\}_m^T \{f\}_m^* + \{q_1\}_m^T \{\bar{f}\}_m^*) \tag{6.1.12}$$

in which

$$[K]_m^* = [\overline{T}]_m^T [K]^e [T]_m , \tag{6.1.13}$$

$$\{f\}_m^* = [\overline{T}]_m^T \{f\}_m . \tag{6.1.14}$$

In equation (6.1.12), the real and imaginary parts of vector $\{q_1\}_m$ (m=1,2,...,n,2n) are independent variables. Let

$$\delta\Pi = 0 . \tag{6.1.15}$$

The symbol $\delta\Pi$ denotes the variation of the first order of the potential energy. Substituting equation (6.1.12) into the variation equation (6.1.15) results in

$$[K]_m^* \{q_1\}_m = \{f\}_m^* \qquad m=1,2,....,2n \tag{6.1.16}$$

This equation is made up of 2n independent equations where each equation includes two variables only. The solutions for $\{q_1\}_m$ (m=1,2,...,2n) of equation (6.7.16) can be found easily for specific loading condition and element. Then the nodal displacements can be obtained by substituting the solutions $\{q_1\}_m$ into the U-transformation (6.1.5), *i.e.*,

$$\begin{Bmatrix} w_{j1} \\ \theta_{j1} \end{Bmatrix} = \frac{1}{\sqrt{2n}} \sum_{m=1}^{2n} e^{i(j-1)m\psi} \{q_1\}_m \tag{6.1.17}$$

6.1.1 Static Problem

A simply supported uniform beam with uniform subdivisions and uniform loading is considered. A cubic interpolation function is used for the finite element idealization. For the present case, the element loading vectors may be expressed as

$$\{F\}_j = \frac{p_0 a}{2} [1, a/6, 1, -a/6]^T \qquad j=1,2,...,n \tag{6.1.18}$$

where p_0 denotes the loading per unit length and "a" denotes the length of the beam element. Other element loading vectors of the equivalent system shown in Figure 6.1.1(b) may be obtained from the anti-symmetrical load condition, *i.e.*,

$$\{F\}_j = -\frac{p_0 a}{2}\left[1, a/6, 1, -a/6\right]^T \qquad j=n+1, n+2, \ldots 2n. \tag{6.1.19}$$

Substituting equations (6.1.18) and (6.1.19) into equation (6.1.9), results in

$$\{f\}_m = \frac{2p_0 a}{\sqrt{2n}}\frac{1}{1-e^{-im\psi}}\left[1, a/6, 1, -a/6\right]^T \qquad m=1,3,\ldots,2n\text{-}1$$

$$\{f\}_m = 0 \qquad m=2,4,\ldots,2n. \tag{6.1.20}$$

Then substituting equations (6.1.20) and (6.1.11) into equation (6.1.14) yields

$$\{f\}^*_m = \frac{2p_0 a}{\sqrt{2n}}\left[-\frac{i\sin m\psi}{1-\cos m\psi},\ \frac{a}{6}\right]^T \qquad m=1,3,\ldots,2n\text{-}1$$

$$\{f\}^*_m = 0 \qquad m=2,4,\ldots,2n. \tag{6.1.21}$$

The stiffness matrix for beam element may be written as

$$[K]^e = \frac{EI}{a^3}\begin{bmatrix} K_{11} & K_{12} \\ K_{21} & K_{22} \end{bmatrix} \tag{6.1.22}$$

with

$$[K_{11}] = \begin{bmatrix} 12 & 6a \\ 6a & 4a^2 \end{bmatrix}, \qquad [K_{22}] = \begin{bmatrix} 12 & -6a \\ -6a & 4a^2 \end{bmatrix},$$

$$[K_{12}]^T = [K_{21}] = \begin{bmatrix} -12 & -6a \\ 6a & 2a^2 \end{bmatrix}, \tag{6.1.23}$$

where EI denotes the flexural rigidity of the beam.

Substituting equations (6.1.11) and (6.1.22) into equation (6.1.13) yields

$$[K]_m^* = \frac{EI}{a^3}\left[K_{11} + K_{22} + e^{im\psi} K_{12} + e^{-im\psi} K_{21}\right] \tag{6.1.24}$$

and then inserting equation (6.1.23) into equation (6.1.24) results in

$$[K]_m^* = \frac{4EI}{a^3}\begin{bmatrix} 6(1-\cos m\psi) & (3a\sin m\psi)i \\ -(3a\sin m\psi)i & a^2(2+\cos m\psi) \end{bmatrix} \tag{6.1.25}$$

with $\psi = \pi/n$ and $i = \sqrt{-1}$.

Noting equations (6.1.21) and (6.1.25), the uncoupling equation (6.1.16) can be written specifically as

$$\begin{bmatrix} 6(1-\cos m\psi) & (3a\sin m\psi)i \\ -(3a\sin m\psi)i & a^2(2+\cos m\psi) \end{bmatrix}\begin{Bmatrix} q_{m1} \\ q_{m2} \end{Bmatrix} = \frac{p_0 a^4}{2EI\sqrt{2n}}\begin{Bmatrix} -\dfrac{i\sin m\psi}{1-\cos m\psi} \\ \dfrac{a}{6} \end{Bmatrix}$$

$$m=1,3,\ldots,2n-1 \tag{6.1.26}$$

and

$$q_{m1} = q_{m2} = 0 \qquad m=2,4,\ldots,2n \tag{6.1.27}$$

where q_{m1} and q_{m2} denote two components of vector $\{q_1\}_m$.

The solution for $\{q_1\}_m$ of equation (6.1.26) may be expressed as

$$q_{m1} = -\frac{p_0 a^4}{12EI\sqrt{2n}} i\sin m\psi \frac{5+\cos m\psi}{(1-\cos m\psi)^3}$$

$$q_{m2} = \frac{p_0 a^3}{3EI\sqrt{2n}}\frac{2+\cos m\psi}{(1-\cos m\psi)^2} \qquad m=1,3,\ldots,2n-1 \tag{6.1.28}$$

The nodal displacements can be found by substituting equation (6.1.28) into the U-transformation (6.1.17) as

$$w_{j1} = -\frac{p_0 L^4}{24EI}\frac{1}{n^5}\sum_{m=1,3..}^{2n-1} e^{i(j-1)m\psi}\, i \sin m\psi \frac{5+\cos m\psi}{(1-\cos m\psi)^3}$$

$$j=1,2,...,2n \qquad (6.1.29)$$

$$\theta_{j1} = \frac{p_0 L^3}{6EI}\frac{1}{n^4}\sum_{m=1,3..}^{2n-1} e^{i(j-1)m\psi}\frac{2+\cos m\psi}{(1-\cos m\psi)^2} \qquad j=1,2,...,2n \qquad (6.1.30)$$

where L denotes the span length of the original beam, *i.e.*, L=na.

In order to make the mid-span a node, the number of elements for the actual beam must be even. Let n= 2s and s is a positive integer. The maximum deflection can be obtained conveniently from equation (6.1.29) as

$$W_{max} = w_{s+1,1} = \frac{p_0 L^4}{12EI}\frac{1}{n^5}\sum_{m=1,3}^{n-1} \sin\frac{m\pi}{2}\sin m\psi \frac{5+\cos m\psi}{(1-\cos m\psi)^3} \qquad (6.1.31)$$

The maximum slope occurs at supports with magnitude

$$\theta_{max} = \theta_{1,1} = \frac{p_0 L^3}{3EI}\frac{1}{n^4}\sum_{m=1,3}^{n-1}\frac{2+\cos m\psi}{(1-\cos m\psi)^2}. \qquad (6.1.32)$$

According to the cubic interpolation formula, the maximum bending moment can be expressed as

$$M_{max} = -EIw''_{s+1}(0) = -2EI(3w_{s+2,1} - 3w_{s+1,1} - a\,\theta_{s+2,1} - 2a\,\theta_{s+1,1})/a^2$$

$$(6.1.33)$$

Substituting equations (6.1.29) and (6.1.30) into equation (6.1.33) yields

$$M_{max} = \frac{p_0 L^2}{6}\frac{1}{n^3}\sum_{m=1,3}^{n-1}\sin\frac{m\pi}{2}\sin m\psi\frac{7-\cos m\psi}{(1-\cos m\psi)^2}. \qquad (6.1.34)$$

Consider now the convergence rates of the solutions obtained by finite element method with cubic interpolation formula. Expansion of the trigonometric function shown in equation (6.1.31) into power series of ψ results in

$$W_{max} = \frac{p_0 L^4}{12EI}\left[\frac{48}{\pi^5}\sum_{m=1,3}^{n-1}(-1)^{\frac{m-1}{2}} m^{-5} + O(n^{-4})\right]$$

$$= \frac{p_0 L^4}{12EI}\left[\frac{48}{\pi^5}\sum_{m=1,3}^{\infty}(-1)^{\frac{m-1}{2}} m^{-5} + O(n^{-4})\right] \qquad (6.1.35)$$

Noting

$$\sum_{m=1,3,5}^{\infty}(-1)^{\frac{m-1}{2}} m^{-5} = \frac{5}{1536}\pi^5,$$

the equation (6.1.35) may be rewritten as

$$W_{max} = \frac{5p_0 L^4}{384EI}\left[1 + O(n^{-4})\right]. \qquad (6.1.36)$$

The maximum defection obtained by beam element converges to the analytical solution and the order of convergence rate is n^{-4} where n is the total number of elements for actual beam.

In a similar manner used to derive the equation (6.1.36), the maximum slope shown in equation (6.1.32) may be expressed as

$$\theta_{max} = \frac{p_0 L^3}{3EI}\left[\frac{12}{\pi^4}\sum_{m=1,3,5}^{n-1} m^{-4} + O(n^{-4})\right] \qquad (6.1.37)$$

When n approaches infinity, the series in equation (6.1.37) approaches to $\pi^4 / 96$ as a limit. Therefore, the equation (6.1.37) may be rewritten as

$$\theta_{max} = \frac{p_0 L^3}{24EI}\left[1 + O(n^{-3})\right]. \qquad (6.1.38)$$

The maximum slope found by beam element with cubic interpolation formula converges also to the analytical solution with convergence rate n^{-3}.

Similar expression for maximum bending moment can be obtained from equation (6.1.34) as

$$M_{max} = \frac{p_0 L^2}{6}\left[\frac{24}{\pi^3}\sum_{m=1,3}^{n-1}(-1)^{\frac{m-1}{2}} m^{-3} + \frac{2}{\pi n^2}\sum_{m=1,3}^{n-1}(-1)^{\frac{m-1}{2}} m^{-1} + O(n^{-3})\right]$$

$$= \frac{p_0 L^2}{8}\left[1 + \frac{2}{3}n^{-2} + O(n^{-3})\right]. \qquad (6.1.39)$$

It is obvious that the maximum moment found by beam element converges to the analytical solution with convergence rate of n^{-2}.

6.1.2 Buckling Load

Consider now the simply supported beam subjected to the axial pressure with magnitude P at two ends. For this case, in the expression of potential energy, the $[K]^e$ should be replaced by $([K]^e - [K_G]^e)$ in which $[K_G]^e$ denotes the geometric stiffness matrix for the element and loading vectors vanish. The uncoupled equation with two degrees of freedom may be expressed as

$$([K]_m^* - [K_G]_m^*)\{q_1\}_m = \{0\} \qquad (6.1.40)$$

where

$$[K_G]_m^* = [\overline{T}]_m^T [K_G]^e [T]_m \qquad (6.1.41)$$

and $[K]_m^*, [T]_m$ have been defined by equations (6.1.13) and (6.1.11) respectively. For a beam element with cubic interpolation formula, the geometric stiffness matrix of element may be expressed as

$$[K_G]^e = \frac{P}{30a}\begin{bmatrix} 36 & 3a & -36 & 3a \\ & 4a^2 & -3a & -a^2 \\ & & 36 & -3a \\ \text{symmetric} & & & 4a^2 \end{bmatrix}. \qquad (6.1.42)$$

Substituting equations (6.1.11) and (6.1.42) into equation (6.1.41) gives

$$[K_G]_m^* = \frac{P}{15a}\begin{bmatrix} 36(1-\cos m\psi) & (3a\sin m\psi)i \\ -(3a\sin m\psi)i & a^2(4-\cos m\psi) \end{bmatrix}. \tag{6.1.43}$$

Let the coefficient determinant of the simultaneous equations (6.1.40) be equal to zero , *i.e.*

$$\left|[K]_m^* - [K_G]_m^*\right| = 0 . \tag{6.1.44}$$

Substituting equations (6.1.25) and (6.1.43) into equation (6.1.44) yields

$$\begin{vmatrix} 6(1-\cos m\psi)(1-6\sigma) & 3ai\sin m\psi(1-\sigma) \\ -3ai\sin m\psi(1-\sigma) & a^2[(2+\cos m\psi)-(4-\cos m\psi)\sigma] \end{vmatrix} = 0 \tag{6.1.45}$$

where

$$\sigma \equiv \frac{Pa^2}{60EI} . \tag{6.1.46}$$

σ is a non-dimensional parameter of the buckling load. Expanding this determinant yields the buckling load equation

$$15(3-\cos m\psi)\sigma^2 - (26+4\cos m\psi)\sigma + 1 - \cos m\psi = 0$$

$$m=1,2,...,n \tag{6.1.47}$$

The smaller root of equation (6.1.47) can be expressed as

$$\sigma_m = \frac{13+2\cos m\psi - \sqrt{124+112\cos m\psi - 11\cos^2 m\psi}}{15(3-\cos m\psi)}$$

$$m=1,2,...,n. \tag{6.1.48}$$

The minimum root σ_1 is corresponding to critical load. Noting equations (6.1.46) and (6.1.48) the critical load found by the beam element with cubic interpolation can be expressed as

$$P_{cr} = \frac{60EI}{a^2}\sigma_1 = \frac{4EI}{a^2}\frac{13 + 2\cos\psi - \sqrt{124 + 112\cos\psi - 11\cos^2\psi}}{3 - \cos\psi} \tag{6.1.49}$$

with $\psi = \pi/n$ and $a = L/n$.

Some numerical results are shown in Table 6.1.1.

Table 6.1.1 Buckling load

Number of elements n	2	4	8	16	32	∞
Buckling load P_{cr}	1.0075223	1.0005121	1.0000328	1.0000021	1.0000001	1
Multiplier	$EI\pi^2 / L^2$					

Expansion of the right hand side of equation (6.1.49) into a power series of ψ results in

$$P_{cr} = \frac{EI\pi^2}{L^2}\left[1 + \frac{\pi^4}{720}n^{-4} + O(n^{-6})\right] \tag{6.1.50}$$

It is well known that the multiplier before the square brackets represents the classical Euler critical load for simply supported beam subjected to axial pressure. The buckling load found by beam element with four degrees of freedom converges from above the Euler critical load at an asymptotic rate of n^{-4}.

6.1.3. Natural Frequency

When we consider dynamic problems, the inertia force $-[M]^e\{\ddot{\delta}\}_j$ must be added into the element loading vector $\{F\}_j$ and then equation (6.1.16) becomes

$$[K]^*_m\{q_1\}_m + [M]^*_m\{\ddot{q}_1\}_m = \{f\}^*_m \qquad m=1,2,...,2n \tag{6.1.51}$$

where

$$[M]^*_m = [\overline{T}]^T_m [M]^e [T]_m \tag{6.1.52}$$

and $[M]^e$ denotes the element mass matrix . The superior double dots denotes the second derivative with respect to the time variable.

The eigenvalue equation for natural vibration may be obtained from equation (6.1.51) as

$$([K]^*_m - \tilde{\omega}^2_m [M]^*_m)\{q_1\}_m = \{0\} \qquad m=1,2,...,n \tag{6.1.53}$$

where $\tilde{\omega}_m$ denotes the natural frequency found by finite element method.

In this section the beam element with four degrees of freedom is studied only. Substituting equation (6.1.11) and the consistent mass matrix of the beam element into equation (6.1.52) yields

$$[M]^*_m = \frac{\rho a}{210}\begin{bmatrix} 156 + 54\cos m\psi & -(13a\sin m\psi)i \\ (13a\sin m\psi)i & a^2(4 - 3\cos m\psi) \end{bmatrix} \tag{6.1.54}$$

where ρ and "a" denote the mass per unit length and length of the element respectively.

As a general rule, the generalized mass matrix $[M]^*_m$ is a Hermitian matrix as $[K]^*_m$, therefore the eigenvalues of equation (6.1.53) are real .

Let

$$\left|[K]^*_m - \tilde{\omega}^2_m [M]^*_m\right| = 0\,. \tag{6.1.55}$$

Substituting equations (6.1.25) and (6.1.54) into equation (6.1.55) and expanding the determinant yields the frequency equation

$$7(65 - 36\cos m\psi + \cos^2 m\psi)\,\lambda^2_m - 6(69 + 37\cos m\psi - \cos^2 m\psi)\,\lambda_m$$
$$+ 3(1 - \cos m\psi)^2 = 0 \tag{6.1.56}$$

where

$$\lambda_m = \frac{\rho a^4 \tilde{\omega}_m^2}{840EI} \tag{6.1.57}$$

which is a non-dimensional parameter of frequency.

Two roots of equation (6.1.56) may be expressed as

$$\lambda_{m2}^{m1} = (B_m \mp \sqrt{C_m})/A_m \qquad m=1,2,...,n \tag{6.1.58}$$

in which

$$A_m = 7(65 - 36\cos m\psi + \cos^2 m\psi)$$

$$B_m = 3(69 + 37\cos m\psi - \cos^2 m\psi)$$

$$C_m = 3(13828 + 16480\cos m\psi + 2727\cos^2 m\psi + 44\cos^3 m\psi - 4\cos^4 m\psi) \tag{6.1.59}$$

The first n natural frequencies found by beam element with four degrees of freedom can be obtained by substituting equation (6.1.58) into equation (6.1.57) as

$$\tilde{\omega}_{m1}^2 = \frac{840EI}{\rho a^4}\lambda_{m1} \qquad m=1,2,...,n \tag{6.1.60}$$

and the last n natural frequencies are

$$\tilde{\omega}_{m2}^2 = \frac{840EI}{\rho a^4}\lambda_{m2} \qquad m=1,2,...,n \tag{6.1.61}$$

Noting that $\tilde{\omega}_{m1}^2$ is the function of $m\psi$, expansion of right hand side of equation (6.1.60) into power series of $m\psi$, results in[23]

$$\tilde{\omega}_{m1}^2 = \omega_m^2[1 + \frac{1}{720}(\frac{m\pi}{n})^4 + O(n^{-6})] \tag{6.1.62}$$

where ω_m denotes the analytical solution for the m-th natural frequency of the simply supported beam, *i.e.*,

$$\omega_m^2 = m^4 \pi^4 \frac{EI}{\rho L^4} \tag{6.1.63}$$

in which L=na.

Equation (6.1.62) shows that the natural frequencies (for fixed m) found by using beam element with four degrees of freedom converge from above the exact answer at an asymptotic rate of n^{-4}.

Similar to the derivation of equation (6.1.62), another set of frequencies $\tilde{\omega}_{m2}$ obtained by using the beam element and consistent mass matrix may be expressed as

$$\tilde{\omega}_{m2}^2 = \frac{840EIn^4}{\rho L^4}\left[3 + O(n^{-2})\right]. \tag{6.1.64}$$

It is obvious that $\tilde{\omega}_{m2}$ (m=1,2,...,n) are not convergent. Therefore the last half of the eigenvalues obtained by the beam element with consistent mass matrix cannot be regarded as approximate frequencies.

Consider now the lumped mass matrix which does not include any rotary inertia, *i.e.*,

$$[M]^e = \frac{\rho a}{2}\begin{bmatrix} 1 & & & 0 \\ & 0 & & \\ & & 1 & \\ 0 & & & 0 \end{bmatrix}. \tag{6.1.65}$$

Substituting equations (6.1.11) and (6.1.65) into equation (6.1.52) yields

$$[M]_m^* = \rho a\begin{bmatrix} 1 & 0 \\ 0 & 0 \end{bmatrix}. \tag{6.1.66}$$

Substituting equations (6.1.25) and (6.1.66) into equation (6.1.55) and expanding the characteristic determinant yields

$$\tilde{\omega}_m^2 = \frac{12n^4(1-\cos m\psi)^2}{2+\cos m\psi}\frac{EI}{\rho L^4} \qquad m=1,2,...,n. \tag{6.1.67}$$

The first five natural frequencies found by using beam element with lumped mass formulation are given in Table 6.1.2.

Table 6.1.2 Natural frequencies $\tilde{\omega}_m^2$

n / m	2	4	8	16	32	∞
1st	96.00000	97.34966	97.40575	97.40889	97.40908	97.40909
	(-1.40909)*	(-0.05943)	(-0.00334)	(-0.00020)	(-0.00001)	
2nd		1536.000	1557.595	1558.492	1558.542	1558.545
		(-22.545)	(-0.950)	(-0.053)	(-0.003)	
3rd		6924.365	7861.234	7888.705	7890.052	7890.136
		(-965.771)	(-28.902)	(-1.431)	(-0.084)	
4th			24576.00	24921.51	24935.87	24936.73
			(-360.73)	(-15.22)	(-0.86)	
5th			58102.08	60782.65	60875.49	60880.68
			(-2778.60)	(-98.03)	(-5.19)	
Multiplier			$EI/\rho L^4$			

* The numbers in the round brackets denote the error.

Upon expansion of the trigonometric functions on the right hand side of equation (6.1.67), we obtain , in the case of the lumped mass matrix , as

$$\tilde{\omega}_m^2 = \omega_m^2[1 - \frac{1}{720}(\frac{m\pi}{n})^4 + O(n^{-6})] \qquad m=1,2,...,n \tag{6.1.68}$$

By comparison with equation (6.1.62) , it is obvious that when n approaches infinity, the eigenvalues $\tilde{\omega}_m^2$ by the lumped mass matrix will have the same rate of convergence as those by the consistent mass matrix, and former converges from below the analytical solution at an asymptotic rate of n^{-4} .

6.1.4 Dynamic Response

Let us consider a concentrated load of magnitude P(t) acting at the midpoint of a simply supported beam[23]. It is convenient to assume that the number of elements is even, *i.e.*, n=2s and s denotes a positive integer. For this case, the loading vectors of elements for the equivalent structure may be expressed as

$$\{F\}_{s+1} = [P(t),\ 0,\ 0,\ 0]^T, \qquad \{F\}_{3s+1} = [-P(t),\ 0,\ 0,\ 0]^T,$$

$$\{F\}_j = \{0\}, \qquad j \neq s+1, 3s+1. \tag{6.1.69}$$

Substituting equation (6.1.69) into equation (6.1.9), yields

$$\{f\}_m = \left[-\frac{2i}{\sqrt{2n}}\sin\frac{m\pi}{2}P(t),\ 0,\ 0,\ 0\right]^T \tag{6.1.70}$$

and then substituting equations (6.1.11) and (6.1.70) into equation (6.1.14) gives

$$\{f\}_m^* = \left[-\frac{2P(t)}{\sqrt{2n}}\mathrm{i}\sin\frac{m\pi}{2},\ 0\right]^T. \tag{6.1.71}$$

The dynamic equation (6.1.51) may be written specifically as

$$\frac{4EI}{a^3}\begin{bmatrix} 6(1-\cos m\psi) & 3ai\sin m\psi \\ -3ai\sin m\psi & a^2(2+\cos m\psi) \end{bmatrix}\begin{Bmatrix} q_{m1} \\ q_{m2} \end{Bmatrix} + \rho a\begin{bmatrix} 1 & 0 \\ 0 & 0 \end{bmatrix}\begin{Bmatrix} \ddot{q}_{m1} \\ \ddot{q}_{m2} \end{Bmatrix}$$

$$= \begin{Bmatrix} -\dfrac{2P(t)}{\sqrt{2n}}\mathrm{i}\sin\dfrac{m\pi}{2} \\ 0 \end{Bmatrix} \tag{6.1.72}$$

where the lumped mass matrix is used and q_{m1}, q_{m2} denote two components of vector $\{q_1\}_m$. Then eliminating q_{m2} from equation (6.1.72), yields

$$\ddot{q}_{m1} + \tilde{\omega}_m^2 q_{m1} = -\frac{2P(t)}{\rho a\sqrt{2n}} i \sin\frac{m\pi}{2} \tag{6.1.73}$$

where $\tilde{\omega}_m^2$ has been giving by equation (6.1.67).

The solution for q_{m1} of equation (6.1.73) with zero initial condition can be expressed as a Duhamel integral

$$q_{m1}(t) = -\frac{2i\sin(m\pi/2)}{\rho a\sqrt{2n}}\frac{1}{\tilde{\omega}_m}\int_0^t P(\tau)\sin\tilde{\omega}_m(t-\tau)d\tau$$

$$m=1,2,\ldots,2n. \tag{6.1.74}$$

The first expression of the U-transformation (6.1.17) may be written as

$$w_{j1} = \frac{1}{\sqrt{2n}}\sum_{m=1}^{2n} e^{i(j-1)m\psi} q_{m1} \tag{6.1.75}$$

where w_{j1} represent the deflection at left node of the j-th element. By substituting equation (6.1.74) and j=s+1 into equation (6.1.75), the deflection response function at midpoint of the simply supported beam may be found as

$$\tilde{w}_{mid}(t) = w_{s+1,1}(t) = \frac{2}{\rho L}\sum_{m=1,3..}^{n-1}\frac{1}{\tilde{\omega}_m}\int_0^t P(\tau)\sin\tilde{\omega}_m(t-\tau)d\tau\,. \tag{6.1.76}$$

The analytical solution by using mode method is given as

$$w_{mid}(t) = \frac{2}{\rho L}\sum_{m=1,3..}^{\infty}\frac{1}{\omega_m}\int_0^t P(\tau)\sin\omega_m(t-\tau)d\tau\,. \tag{6.1.77}$$

The finite element solution shown in equation (6.1.76) converges but does not converge uniformly to the analytical solution when the number of elements approaches infinity. The convergence rate is dependent on the characteristic of the loading function.

6.2 Plate Element

A rectangular plate with all edges simply supported is considered. A rectangle plate may be divided into $m \times n$ equal rectangular elements as shown in Figure 6.2.1(a). Now the plate may be regarded as a periodic structure in two directions. The equivalent system with cyclic periodicity in two directions (Figure 6.2.1c) can be produced in a similar manner as described in section 5.2. Similar to equation (6.1.1) the potential energy of the extended plate may be expressed as[24]

$$\Pi = \sum_{j=1}^{2m} \sum_{k=1}^{2n} \pi_{jk} \tag{6.2.1}$$

where π_{jk} denotes the potential energy of element (j,k) and j,k are the serial numbers of the element in x and y directions respectively. Also similar to equation (6.1.2), the potential energy for element(j,k) can be expressed as

$$\pi_{jk} = \frac{1}{2}\{\bar{\delta}\}^T_{(j,k)}[K]^e\{\delta\}_{(j,k)} - \frac{1}{2}\left(\{\bar{\delta}\}^T_{(j,k)}\{F\}_{(j,k)} + \{\delta\}^T_{(j,k)}\{\bar{F}\}_{(j,k)}\right) \tag{6.2.2}$$

where the element displacement vector $\{\bar{\delta}\}_{(j,k)}$ is made up of four nodal displacement vectors, *i.e.*,

$$\{\delta\}_{(j,k)} = \left[\{\delta_1\}^T_{(j,k)} \ \{\delta_2\}^T_{(j,k)} \ \{\delta_3\}^T_{(j,k)} \ \{\delta_4\}^T_{(j,k)}\right]^T \tag{6.2.3}$$

with vector components as

$$\{\delta_i\} = \left[w_i, \theta_{xi}, \theta_{yi}\right]^T \qquad i=1,2,3,4 \tag{6.2.4}$$

where w, θ_x, θ_y denote the deflection and two angular rotations about the x-and y-axis. The node numbers of an element are shown in Figure 6.2.1(b).

The continuity across the nodes requires the following conditions to be satisfied:

$$\{\delta_2\}_{(j,k)} = \{\delta_1\}_{(j+1,k)}, \ \{\delta_3\}_{(j,k)} = \{\delta_1\}_{(j+1,k+1)}, \ \{\delta_4\}_{(j,k)} = \{\delta_1\}_{(j,k+1)}$$
$$j=1,2,\ldots,2m, \quad k=1,2,\ldots,2n \tag{6.2.5}$$

where

$$\{\delta_1\}_{(2m+1,k)} = \{\delta_1\}_{(1,k)}, \{\delta_1\}_{(2m+1,2n+1)} = \{\delta_1\}_{(1,1)}, \; \{\delta_1\}_{(j,2n+1)} = \{\delta_1\}_{(j,1)} \quad (6.2.6)$$

which result is obtained from the cyclic periodicity in two directions.

Applying the double U-transformation to equations (6.2.1), (6.2.2) and (6.2.5), *i.e.*, substituting the following equation

$$\{\delta\}_{(j,k)} = \frac{1}{\sqrt{2m}\sqrt{2n}} \sum_{r=1}^{2m} \sum_{s=1}^{2n} e^{i(j-1)r\psi_1} e^{i(k-1)s\psi_2} \{q\}_{(r,s)} \quad (6.2.7)$$

with $i = \sqrt{-1}, \psi_1 = \pi/m$ and $\psi_2 = \pi/m$ into equation (6.2.1), (6.2.2) and (6.2.5), results in

$$\Pi = \sum_{r=1}^{2m} \sum_{s=1}^{2n} \frac{1}{2} \{\bar{q}\}^T_{(r,s)} [K]^e \{q\}_{(r,s)} - \sum_{r=1}^{2m} \sum_{s=1}^{2n} \frac{1}{2} \left(\{\bar{q}\}^T_{(r,s)} \{f\}_{(r,s)} + \{q\}^T_{(r,s)} \{\bar{f}\}_{(r,s)} \right)$$

(6.2.8)

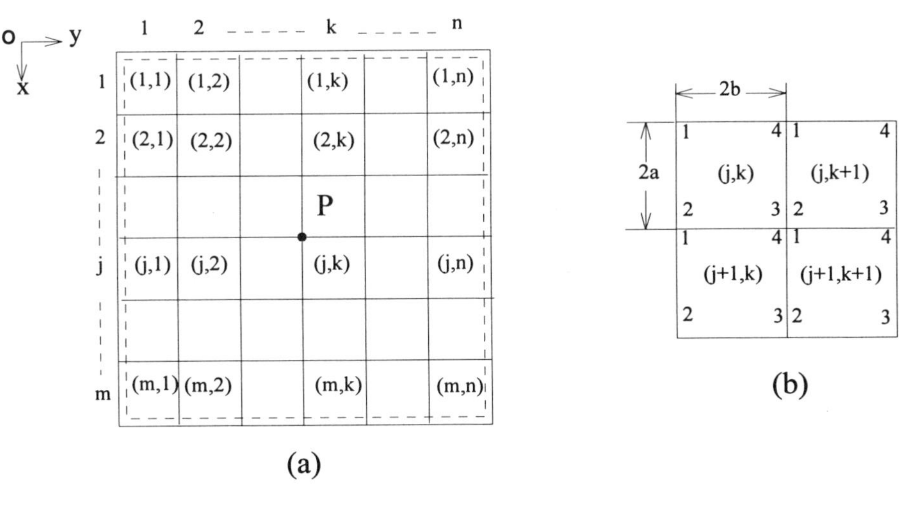

Figure 6.2.1 Rectangular plate with simply supported edges
(a) Actual system with $m \times n$ elements; (b) Node numbers of element

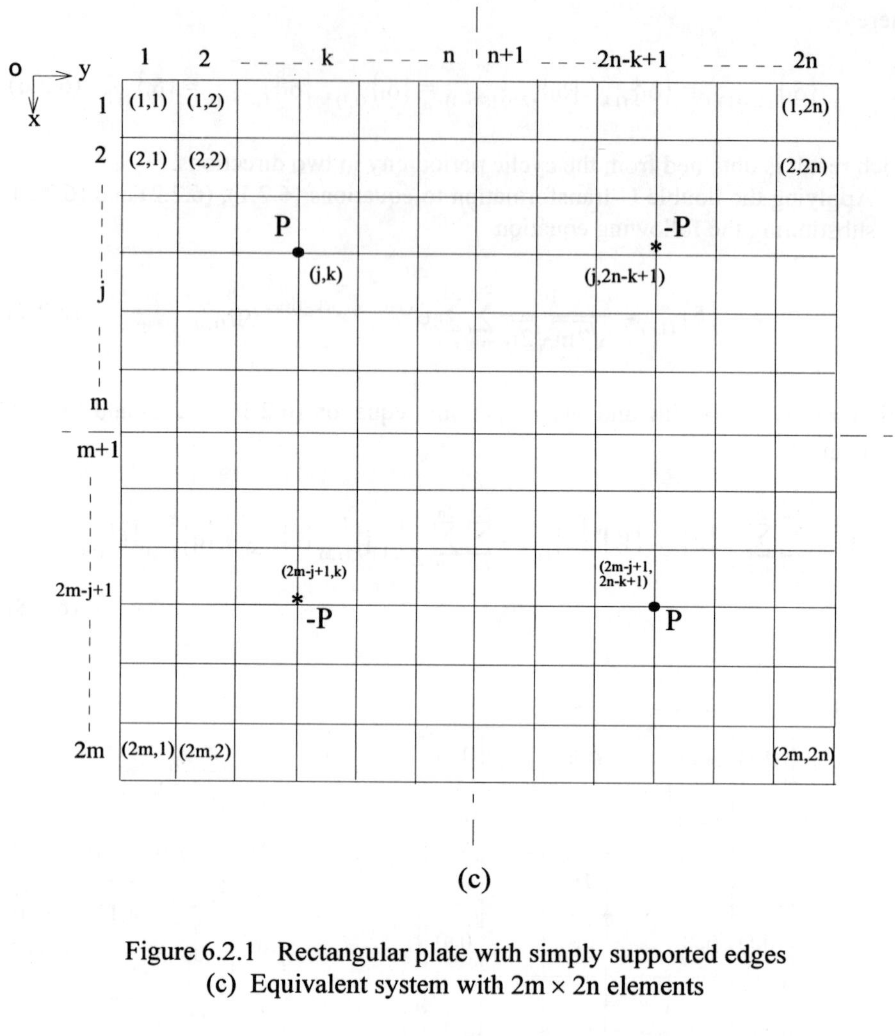

(c)

Figure 6.2.1 Rectangular plate with simply supported edges
(c) Equivalent system with 2m × 2n elements

and

$$\{q_2\}_{(r,s)} = e^{ir\psi_1}\{q_1\}_{(r,s)}, \ \{q_3\}_{(r,s)} = e^{i(r\psi_1+s\psi_2)}\{q_1\}_{(r,s)}, \ \{q_4\}_{(r,s)} = e^{is\psi_2}\{q_1\}_{(r,s)} \tag{6.2.9}$$

in which

$$\{f\}_{(r,s)} = \frac{1}{\sqrt{2m}\sqrt{2n}} \sum_{j=1}^{2m} \sum_{k=1}^{2n} e^{-i(j-1)r\psi_1} e^{-i(k-1)s\psi_2} \{F\}_{(j,k)} \tag{6.2.10}$$

and $\{q_k\}_{(r,s)}$ (k=1,2,3,4) denotes the k-th vector components of $\{q\}_{(r,s)}$.

The continuity condition (6.2.9) may be rewritten as

$$\{q\}_{(r,s)} = [T]_{(r,s)} \{q_1\}_{(r,s)} \tag{6.2.11}$$

where

$$[T]_{(r,s)} = \begin{bmatrix} I_3 & e^{ir\psi_1} I_3 & e^{i(r\psi_1 + s\psi_2)} I_3 & e^{is\psi_2} I_3 \end{bmatrix}^T \tag{6.2.12}$$

and I_3 represents an unit matrix of three order. Substituting equation (6.2.11) into equation (6.2.8), yields

$$\Pi = \sum_{r=1}^{2m} \sum_{s=1}^{2n} \left(\frac{1}{2} \{\bar{q}_1\}_{(r,s)}^T [K]_{(r,s)}^* \{q_1\}_{(r,s)} - \frac{1}{2} \{\bar{q}_1\}_{(r,s)}^T \{f\}_{(r,s)}^* - \frac{1}{2} \{q_1\}_{(r,s)}^T \{\bar{f}\}_{(r,s)}^* \right) \tag{6.2.13}$$

in which

$$[K]_{(r,s)}^* = [\bar{T}]_{(r,s)}^T [K]^e [T]_{(r,s)} \tag{6.2.14}$$

$$\{f\}_{(r,s)}^* = [\bar{T}]_{(r,s)}^T \{f\}_{(r,s)} \tag{6.2.15}$$

Similar to equation (6.1.16), the necessary condition to make the potential energy minimum may be expressed as

$$[K]_{(r,s)}^* \{q_1\}_{(r,s)} = \{f\}_{(r,s)}^* \qquad r=1,2,\ldots,2m \qquad s=1,2,\ldots,2n \tag{6.2.16}$$

This equation is made up of $2m \times 2n$ independent equations, in which each includes three variables. Therefore, the closed form solution for $\{q_1\}_{(r,s)}$ of equation

(6.2.16) can be found for specific loading condition and element. Then the nodal displacements can be obtained in terms of $\{q_1\}_{(r,s)}$ as

$$\begin{Bmatrix} w_1 \\ \theta_{x1} \\ \theta_{y1} \end{Bmatrix}_{(j,k)} = \frac{1}{\sqrt{2m}\sqrt{2n}} \sum_{r=1}^{2m} \sum_{s=1}^{2n} e^{i(j-1)r\psi_1} e^{i(k-1)s\psi_2} \{q_1\}_{(r,s)}$$

$$j=1,2,..,2m \quad k=1,2,..,2n \tag{6.2.17}$$

6.2.1 Static Displacements

Now consider a square uniform plate with $n \times n$ uniform meshes subjected to an uniform load with magnitude p_0. Noting that the loading function must satisfy the anti-symmetric condition for the equivalent system, the elements (j,k) (j,k=1,2,...,n and j,k=n+1,n+2,...,2n) are subjected to uniform load p_0 but other elements must be subjected to equal and opposite loading $-p_0$. The well known 12 degrees of freedom non-conforming rectangular element will be studied in this section. The consistent loading vectors of the plate element may be expressed as

$$\{F\}_{(j,k)} = 4a^2 p_0 \begin{bmatrix} \frac{1}{4} & \frac{a}{12} & -\frac{a}{12} & \frac{1}{4} & \frac{a}{12} & \frac{a}{12} & \frac{1}{4} & -\frac{a}{12} & \frac{a}{12} & \frac{1}{4} & -\frac{a}{12} & -\frac{a}{12} \end{bmatrix}^T$$

$$j,k=1,2,...,n \quad \text{or} \quad j,k=n+1,n+2,...,2n$$

$$\{F\}_{(j,k)} = -4a^2 p_0 \begin{bmatrix} \frac{1}{4} & \frac{a}{12} & -\frac{a}{12} & \frac{1}{4} & \frac{a}{12} & \frac{a}{12} & \frac{1}{4} & -\frac{a}{12} & \frac{a}{12} & \frac{1}{4} & -\frac{a}{12} & -\frac{a}{12} \end{bmatrix}^T$$

$$j=1,2,...,n \quad k=n+1,n+2,...,2n \quad \text{or} \quad j=n+1,n+2,...,2n \quad k=1,2,...,n \tag{6.2.18}$$

where "a" denotes one half of the side length of the square element.

Substituting equation (6.2.18) into equation (6.2.10) yields

$$\{f\}_{(r,s)} = 4a^2 f_{rs} \begin{bmatrix} \frac{1}{4} & \frac{a}{12} & -\frac{a}{12} & \frac{1}{4} & \frac{a}{12} & \frac{a}{12} & \frac{1}{4} & -\frac{a}{12} & \frac{a}{12} & \frac{1}{4} & -\frac{a}{12} & -\frac{a}{12} \end{bmatrix}^T \tag{6.2.19}$$

in which

$$f_{rs} = \frac{8p_0}{n}\frac{1}{(1-e^{-ir\psi})(1-e^{-is\psi})} \qquad r,s=1,3,...,2n-1$$

$$f_{rs} = 0 \qquad r \text{ or } s = 2,4,...,2n \tag{6.2.20}$$

with $\psi = \pi/n$.

Then substituting equations (6.2.19), (6.2.20) and (6.2.12) into equation (6.2.15) results in

$$\{f\}^*_{(r,s)} = \frac{8p_0a^2}{n}\begin{Bmatrix} -\operatorname{ctg}(r\psi/2)\operatorname{ctg}(s\psi/2) \\ -(a/3)\, i\, \operatorname{ctg}(r\psi/2) \\ (a/3)\, i\, \operatorname{ctg}(s\psi/2) \end{Bmatrix} \qquad r,s=1,3,...,2n-1$$

$$\{f\}^*_{rs} = \{0\} \qquad r \text{ or } s = 2,4,...,2n. \tag{6.2.21}$$

Substituting the stiffness matrix $[K]^e$ for 12 degrees of freedom non-conforming square element and equation (6.2.12) into equation (6.2.14) gives

$$[K]^*_{(r,s)} = \frac{D}{30a^2}\begin{bmatrix} k_{11} & k_{12} & k_{13} \\ k_{21} & k_{22} & k_{23} \\ k_{31} & k_{32} & k_{33} \end{bmatrix} \tag{6.2.22}$$

where D denotes the flexural rigidity and

$$k_{11} = 12[27 - 2\mu - 2(6-\mu)(\cos r\psi + \cos s\psi) - (3+2\mu)\cos r\psi \ \cos s\psi]$$

$$k_{22} = 8a^2[4(6-\mu) + 2(3+2\mu)\cos r\psi + (9+\mu)\cos s\psi + (6-\mu)\cos r\psi \ \cos s\psi]$$

$$k_{33} = 8a^2[4(6-\mu) + (9+\mu)\cos r\psi + 2(3+2\mu)\cos s\psi + (6-\mu)\cos r\psi \ \cos s\psi]$$

$$k_{12} = \bar{k}_{21} = 12ai\sin s\psi[11 - \mu + (4+\mu)\cos r\psi]$$

$$k_{13} = \bar{k}_{31} = -12ai\sin r\psi[11 - \mu + (4+\mu)\cos s\psi]$$

$$k_{23} = \bar{k}_{32} = 0 \tag{6.2.23}$$

in which μ denotes the Poisson's ratio.

Substituting equations (6.2.21) and (6.2.22) into equation (6.2.16) and solving this equation for $\{q_1\}_{(r,s)}$, results in

$$q_{(r,s)1} = \frac{\Delta_1}{\Delta},\ q_{(r,s)2} = \frac{\Delta_2}{\Delta},\ q_{(r,s)3} = \frac{\Delta_3}{\Delta} \qquad r,s=1,3,...,2n\text{-}1$$

$$q_{(r,s)1} = q_{(r,s)2} = q_{(r,s)3} = 0 \qquad r \text{ or } s=2,4,...,2n \tag{6.2.24}$$

where

$$\Delta = k_{11}k_{22}k_{33} - k_{13}k_{22}k_{31} - k_{12}k_{21}k_{33}$$

$$\Delta_1 = b_1k_{22}k_{33} - b_2k_{12}k_{33} - b_3k_{13}k_{22}$$

$$\Delta_2 = -b_1k_{21}k_{33} + b_2(k_{11}k_{33} - k_{13}k_{31}) + b_3k_{21}k_{13}$$

$$\Delta_3 = -b_1k_{22}k_{31} + b_2k_{12}k_{31} + b_3(k_{11}k_{22} - k_{12}k_{21}) \tag{6.2.25}$$

in which

$$b_1 = -\frac{240p_0a^4}{D\,n}\mathrm{ctg}\frac{r\psi}{2}\mathrm{ctg}\frac{s\psi}{2}, \qquad b_2 = -\frac{80p_0a^5}{D\,n}i\,\mathrm{ctg}\frac{r\psi}{2},$$

$$b_3 = \frac{80p_0a^5}{D\,n}i\,\mathrm{ctg}\frac{s\psi}{2} \tag{6.2.26}$$

$q_{(r,s)k}$ (k=1,2,3) denotes the k-th component of vector $\{q_1\}_{(r,s)}$. The nodal deflections can be obtained from the double U-transformation (6.2.17) as

$$w_{(j,k)1} = \frac{1}{\sqrt{2n}\sqrt{2n}}\sum_{r=1}^{2n}\sum_{s=1}^{2n} e^{i(j-1)r\psi}e^{i(k-1)s\psi}q_{(r,s)1} \tag{6.2.27}$$

where $w_{(j,k)1}$ denotes the deflection at the first node of element (j,k). The maximum deflection occurs at center of the square plate and its magnitude can be expressed as

$$W_{max} = w_{(\frac{n}{2}+1,\frac{n}{2}+1)1} = -\frac{2}{n}\sum_{r=1,3}^{n-1}\sum_{s=1,3}^{n-1}\sin\frac{r\pi}{2}\sin\frac{s\pi}{2}q_{(r,s)1} \tag{6.2.28}$$

where n is an even number.

In order to discuss the convergence of the series on the right hand side of equation (6.2.28) when n approaches infinity, expansion of $q_{(r,s)1}$ into a power series of ψ, equation (6.2.28) becomes[24]

$$W_{max} =$$

$$\frac{16p_0L^4}{D\pi^6}\Big[\sum_{r=1,3}^{n-1}\sum_{s=1,3}^{n-1}\frac{\sin\frac{r\pi}{2}\sin\frac{s\pi}{2}}{rs(r^2+s^2)^2}+\frac{(1+\mu)\pi^2}{6n^2}\sum_{r=1,3}^{n-1}\sum_{s=1,3}^{n-1}\frac{rs\sin\frac{r\pi}{2}\sin\frac{s\pi}{2}}{(r^2+s^2)^3}+O(n^{-4})\Big]=$$

$$\frac{16p_0L^4}{D\pi^6}\sum_{r=1,3}^{\infty}\sum_{s=1,3}^{\infty}\frac{\sin\frac{r\pi}{2}\sin\frac{s\pi}{2}}{rs(r^2+s^2)^2}+\frac{8(1+\mu)p_0L^4}{3D\pi^4}\sum_{r=1,3}^{\infty}\sum_{s=1,3}^{\infty}\frac{rs\sin\frac{r\pi}{2}\sin\frac{s\pi}{2}}{(r^2+s^2)^3}n^{-2}+O(n^{-4})$$

(6.2.29)

where L denotes the side length of the square plate , *i.e.*, L=2na.

The first term on the right hand side of equation (6.2.29) is in agreement with the analytical solution. The second term represents the main error of the finite element solution. The maximum deflection found by the plate element with 12 degrees-of-freedom converges from above the analytical solution at an asymptotic rate of n^{-2} (n is even).

6.2.2 Buckling Load

Simply supported square plates subjected to uniformly distributed compression in x- or x- and y- directions are considered. p_x and p_y denote the pressure per unit length in x- and y- directions as shown in Figure 6.2.2. Similar to equation (6.1.40), the buckling equation for the present case can be expressed as

$$\left([K]^*_{(r,s)} - [K_G]^*_{(r,s)}\right)\{q_1\}_{(r,s)} = \{0\} \tag{6.2.30}$$

where

$$[K_G]^*_{(r,s)} = \left[\overline{T}\right]^T_{(r,s)} [K_G]^e [T]_{(r,s)} \tag{6.2.31}$$

and $[K_G]^e$ denotes the geometric stiffness matrix of the plate element.

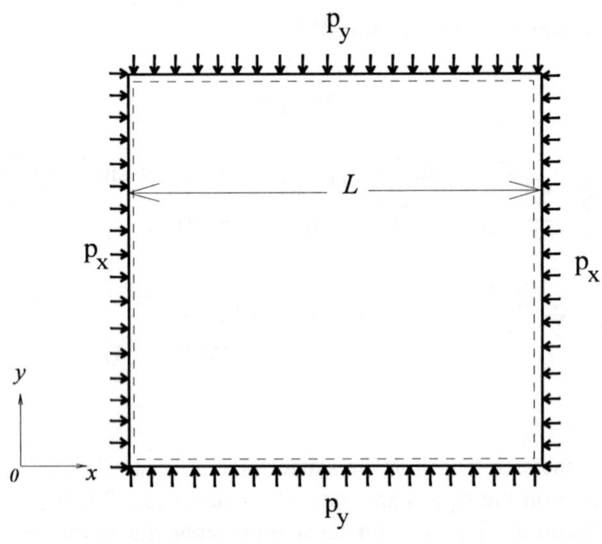

Figure 6.2.2 Simply supported square plate subjected to pressure in middle plane

Substituting the geometric stiffness matrix $[K_G]^e$ for 12 degrees of freedom non-conforming square plate element and equation (6.2.12) into equation (6.2.31) results in[24]

$$[K_G]^*_{(r,s)} = \frac{p_x}{420}\begin{bmatrix} G_{x11} & G_{x12} & G_{x13} \\ G_{x21} & G_{x22} & G_{x23} \\ G_{x31} & G_{x32} & G_{x33} \end{bmatrix} + \frac{p_y}{420}\begin{bmatrix} G_{y11} & G_{y12} & G_{y13} \\ G_{y21} & G_{y22} & G_{y23} \\ G_{y31} & G_{y32} & G_{y33} \end{bmatrix} \tag{6.2.32}$$

in which

$$\begin{aligned}
G_{x11} &= 16(1-\cos r\psi)(46+17\cos s\psi) \\
G_{x22} &= 16a^2(1-\cos r\psi)(4-3\cos s\psi) \\
G_{x33} &= (112/3)a^2(4-\cos r\psi)(2+\cos s\psi) \\
G_{x12} &= \overline{G}_{x21} = -104ai\sin s\psi(1-\cos r\psi) \\
G_{x13} &= \overline{G}_{x31} = -56ai\sin r\psi(2+\cos s\psi) \\
G_{x23} &= \overline{G}_{x32} = 0 \\
G_{y11} &= 16(1-\cos s\psi)(46+17\cos r\psi) \\
G_{y22} &= (112/3)a^2(4-\cos s\psi)(2+\cos r\psi) \\
G_{y33} &= 16a^2(1-\cos s\psi)(4-3\cos r\psi) \\
G_{y12} &= \overline{G}_{y21} = 56ai\sin s\psi(2+\cos r\psi) \\
G_{y13} &= \overline{G}_{y31} = 104ai\sin r\psi(1-\cos s\psi) \\
G_{y23} &= \overline{G}_{y32} = 0
\end{aligned} \tag{6.2.33}$$

with $\psi = \pi/n$ and $a = L/2n$.

The buckling load must satisfy the following equation:

$$\left| [K]^*_{(r,s)} - [K_G]^*_{(r,s)} \right| = 0 . \tag{6.2.34}$$

Obviously the buckling load depends on the ratios between p_x and p_y. Firstly consider the case of bi-directional uniform compression, *i.e.*, $p_x = p_y = p$. Substitution equations (6.2.32) and (6.2.22) into equation (6.2.34) results in

$$-C_1\sigma^3+C_2\sigma^2-C_3\sigma+C_4=0 \tag{6.2.35}$$

where

$$\sigma=\frac{pL^2}{56Dn^2} \tag{6.2.36}$$

which is a non-dimensional parameter of the buckling load and the coefficients of the cubic equation (6.2.35) are defined as

$$C_1=\left\|[G_{ij}]\right\|,\qquad C_4=\left\|[k_{ij}]\right\|$$

$$C_2=\begin{vmatrix}k_{11}&k_{12}&k_{13}\\G_{21}&G_{22}&G_{23}\\G_{31}&G_{32}&G_{33}\end{vmatrix}+\begin{vmatrix}G_{11}&G_{12}&G_{13}\\k_{21}&k_{22}&k_{23}\\G_{31}&G_{32}&G_{33}\end{vmatrix}+\begin{vmatrix}G_{11}&G_{12}&G_{13}\\G_{21}&G_{22}&G_{23}\\k_{31}&k_{32}&k_{33}\end{vmatrix}$$

$$C_3=\begin{vmatrix}G_{11}&G_{12}&G_{13}\\k_{21}&k_{22}&k_{23}\\k_{31}&k_{32}&k_{33}\end{vmatrix}+\begin{vmatrix}k_{11}&k_{12}&k_{13}\\G_{21}&G_{22}&G_{23}\\k_{31}&k_{32}&k_{33}\end{vmatrix}+\begin{vmatrix}k_{11}&k_{12}&k_{13}\\k_{21}&k_{22}&k_{23}\\G_{31}&G_{32}&G_{33}\end{vmatrix} \tag{6.2.37}$$

where $G_{ij}=G_{xij}+G_{yij}$ (i,j=1,2,3) and G_{xij}, G_{yij}, and k_{ij} have been defined by equations (6.2.33) and (6.2.23) respectively.

Consider now the asymptotic characteristic of the finite element solution for buckling load. Noting equation (6.2.36), it may be assumed that

$$\sigma=\lambda_1 n^{-2}+\lambda_2 n^{-4}+O(n^{-6}) \tag{6.2.38}$$

Expansion of the coefficients C_i $(i=1,2,3,4)$ into the power series of ψ, result in

$$C_1=O(n^{-6}),$$

$$C_2=32400a^4\times 3136\,(r^2+s^2)\pi^2 n^{-2}\left[1+O(n^{-2})\right],$$

$$C_3=32400a^4\times 1680\,(r^2+s^2)\pi^2 n^{-2}\left[1-\frac{8+\mu}{30}(r^2+s^2)\pi^2 n^{-2}+O(n^{-4})\right],$$

$$C_4 = 32400a^4 \times 30\,(r^2+s^2)^2\,\pi^4 n^{-4}$$

$$\times\Big[1-\frac{9+\mu}{30}(r^2+s^2)\,\pi^2 n^{-2}-\frac{1+\mu}{6}\frac{r^2 s^2}{r^2+s^2}\pi^2 n^{-2}+O(n^{-4})\Big] \tag{6.2.39}$$

Substituting equations (6.2.38) and (6.2.39) into equation (6.2.35) and letting the coefficients of terms n^{-4} and n^{-6} vanishing , yields

$$30(r^2+s^2)\,\pi^2-1680\,\lambda_1=0\,,$$

$$-(9+\mu)(r^2+s^2)^2\,\pi^4-5(1+\mu)\,r^2 s^2\pi^4-1680\,\lambda_2+56(8+\mu)(r^2+s^2)\,\pi^2\lambda_1$$
$$+3160\,\lambda_1^2=0 \tag{6.2.40}$$

The solution for λ_1 and λ_2 of equation (6.2.40) can be found as

$$\lambda_1=\frac{1}{56}(r^2+s^2)\,\pi^2\,,$$

$$\lambda_2=-\frac{1+\mu}{336}r^2 s^2\pi^4\,. \tag{6.2.41}$$

The minimum buckling parameter σ_{min} can be expressed , by substituting equation (6.2.41) with r=1 and s=1 into equation (6.2.38), as

$$\sigma_{min}=\frac{1}{28}\pi^2 n^{-2}\left[1-\frac{1+\mu}{12}\pi^2 n^{-2}+O(n^{-4})\right]. \tag{6.2.42a}$$

The magnitude of buckling load is

$$P_{cr}=\frac{56Dn^2}{L^2}\sigma_{min}=\frac{2\pi^2 D}{L^2}-\frac{1+\mu}{6L^2}\pi^4 Dn^{-2}+O(n^{-4}) \tag{6.2.42b}$$

The first term on the right hand side of equation (6.2.42b) is in agreement with the analytical solution and then the second term represents the main error of the buckling load found by non-conforming element with 12 degrees of freedom. When n approaches infinity, the finite element solution for buckling load converges from below the analytical solution at an asymptotic rate of n^{-2}.

Secondly, consider the case of unidirectional compression, *i.e.*, $P_x = P$ and $P_y = 0$. In a similar way to derive the buckling load equation for the case of bi-directional compression , the equations (6.2.34)-(6.2.38) are applicable to the present case but G_{ij} in equation (6.2.37) should be replaced by G_{xij}. Then the expanded forms of the coefficients C_i (i=1,2,3,4) may be expressed as

$$C_1 = O(n^{-6}),$$

$$C_2 = 32400a^4 \times 1568 r^2 \pi^2 n^{-2}\left[1 + O(n^{-2})\right],$$

$$C_3 = 32400a^4 \times 1680 \pi^2 n^{-2}\left\{r^2 - \frac{r^2+s^2}{60}\left[2(9+\mu)r^2 - r^2 - s^2\right]\pi^2 n^{-2} + O(n^{-4})\right\},$$

$$C_4 = 32400a^4 \times 30(r^2+s^2)^2 \pi^4 n^{-4}$$

$$\times\left[1 - \frac{9+\mu}{30}(r^2+s^2)\pi^2 n^{-2} - \frac{1+\mu}{6}\frac{r^2 s^2}{r^2+s^2}\pi^2 n^{-2} + O(n^{-4})\right] \tag{6.2.43}$$

Inserting equations (6.2.38) and (6.2.43) into equation (6.2.35) results in

$$30(r^2+s^2)^2\pi^2 - 1680 r^2 \lambda_1 = 0,$$

$$30(r^2+s^2)^2\pi^4\left[-\frac{9+\mu}{30}(r^2+s^2) - \frac{1+\mu}{6}\frac{r^2 s^2}{r^2+s^2}\right]$$

$$+1680\frac{r^2+s^2}{60}\left[2(9+\mu)r^2 - r^2 - s^2\right]\pi^2\lambda_1 - 1680 r^2\lambda_2 + 1568 r^2 \lambda_1^2 = 0. \tag{6.2.44}$$

The solution for λ_1 and λ_2 of equation (6.2.44) can be found as

$$\lambda_1 = \frac{(r^2+s^2)^2}{56r^2}\pi^2,$$

$$\lambda_2 = -\frac{1+\mu}{336} s^2 (r^2 + s^2) \pi^4 . \tag{6.2.45}$$

Substituting equation (6.2.45) into equation (6.2.38) yields

$$\sigma = \frac{(r^2 + s^2)^2}{56 r^2} \pi^2 n^{-2} \left[1 - \frac{1+\mu}{6} \frac{r^2 s^2}{r^2 + s^2} \pi^2 n^{-2} + O(n^{-4}) \right] \tag{6.2.46}$$

and

$$\sigma_{min} = \frac{1}{14} \pi^2 n^{-2} \left[1 - \frac{1+\mu}{12} \pi^2 n^{-2} + O(n^{-4}) \right] . \tag{6.2.47}$$

The buckling load can be obtained from equation (6.2.36) as

$$P_{cr} = \frac{4\pi^2 D}{L^2} - \frac{1+\mu}{3L^2} \pi^4 D n^{-2} + O(n^{-4}) . \tag{6.2.48}$$

The first term on the right hand side of equation (6.2.48) represents the limiting solution which is in agreement with the analytical solution. Then the second term represents the main error . Making a comparison between the equations (6.2.47) and (6.2.42b), leads to the conclusion that the rates of convergence of finite element solutions for both unidirectional and bi-directional compression are the same.

6.2.3 Natural Frequency

The equation of natural vibration in terms of the generalized displacements is similar to equation (6.1.53), *i.e.*,

$$\left([K]^*_{(r,s)} - \tilde{\omega}^2_{rs} [M]^*_{(r,s)} \right) \{q_1\}_{(r,s)} = \{0\} \tag{6.2.49}$$

where $\tilde{\omega}_{rs}$ denotes the natural frequency found by plate element and

$$[M]^*_{(r,s)} = [\overline{T}]^T_{(r,s)} [M]^e [T]_{(r,s)} \tag{6.2.50}$$

in which $[M]^e$ denotes the mass matrix of element. For the non-conforming rectangular plate element with 12 degrees-of-freedom, the explicit formulation of

$[M]^e$ may be obtained from any textbook about Finite Element Methods. Substituting the consistent mass matrix of square plate element and equation (6.2.12) into equation (6.2.50) yields[23]

$$[M]^*_{(r,s)} = \frac{\rho a^2}{6300}\begin{bmatrix} m_{11} & m_{12} & m_{13} \\ m_{21} & m_{22} & m_{23} \\ m_{31} & m_{32} & m_{33} \end{bmatrix} \tag{6.2.51}$$

where ρ and "a" denote the mass per unit area of plate and one half of the side length of the square element and

$$m_{11} = 8[1727 + 613(\cos r\psi + \cos s\psi) + 197\cos r\psi\ \cos s\psi]$$

$$m_{22} = 160a^2(8 + 4\cos r\psi - 6\cos s\psi - 3\cos r\psi\ \cos s\psi)$$

$$m_{33} = 160a^2(8 - 6\cos r\psi + 4\cos s\psi - 3\cos r\psi\ \cos s\psi)$$

$$m_{12} = \overline{m}_{21} = -16ai\sin s\psi(137 + 58\cos r\psi)$$

$$m_{13} = \overline{m}_{31} = 16ai\sin r\psi(137 + 58\cos s\psi)$$

$$m_{23} = \overline{m}_{32} = -448a^2\sin s\psi\sin r\psi\ . \tag{6.2.52}$$

Substituting equations (6.2.22) and (6.2.51) into the frequency equation

$$\left|[K]^*_{(r,s)} - \tilde{\omega}^2_{rs}[M]^*_{(r,s)}\right| = 0 \tag{6.2.53}$$

and expanding the above determinant results in

$$-C_1\lambda^3_{rs} + C_2\lambda^2_{rs} - C_3\lambda_{rs} + C_4 = 0 \tag{6.2.54}$$

where λ_{rs} denotes the frequency parameter defined as

$$\lambda_{rs} = \frac{1}{3360D}\rho L^4\tilde{\omega}^2_{rs}n^{-4} \tag{6.2.55}$$

and the coefficients of frequency equation (6.2.54) are defined as

$$C_1 = \left\|[m_{ij}]\right\|, \qquad C_4 = \left\|[k_{ij}]\right\|,$$

$$C_2 = \begin{bmatrix} k_{11} & k_{12} & k_{13} \\ m_{21} & m_{22} & m_{23} \\ m_{31} & m_{32} & m_{33} \end{bmatrix} + \begin{bmatrix} m_{11} & m_{12} & m_{13} \\ k_{21} & k_{22} & k_{23} \\ m_{31} & m_{32} & m_{33} \end{bmatrix} + \begin{bmatrix} m_{11} & m_{12} & m_{13} \\ m_{21} & m_{22} & m_{23} \\ k_{31} & k_{32} & k_{33} \end{bmatrix},$$

$$C_3 = \begin{bmatrix} m_{11} & m_{12} & m_{13} \\ k_{21} & k_{22} & k_{23} \\ k_{31} & k_{32} & k_{33} \end{bmatrix} + \begin{bmatrix} k_{11} & k_{12} & k_{13} \\ m_{21} & m_{22} & m_{23} \\ k_{31} & k_{32} & k_{33} \end{bmatrix} + \begin{bmatrix} k_{11} & k_{12} & k_{13} \\ k_{21} & k_{22} & k_{23} \\ m_{31} & m_{32} & m_{33} \end{bmatrix}. \tag{6.2.56}$$

Explicit solution for λ_{rs} of cubic equation (6.2.54) cannot be obtained. But the convergence of λ_{rs} can be analyzed when n approaches infinity. Expansion of the coefficients of the cubic equation (6.2.54) into power series of ψ results in

$$C_1 = O(1), \qquad C_2 = O(1),$$

$$C_3 = 25200\left[1 - \frac{9+\mu}{30}(r^2+s^2)\pi^2 n^{-2} + O(n^{-4})\right] \times 129600a^4,$$

$$C_4 = \frac{15}{2}\Big[(r^2+s^2)^2\pi^4 n^{-4} - \frac{9+\mu}{30}(r^2+s^2)^3\pi^6 n^{-6}$$

$$- \frac{1+\mu}{6} r^2 s^2 (r^2+s^2)\pi^6 n^{-6} + O(n^{-8})\Big] \times 129600a^4 \tag{6.2.57}$$

Noting equation (6.2.55) the expansion of λ_{rs} can be expressed as

$$\lambda_{rs} = \lambda_1 n^{-4} + \lambda_2 n^{-6} + O(n^{-8}). \tag{6.2.58}$$

Substituting equations (6.2.57) and (6.2.58) into equation (6.2.54) let the coefficients of n^{-4} and n^{-6} terms be equal to zero respectively, *i.e.*,

$$-25200\lambda_1+\frac{15}{2}(r^2+s^2)^2\pi^4=0\,,$$

$$-25200\lambda_2+840(9+\mu)(r^2+s^2)\pi^2\lambda_1-\frac{9+\mu}{4}(1+\mu)(r^2+s^2)^3\pi^6\,.$$

$$-\frac{5(1+\mu)}{4}r^2s^2(r^2+s^2)\pi^6=0 \tag{6.2.59}$$

The solution for λ_1 and λ_2 of equation (6.2.59) can be found as

$$\lambda_1=\frac{1}{3360}(r^2+s^2)^2\pi^4\,,$$

$$\lambda_2=-\frac{1+\mu}{20160}r^2s^2(r^2+s^2)\pi^6\,. \tag{6.2.60}$$

Substituting equation (6.2.60) into equation (6.2.58) yields

$$\lambda_{rs}=\frac{1}{3360}(r^2+s^2)^2\pi^4n^{-4}\Big[1-\frac{1+\mu}{6}\frac{r^2s^2}{r^2+s^2}\pi^2n^{-2}+O(n^{-4})\Big] \tag{6.2.61}$$

and then substituting the above expansion into the left hand side of equation (6.2.55) results in

$$\tilde{\omega}^2_{rs}=\omega^2_{rs}\Big[1-\frac{1+\mu}{6}\frac{r^2s^2}{r^2+s^2}\pi^2n^{-2}+O(n^{-4})\Big] \tag{6.2.62}$$

where ω_{rs} denotes the analytical solution for natural frequencies defined as

$$\omega^2_{rs}=(r^2+s^2)^2\pi^4\frac{D}{\rho L^4} \tag{6.2.63}$$

in which the subscripts r,s are in agreement with the numbers of the half wave of natural mode in x-and y-directions respectively.

The eigenvalue solutions found by the 12 degrees-of -freedom non-conforming element with the consistent mass matrix converge from below the exact answers at an asymptotic rate of n^{-2} when n approaches infinity. It must be pointed out that the other two sets of eigenvalue solutions corresponding to two bigger roots of equation (6.2.54) are divergent with the order of n^4. Therefore, the last two thirds of the natural frequencies obtained by the plate element with consistent mass matrix cannot be regarded as the approximate frequencies.

Moreover, consider the lumped mass matrix neglecting the rotary inertia, *i.e.*, let

$$[M]^e \equiv [M_{ij}] \tag{6.2.64}$$

with

$$M_{11} = M_{44} = M_{77} = M_{10} = \rho a^2 \tag{6.2.65}$$

and other elements of mass matrix $[M]^e$ being equal to zero. For this case

$$[M]^*_{(r,s)} = 4\rho a^2 \begin{bmatrix} 1 & 0 & 0 \\ 0 & 0 & 0 \\ 0 & 0 & 0 \end{bmatrix}. \tag{6.2.66}$$

Substituting equations (6.2.22) and (6.2.66) into equation (6.2.53) yields

$$\begin{vmatrix} k_{11} - \beta\tilde{\omega}^2_{rs} & k_{12} & k_{13} \\ k_{21} & k_{22} & k_{23} \\ k_{31} & k_{32} & k_{33} \end{vmatrix} = 0 \tag{6.2.67}$$

where

$$\beta = \frac{15\rho L^4}{2D\,n^4}. \tag{6.2.68}$$

Expanding the characteristic determinant on the left hand side of equation (6.2.67) results in

$$\tilde{\omega}_{rs}^2 = \frac{1}{\beta} \frac{\left|[k_{ij}]\right|}{k_{22} k_{33}} . \tag{6.2.69}$$

where the elements of matrix $[k_{ij}]$ and β have been defined as equations (6.2.23) and (6.2.68) respectively. Upon expanding the expression on the right hand side of equation (6.2.69) into the power series of ψ results in

$$\tilde{\omega}_{rs}^2 = \omega_{rs}^2 \left[1 - \frac{1+\mu}{6} \frac{r^2 s^2}{r^2+s^2} \pi^2 n^{-2} + O(n^{-4}) \right] . \tag{6.2.70}$$

By comparison with equation (6.2.62), it may be concluded that the lumped mass system will suffer no loss in rate of convergence. The numerical results shown in equation (6.2.69) obtained by the plate element with lumped mass matrix are given in Table 6.2.1.

Table 6.2.1 $\tilde{\omega}_{rs}^2$ for square plate with all edges simply supported ($\mu = 0.3$)

n (r,s)	2	4	8	16	32	∞
(1,1)	290.1895 (-0.255)*	364.0791 (-0.0656)	383.1605 (-0.0166)	388.0110 (-0.00417)	389.2295 (-0.00104)	389.6364
(1,2),(2,1)		2155.328 (-0.115)	2369.572 (-0.0270)	2418.928 (-0.00669)	2431.156 (-0.00167)	2435.227
(2,2)		4643.032 (-0.255)	5825.265 (-0.0656)	6130.568 (-0.0166)	6208.176 (-0.00417)	6234.182
Multiplier	$D/(\rho L^4)$					

* The numbers in the round brackets denote the relative errors

It is of interest to note that the rates of convergence shown in equations (6.2.29), (6.2.42), (6.2.48), (6.2.62) and (6.2.70) are dependent on μ and the main errors vary directly as $(1+\mu)$. The μ in the stiffness matrix except the implicit one in flexural rigidity D may be regarded as a regulative factor for the stiffness of plate element if the boundary condition is independent of μ. Let $\mu = -1$ in which μ does not represent the real Poisson's ratio and then the convergence rates will be improved and the errors are of the order of n^{-4} instead of n^{-2}. The numerical results of equation (6.2.69) for $\mu=-1$ are given in Table 6.2.2. However , if the boundary conditions of the plate considered have relation with μ , the μ in the stiffness matrix must be equal to the value of the real Poisson's ratio.

Table 6.2.2 $\tilde{\omega}_{rs}^{2}$ for square plate with all edges simply supported (μ =-1)

n (r,s)	2	4	8	16	32	∞
(1,1)	347.4286 (-0.108)*	387.4974 (-0.00549)	389.5159 (-0.000309)	389.6289 (-0.0000192)	389.6358 (-0.00000154)	389.6364
(1,2),(2,1)		2343.936 (-0.0375)	2430.816 (0.00181)	2434.972 (-0.000105)	2435.211 (-0.00000657)	2435.227
(2,2)		5558.857 (-0.108)	6199.959 (-0.00549)	6232.254 (-0.000309)	6234.062 (-0.0000192)	6234.182
Multiplier			$D/(\rho L^4)$			

* The numbers in the round brackets denote the relative errors

6.2.4 Dynamic Response

Let a concentrated load of magnitude P being time dependent act at the center of the simply supported square plate with $n \times n$ uniform meshes where n is an even number , say n=2m . For this case[23], the element loading vectors for equivalent system may be expressed as

$$\{F\}_{(m+1,m+1)} = \{F\}_{(3m+1,3m+1)} = [P(t),0,0,\cdots,0]_{12\times1}^{T}$$

$$\{F\}_{(3m+1,m+1)} = \{F\}_{(m+1,3m+1)} = [-P(t),0,0,\cdots,0]_{12\times1}^{T} \tag{6.2.71}$$

with other element loading vectors being equal to zero.

Substituting equation (6.2.71) into equation (6.2.10) yields

$$\{f\}_{(r,s)} = \left[-\frac{2P(t)}{n}\sin\frac{r\pi}{2}\sin\frac{s\pi}{2},0,0,\ldots,0\right]_{12\times1}^{T} \tag{6.2.72}$$

and then inserting equations (6.2.72) with (6.2.12) in equation (6.2.15) results in

$$\{f\}^*_{(r,s)} = \left[-\frac{2P(t)}{n} \sin\frac{r\pi}{2} \sin\frac{s\pi}{2}, 0, 0 \right]^T . \tag{6.2.73}$$

In a similar way to derive the dynamic equation (6.1.51) for beam, the dynamic equation with three degrees of freedom can be expressed as

$$\{K\}^*_{(r,s)} \{q_1\}_{(r,s)} + \{M\}^*_{(r,s)} \{\ddot{q}_1\}_{(r,s)} = \{f\}^*_{(r,s)} \qquad r,s=1,2,...,2n \tag{6.2.74}$$

The lumped mass matrix will be adopted once again for dynamic equation (6.2.74). Substituting equations (6.2.22), (6.2.66) and (6.2.73) into equation (6.2.74) yields

$$\ddot{q}_{(r,s)1} + \tilde{\omega}^2_{rs} q_{(r,s)1} = -\frac{2nP}{\rho L^2} \sin\frac{r\pi}{2} \sin\frac{s\pi}{2} \tag{6.2.75}$$

$$q_{(r,s)2} = -\frac{k_{21}}{k_{22}} q_{(r,s)1}, \qquad q_{(r,s)3} = -\frac{k_{31}}{k_{33}} q_{(r,s)1} \tag{6.2.76}$$

where $q_{(r,s)j}$ (j=1,2,3) denote the components of vectors $\{q_1\}_{(r,s)}$ and $\tilde{\omega}^2_{rs}$, k_{ij} have been defined by equations (6.2.69) and (6.2.23). The dynamic equation (6.2.75) with single degree of freedom can be solved by means of Duhamel integral, *i.e.*,

$$q_{(r,s)1} = -\frac{2n}{\rho L^2} \frac{1}{\tilde{\omega}_{rs}} \sin\frac{r\pi}{2} \sin\frac{s\pi}{2} \int_0^t P(\tau) \sin\tilde{\omega}_{rs}(t-\tau) d\tau . \tag{6.2.77}$$

Then $q_{(r,s)2}$, $q_{(r,s)3}$ can be found by substituting equation (6.2.77) into equation (6.2.76). Now every nodal displacement vector can be obtained from the double U-transformation (6.2.17) with equation (6.2.77), *i.e.*,

$$\begin{Bmatrix} w_1 \\ \theta_{x1} \\ \theta_{y1} \end{Bmatrix}_{(j,k)} = \frac{1}{2n} \sum_{r=1,3}^{2n-1} \sum_{s=1,3}^{2n-1} e^{i(j-1)r\psi} e^{i(k-1)s\psi} \begin{Bmatrix} q_{(r,s)1} \\ q_{(r,s)2} \\ q_{(r,s)3} \end{Bmatrix} \tag{6.2.78}$$

The response function of the deflection at center of the plate may be found by using the first equation of the vector equation (6.2.78) in which $j = m+1$ and $k = m+1$, *i.e.*,

$$\widetilde{W}_c(t) = \frac{4}{\rho L^2} \sum_{r=1,3}^{n-1} \sum_{s=1,3}^{n-1} \frac{1}{\widetilde{\omega}_{rs}} \int_0^t P(\tau) \sin\widetilde{\omega}_{rs}(t-\tau) d\tau \,. \tag{6.2.79}$$

The analytical solution may be obtained by the double sine series method. The center deflection can be expressed as

$$W_c(t) = \frac{4}{\rho L^2} \sum_{r=1,3}^{\infty} \sum_{s=1,3}^{\infty} \frac{1}{\omega_{rs}} \int_0^t P(\tau) \sin\omega_{rs}(t-\tau) d\tau \tag{6.2.80}$$

The error $(\widetilde{W}_c - W_c)$ is caused by both the errors of frequencies and cut-out of the modes for the discrete system by plate element . In a general way, the response function found by the non-conforming rectangular element and lumped mass matrix will converge non-uniformly to the analytical solution at an asymptotic rate of n^{-2} .

Chapter 7

NEARLY PERIODIC STRUCTURES

In engineering, periodic structures may not always be perfect, *i.e.*, sometimes a very few number of substructures are different from other substructures in geometric and mechanical properties. Such a kind of structures will be considered in this chapter. The dynamic behavior of nearly periodic structures may be more different from that of the periodic structures. If we regard the nearly periodic structure as a perfect one, the results of the dynamic analysis may be erroneous. Therefore, an exact method is need to analyze nearly periodic structures.

7.1 One Dimensional System

The essential behavior of nearly periodic structures is best demonstrated by using a simple model. A general model is illustrated in Figure 7.1.1(a), where each substructure is modeled with one degree of freedom and only one substructure is departing from the regularity in both stiffness and mass. In order to apply the U-transformation to uncouple the governing equation, an equivalent system with cyclic periodicity except two substructures must be created. In a similar manner to that described in section 3.1, the original system with linear periodicity and n substructures can be replaced by its equivalent system with cyclic periodicity and 2n substructures, where the structure is symmetrical but loads are anti-symmetrical about the center line as shown in Figure 7.1.1(b) where the first and last (2n-th) masses are imaginarily jointed by a spring with stiffness k_1 which is not subjected to any load for anti-symmetric vibration modes. Therefore, the anti-symmetrical modes of the equivalent system are identical to the modes of the actual system.

The dynamic equations for all subsystems of the equivalent system may be expressed as

$$M\ddot{x}_j + (K + 2k_1)x_j - k_1(x_{j+1} + x_{j-1}) = F_j, \qquad j \neq j_1, 2n - j_1 + 1 \qquad (7.1.1)$$

$$(M + \Delta M)\ddot{x}_j + (K + \Delta K + 2k_1)x_j - k_1(x_{j+1} + x_{j-1}) = F_j, \quad j = j_1, 2n - j_1 + 1 \qquad (7.1.2)$$

where M, K denote the mass and stiffness of ordered subsystem, $\Delta M, \Delta K$ denote the magnitudes of disorder for mass and stiffness and k_1 denotes the coupling stiffness. j_1 denotes the ordinal number of the disordered subsystem; x_j, F_j denote the displacement and load of the j-th subsystem. They must satisfy the anti-symmetry condition, *i.e.*,

$$x_j = x_{2n-j+1},$$
$$F_j = F_{2n-j+1}, \quad j=1,2,...,n \tag{7.1.3}$$

where n denotes the total number of subsystems for the actual system.

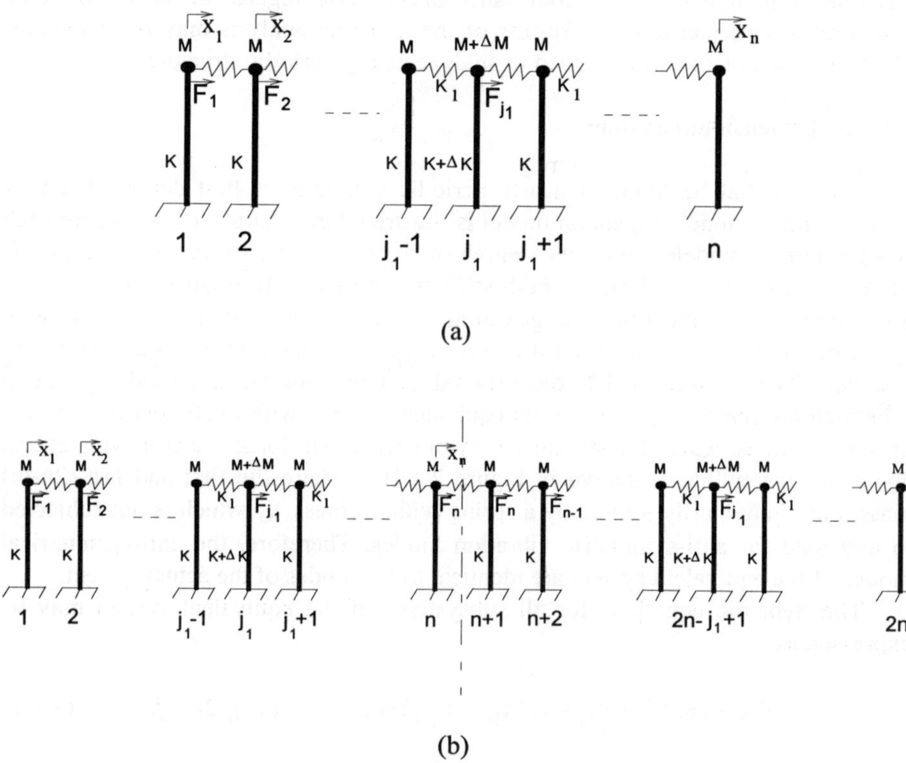

Figure 7.1.1 A nearly periodic one-dimensionnal system
(a) Actual system; (b) Equivalent system

7.1.1 Natural Vibration And Localized Modes

The equations of natural vibration for all subsystems of the equivalent system may be written in the same form[25,26] as:

$$(K+2k_1-M\omega^2)x_j-k_1(x_{j+1}+x_{j-1})=F_j, \qquad j=1,2,\ldots,2n \tag{7.1.4}$$

where ω denotes the natural frequency, x_j denotes the amplitude of j-th particle and

$$F_j=\begin{cases}(\Delta M\omega^2-\Delta K)x_{j_1}, & j=j_1,2n-j_1+1,\\ 0, & j\neq j_1,2n-j_1+1.\end{cases} \tag{7.1.5}$$

In equation (7.1.4), the disordered terms are regarded as the loads.

One can now apply the U-transformation to equations (7.1.4) with (7.1.5). The U-and inverse U-transformation may be defined as

$$x_j=\frac{1}{\sqrt{2n}}\sum_{r=1}^{2n}e^{i(j-1)r\psi}q_r \tag{7.1.6}$$

and

$$q_r=\frac{1}{\sqrt{2n}}\sum_{j=1}^{2n}e^{-i(j-1)r\psi}x_j \tag{7.1.7}$$

where q_r (r=1,2,...,2n) is the image of x_j (j=1,2,...,2n) by U-transformation and $\psi=\pi/n, i=\sqrt{-1}$.

Equation (7.1.4) may be expressed in terms of q_r (r=1,2,...,2n) as

$$(K+2k_1-M\omega^2)q_r-2k_1\cos r\psi\, q_r=f_r \qquad r=1,2,\ldots,2n \tag{7.1.8}$$

in which

$$f_r=\frac{1}{\sqrt{2n}}\sum_{j=1}^{2n}e^{-i(j-1)r\psi}F_j \tag{7.1.9}$$

Substituting equation (7.1.5) into equation (7.1.9) yields

$$f_r = \frac{F_{j_1}}{\sqrt{2n}}(e^{ij_1 r\psi} + e^{-i(j_1-1)r\psi}) . \tag{7.1.10}$$

The q_r in equation (7.1.8) may be formally expressed as

$$q_r = f_r/(K + 2k_1 - M\omega^2 - 2k_1 \cos r\psi) \tag{7.1.11}$$

Substituting equations (7.1.11) and (7.1.10) into the U-transformation (7.1.6), results in

$$x_j = \frac{F_{j_1}}{2n}\sum_{r=1}^{2n}\frac{\cos(j+j_1-1)r\psi + \cos(j-j_1)r\psi}{K+2k_1 - M\omega^2 - 2k_1\cos r\psi}, \quad j=1,2,\ldots,2n. \tag{7.1.12}$$

It is obvious that the displacements shown in equation (7.1.12) satisfy the anti-symmetry condition (7.1.3).

Since the system with infinite number of subsystems (*i.e.*, $n \to \infty$) is to be considered, therefore, with no loss of generality, it is assumed that the disordered subsystem is situated in the middle of the actual system with n number of subsystems, *i.e.*, let n=2 j_1 -1. For this case, equation (7.1.12) becomes:

$$x_j = \frac{F_{j_1}}{n}\sum_{r=2,4,\ldots}^{2n}\frac{\cos(j-j_1)r\psi}{K+2k_1 - M\omega^2 - 2k_1\cos r\psi} \equiv F_{j_1}\beta_{j,j_1},$$

$$j=1,2,\ldots,2n \tag{7.1.13}$$

where

$$\beta_{j,j_1} = \frac{1}{n}\sum_{r=1}^{n}\frac{\cos(j-j_1)2r\psi}{K+2k_1 - M\omega^2 - 2k_1\cos 2r\psi} \tag{7.1.14}$$

with $\psi = \pi/n$. β_{j,j_1} is referred to as the harmonic influence coefficient for the ordered system which means the amplitude of harmonic vibration at j-th particle caused by unit harmonic force acting at j_1-th particle. It is obvious that the reciprocal relation

$$\beta_{j,j_1} = \beta_{j_1,j} \tag{7.1.15}$$

is satisfied.

Now the frequency equation for the nearly periodic system with a single subsystem departing from the regularity may be obtained by substituting equation (7.1.5) and $j = j_1$ into equation (7.1.13) as

$$\Delta M \omega^2 - \Delta K = 1/\beta_{j_1,j_1}, \tag{7.1.16}$$

where

$$\beta_{j_1,j_1} = \frac{1}{n}\sum_{r=1}^{n} \frac{1}{K + 2k_1 - M\omega^2 - 2k_1 \cos 2r\psi}. \tag{7.1.17}$$

For the ordered system, *i.e.*, $\Delta M = \Delta K = 0$, the frequency equation (7.1.16) may be simplified to

$$K + 2k_1 - M\omega^2 - 2k_1 \cos 2r\psi = 0, \quad r=1,2,...,n \tag{7.1.18}$$

The natural frequencies for the ordered system may be expressed as

$$\tilde{\omega}_{r+1}^2 = (K + 2k_1 - 2k_1 \cos\frac{2r\pi}{n})\Big/M, \quad r=0,1,2,...,n-1 \tag{7.1.19}$$

where $\tilde{\omega}_r$, r=1,2,...,(n+1)/2 (n is odd) or n/2+1 (n is even) represent the different frequencies. The lower and upper limits of the frequency band are obtained as

$$\omega_L^2 = K/M, \qquad \omega_U^2 = (K + 4k_1)/M \tag{7.1.20}$$

where ω_L and ω_U denote the lower and upper limits respectively.

Substituting equation (7.1.19) into equation (7.1.17), yields

$$\beta_{j_1,j_1} = \frac{1}{n}\sum_{r=1}^{n} \frac{1}{M(\tilde{\omega}_r^2 - \omega^2)} \tag{7.1.21}$$

and then the frequency equation (7.1.16) may be rewritten as

$$f(\omega^2) = D(\omega^2) \tag{7.1.22}$$

where

$$f(\omega^2) = \Delta M \omega^2 - \Delta K, \qquad D(\omega^2) = nM \Big/ \sum_{r=1}^{n} \frac{1}{\tilde{\omega}_r^2 - \omega^2} . \tag{7.1.23}$$

The functions f and D of ω^2 may be qualitatively expressed as the curves shown in Figure 7.1.2. The longitudinal coordinates of the intersections represent the roots for ω^2 of the frequency equation (7.1.22) with equation (7.1.23). The function $D(\omega^2)$ shown in Figure 7.1.2, is made up of number of j_1 (when $n = 2j_1 - 1$) continuous and monotonic drop curves. By observation from Figure 7.1.2, it may be concluded that if $f(\tilde{\omega}_i^2) > 0, \omega_i < \tilde{\omega}_i$ and if $f(\tilde{\omega}_i^2) < 0, \omega_i > \tilde{\omega}_i$, where ω_i denotes the i-th different frequency for the disordered system. It is well known that if ω lies in the pass band of the ordered system, *i.e.*, $\omega_L \le \omega \le \omega_U$ the corresponding mode will extend through to the whole system and if ω lies in the stop band, *i.e.*, $\omega < \omega_L$ or $\omega > \omega_U$, the corresponding mode is localized. Four different cases may be obtained from Figure 7.1.2.

1. $f(\tilde{\omega}_L^2) > 0$ and $f(\tilde{\omega}_U^2) \ge 0$, there is one localized mode with ω less than ω_L.
2. $f(\tilde{\omega}_L^2) \le 0$ and $f(\tilde{\omega}_U^2) < 0$, there is one localized mode with ω greater than ω_U.
3. $f(\tilde{\omega}_L^2) > 0$ and $f(\tilde{\omega}_U^2) < 0$, there are two localized modes with ω less than ω_L and greater than ω_U respectively.
4. $f(\tilde{\omega}_L^2) \le 0$ and $f(\tilde{\omega}_U^2) \ge 0$, there is no localized mode.

Noting the definition of function $f(\omega^2)$ given in equation (7.1.23), the above four cases are equivalent to four regions for the disordered parameters as shown in Figure 7.1.3 where the non-dimensional parameters $\varepsilon_m, \varepsilon_k$ and ε_l are defined as

$$\varepsilon_m = \frac{\Delta M}{M}, \qquad \varepsilon_k = \frac{\Delta K}{K}, \qquad \varepsilon_l = \frac{k_l}{K} . \tag{7.1.24}$$

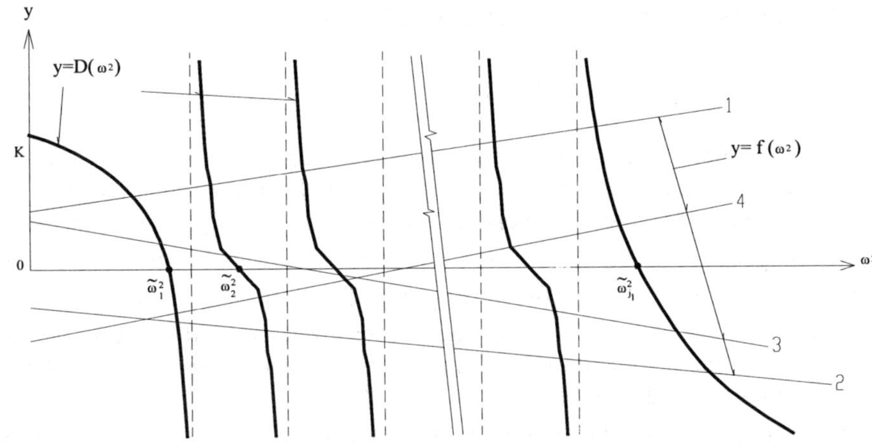

Figure 7.1.2 The functions $f(\omega^2)$ and $D(\omega^2)$

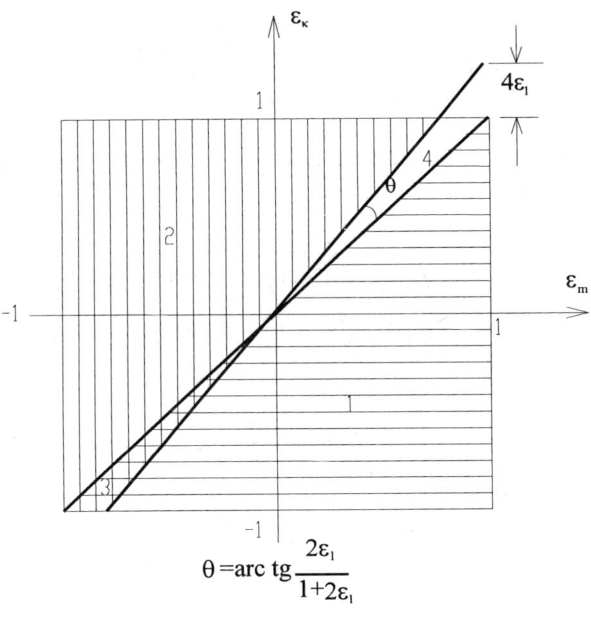

Figure 7.1.3 Four regions for disordered parameters ε_m and ε_k

The region 4 may be expressed in terms of the parameters $\varepsilon_m, \varepsilon_k$ and ε_1 as

$$0 \le \varepsilon_k - \varepsilon_m \le 4\varepsilon_1\varepsilon_m \tag{7.1.25}$$

when the parameters satisfy the above inequality equation, the small level of disorder is not enough to localize one mode. Systems having a strong coupling but with weak disorder fall into this category. However, for the simple case of disorder, *i.e.*, $\varepsilon_k \ne 0, \varepsilon_m = 0$ or $\varepsilon_m \ne 0, \varepsilon_k = 0$ any magnitude of disorder can cause localization of one mode. It is interesting that if $\varepsilon_m < 0$ and $\varepsilon_k = \varepsilon_m$,*i.e.*, the frequencies of all subsystems are the same, the localized mode still occurs, because the behavior of the disordered subsystem in high frequency vibration is different from that of the ordered subsystem.

In order to find the closed form solution for ω^2 corresponding to localized modes consider the limiting case when n is approaching infinity. By letting n approach infinity, the limit of the series summation in the right hand side of equation (7.1.14) becomes the definite integral, *i.e.*,

$$\lim_{n\to\infty} \beta_{j,j_1} = \frac{1}{\pi}\int_0^{\pi} \frac{\cos\left[(j-j_1)\theta\right]d\theta}{K + 2k_1 - M\omega^2 - 2k_1\cos\theta} \tag{7.1.26}$$

If ω lies in the stop bend, the definite integral is in existence and can be expressed by an elementary function. For example:

$$\beta_{j,j} = \begin{cases} 1\Big/\left(2k_1\sqrt{\eta^2-1}\right), & \text{when } \eta > 1 \\ -1\Big/\left(2k_1\sqrt{\eta^2-1}\right), & \text{when } \eta < -1 \end{cases} \tag{7.1.27}$$

$$\beta_{j-1,j} = \beta_{j+1,j} = \begin{cases} \dfrac{1}{2k_1}(-1 + \dfrac{\eta}{\sqrt{\eta^2-1}}), & \text{when } \eta > 1 \\ \dfrac{1}{2k_1}(-1 - \dfrac{\eta}{\sqrt{\eta^2-1}}), & \text{when } \eta < -1 \end{cases} \tag{7.1.28}$$

$$\beta_{j-2,j} = \beta_{j+2,j} = \begin{cases} \dfrac{1}{2k_1}(-2\eta + \dfrac{2\eta^2 - 1}{\sqrt{\eta^2 - 1}}), & \text{when } \eta > 1 \\ \dfrac{1}{2k_1}(-2\eta - \dfrac{2\eta^2 - 1}{\sqrt{\eta^2 - 1}}), & \text{when } \eta < -1 \end{cases} \tag{7.1.29}$$

where η is a non-dimensional frequency parameter defined as

$$\eta = (K + 2k_1 - M\omega^2)/2k_1 . \tag{7.1.30}$$

The domains $\eta > 1$ and $\eta < -1$ in equations (7.1.27)-(7.1.29) are equivalent to $\omega < \omega_L$ and $\omega > \omega_U$ respectively.

Substituting equation (7.1.27) into frequency equation (7.1.16) and eliminating ω^2 by equation (7.1.30) yields the frequency equation in terms of the elementary functions:

$$-\varepsilon_m \eta + (\frac{1}{2\varepsilon_l} + 1)\varepsilon_m - \frac{\varepsilon_k}{2\varepsilon_l} = \text{sgn}(\eta)\sqrt{\eta^2 - 1}, \quad \text{for } |\eta| > 1 \tag{7.1.31}$$

where the symbol sgn denotes the signum. The frequency parameter corresponding to the localized mode, *i.e.*, $|\eta| > 1$, may be found from equation (7.1.31) as

$$\eta = \left[-\varepsilon_m \alpha + \text{sng}(\eta)\sqrt{\alpha^2 + 4(1 - \varepsilon_m^2)}\right] \Big/ 2(1 - \varepsilon_m^2) \tag{7.1.32}$$

where sgn(η) can be determined by using Figure 7.1.3 and

$$\alpha = (\varepsilon_m - \varepsilon_k + 2\varepsilon_m \varepsilon_l)/\varepsilon_l . \tag{7.1.33}$$

The attenuation rate ξ of localized modes is defined as

$$\xi = \frac{x_{j_1+1}}{x_{j_1}} = \frac{\beta_{j_1+1,j_1}}{\beta_{j_1,j_1}} . \tag{7.1.34}$$

Substituting equations (7.1.27), (7.1.28) into equation (7.1.34), results in

$$\xi = \eta - \mathrm{sgn}(\eta)\sqrt{\eta^2 - 1}, \quad \text{for } |\eta| > 1. \tag{7.1.35}$$

It is obvious that ξ and η are of the same sign and $|\xi| < 1$. Equation (7.1.35) may be rewritten by using equation (7.1.31) as

$$\xi = (1+\varepsilon_m)\eta - (\frac{1}{2\varepsilon_1} + 1)\varepsilon_m + \frac{\varepsilon_k}{2\varepsilon_1}, \quad \text{for } |\eta| > 1 \tag{7.1.36}$$

where η has been defined by equation (7.1.32).

Let us consider a simple case, *i.e.*, $\varepsilon_k \neq 0$ and $\varepsilon_m = 0$. For this case, Figure 7.1.3 shows that if $\varepsilon_k < 0$, one localized mode with $\eta > 1$ (*i.e.*, $\omega < \omega_L$) occurs and if $\varepsilon_k > 0$, one localized mode with $\eta < -1$ (*i.e.*, $\omega > \omega_U$) occurs. Substituting $\varepsilon_m = 0$ into equations (7.1.32) and (7.1.33) yields

$$\eta = -\frac{\mathrm{sgn}(\varepsilon_k)}{2}\sqrt{\left(\frac{\varepsilon_k}{\varepsilon_1}\right)^2 + 4} \tag{7.1.37}$$

and then substituting equation (7.1.37) with $\varepsilon_m = 0$ into equation (7.1.36) yields

$$\xi = \frac{1}{2}\left[\frac{\varepsilon_k}{\varepsilon_1} - \mathrm{sgn}(\varepsilon_k)\sqrt{\left(\frac{\varepsilon_k}{\varepsilon_1}\right)^2 + 4}\right]. \tag{7.1.38}$$

The attenuation rate ξ is the odd function of the $\varepsilon_k / \varepsilon_1$ ratio as shown in Figure 7.1.4. Similarly the attenuation rate may be found, for other cases.

Consider now the system with multiple disordered subsystems[27]. Let $j_1, j_2, \ldots, j_m$ denote the ordinal numbers of disordered subsystems and $\Delta M_r, \Delta K_r$ (r=1,2,...,m) denotes the magnitude of disorder in mass and stiffness for the j_r - th subsystem. The governing equation for harmonic vibration can be expressed by means of the harmonic influence coefficients as

$$x_{j_1} = F_{j_1}\beta_{j_1,j_1} + F_{j_2}\beta_{j_1,j_2} + \cdots + F_{j_m}\beta_{j_1,j_m},$$

$$x_{j_2} = F_{j_1}\beta_{j_2,j_1} + F_{j_2}\beta_{j_2,j_2} + \cdots + F_{j_m}\beta_{j_2,j_m} \; ,$$

$$\cdots\cdots\cdots$$

$$x_{j_m} = F_{j_1}\beta_{j_m,j_1} + F_{j_2}\beta_{j_m,j_2} + \cdots + F_{j_m}\beta_{j_m,j_m} \qquad (7.1.39)$$

where

$$F_{j_r} = (\Delta M_r \omega^2 - \Delta K_r)\, x_{j_r} \qquad r=1,2,...,m. \qquad (7.1.40)$$

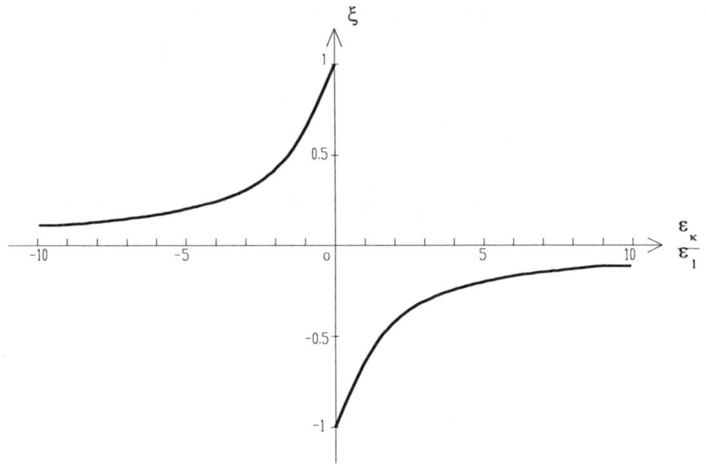

Figure 7.1.4 The attenuation rate of ξ.

When the total number of subsystems approaches infinity, the harmonic influence coefficients can be found from equation (7.1.26). Equation (7.1.39) is simultaneous linear homogeneous equations with m unknowns x_{j_r} (r=1,2,...,m). If and only if the coefficient determinant of the simultaneous equations (7.1.39) with equation (7.1.40) is equal to zero, the nontrivial solution is in existence. Let the coefficient determinant be equal to zero, *i.e.*,

$$\begin{vmatrix} D_1 \beta_{j_1 j_1} - 1 & D_2 \beta_{j_1 j_2} & \cdots & D_m \beta_{j_1 j_m} \\ D_1 \beta_{j_2 j_1} & D_2 \beta_{j_2 j_2} - 1 & \cdots & D_m \beta_{j_2 j_m} \\ \cdots & \cdots & \cdots & \cdots \\ D_1 \beta_{j_m j_1} & D_2 \beta_{j_m j_2} & \cdots & D_m \beta_{j_m j_m} - 1 \end{vmatrix} = 0. \tag{7.1.41}$$

with $D_r = \Delta M_r \omega^2 - \Delta K_r \quad (r = 1, 2, \cdots, m)$.

This is the frequency equation for the nearly periodic system with multiple disordered subsystems. As an example, consider the nearly periodic system with two adjacent and same disordered subsystems, say j-th, (j+1)-th subsystems. The magnitudes of disorder for stiffness and mass are ΔK and zero respectively. For the present case, the frequency equation (7.1.41) becomes

$$\begin{vmatrix} \Delta K \beta_{j,j} + 1 & \Delta K \beta_{j,j+1} \\ \Delta K \beta_{j+1,j} & \Delta K \beta_{j+1,j+1} + 1 \end{vmatrix} = 0. \tag{7.1.42}$$

Noting $\beta_{j,j} = \beta_{j+1,j+1}$ and $\beta_{j,j+1} = \beta_{j+1,j}$, the equation (7.1.42) is equivalent to

$$\Delta K \beta_{j,j} + 1 = \Delta K \beta_{j+1,j} \tag{7.1.43}$$

or

$$\Delta K \beta_{j,j} + 1 = -\Delta K \beta_{j+1,j}. \tag{7.1.44}$$

The following discussion is aimed at analyzing the localized modes, *i.e.*, the roots for ω of equations (7.1.43) and (7.1.44) lie in the stop band $(\omega < \omega_L \text{ or } \omega > \omega_U)$. Substituting equations (7.1.27) and (7.1.28) into equation (7.1.43), results in

$$\left(\frac{2k_1}{\Delta K} + 1\right)\sqrt{\eta^2 - 1} = \operatorname{sgn}(\eta)(\eta - 1), \quad |\eta| > 1 \tag{7.1.45}$$

where the frequency parameter η has been defined as equation (7.1.30). If there is a root for η of equation (7.1.45), the necessary condition is

$$\frac{2k_1}{\Delta K} > -1, \qquad (\Delta K > 0 \ \text{ or } \ \Delta K < -2k_1)\,. \tag{7.1.46}$$

The solution for η of equation (7.1.45) can be found easily as

$$\eta = \left[1 + \left(\frac{2k_1}{\Delta K} + 1\right)^2\right] \Bigg/ \left[1 - \left(\frac{2k_1}{\Delta K} + 1\right)^2\right]. \tag{7.1.47}$$

According to equations (7.1.47) and (7.1.46), it may be concluded that if $\Delta K > 0$, the solution η is less than -1 and if $\Delta K < -2k_1$, η is greater than 1.

Discuss now equation (7.1.44). Substituting equations (7.1.27) and (7.1.28) into equation (7.1.44) yields

$$\left(\frac{2k_1}{\Delta K} - 1\right)\sqrt{\eta^2 - 1} = -\,\mathrm{sgn}(\eta)(\eta + 1)\,, \qquad |\eta| > 1\,. \tag{7.1.48}$$

The necessary condition is

$$\frac{2k_1}{\Delta K} < 1, \qquad (\Delta K < 0 \ \text{ or } \ \Delta K > 2k_1) \tag{7.1.49}$$

instead of equation (7.1.46). The solution for η of equation (7.1.48) can be expressed as

$$\eta = \left[\left(\frac{2k_1}{\Delta K} - 1\right)^2 + 1\right] \Bigg/ \left[\left(\frac{2k_1}{\Delta K} - 1\right)^2 - 1\right] \tag{7.1.50}$$

If $\Delta K < 0$ then $\eta > 1$ and if $\Delta K > 2k_1$ then $\eta < -1$.

According to the above discussion, It can be concluded that

1.when $\Delta K > 2k_1$, there are two localized modes with ω greater than ω_U,(*i.e.*, $\eta < -1$),

2.when $\Delta K < -2k_1$, there are two localized modes with ω less than ω_L,(*i.e.*, $\eta > 1$),

3.when $0 < \Delta K \le 2k_1$, there is one localized mode with ω greater than ω_U,

4.when $-2k_1 \le \Delta K < 0$, there is one localized mode with ω less than ω_L

For the present case, the homogeneous equation (7.1.39) can be written as

$$(\Delta K \beta_{j,j} + 1) x_j + \Delta K \beta_{j,j+1} x_{j+1} = 0$$

$$\Delta K \beta_{j+1,j} x_j + (\Delta K \beta_{j+1,j+1} + 1) x_{j+1} = 0 \tag{7.1.51}$$

Substituting the root for η of equation (7.1.43), as shown in equation (7.1.47), into equation (7.1.51), the non-trivial solution may be expressed as

$$x_j = -x_{j+1} . \tag{7.1.52}$$

For this localized mode, the displacements of two disordered subsystems are of same magnitude and opposite directions. The attenuation rate ξ in ordered subsystems is the function of η as shown in equation (7.1.35), *i.e.*,

$$\xi = \frac{x_{j-1}}{x_j} = \frac{x_{j+2}}{x_{j+1}} = \eta - \mathrm{sgn}(\eta)\sqrt{\eta^2 - 1} , \quad \text{for } |\eta| > 1 . \tag{7.1.53}$$

Substituting equation (7.1.47) with equation (7.1.46) into equation (7.1.53), results in

$$\xi = -\frac{k_1}{k_1 + \Delta K} , \qquad \text{for } \Delta K > 0 \text{ and } \Delta K < -2 k_1 . \tag{7.1.54}$$

In a similar way to discuss above, substituting the solution η of equation (7.1.44) into equation (7.1.51) yields

$$x_j = x_{j+1} \tag{7.1.55}$$

which indicates the displacements of two disordered subsystems are of the same magnitudes and directions in corresponding localized mode. Substituting equation (7.1.50) with equation (7.1.49) into equation (7.1.53) results in

$$\xi = \frac{k_1}{k_1 - \Delta K} , \qquad \text{for } \Delta K < 0 \text{ and } \Delta K > 2 k_1 . \tag{7.1.56}$$

From equations (7.1.54), (7.1.47) and equations (7.1.56), (7.1.50), it can be deduced that when the natural frequency ω is less than ω_L (*i.e.*, $\eta > 1$), $0 < \xi < 1$ and when ω is greater than ω_U (*i.e.*, $\eta < -1$), $-1 < \xi < 0$.

It can be pointed out that the method used in this section may be applied to analysis of some bi-periodic structures[28].

7.1.2 Forced Vibration

Consider now the steady state response of the nearly periodic structures subjected to harmonic forces. The load acting at j-th subsystem is denoted by $F_j^* e^{i\omega t}$ (j=1,2,...,n). The forced vibration equation can also be expressed as equation (7.1.4) in which

$$F_{2n-j+1} = F_j = \begin{cases} F_{j_r}^* + (\Delta M_r \omega^2 - \Delta K_r) x_{j_r}, & j = j_r,\ r = 1,2,\ldots,m \\ F_j^*, & j \neq j_1, j_2, \ldots, j_m \end{cases} \tag{7.1.57}$$

instead of equation (7.1.5). The governing equation [27] can be simplified by means of the harmonic influence coefficients as

$$\begin{aligned} x_{j_1} &= \left(\Delta M_1 \omega^2 - \Delta K_1\right)\beta_{j_1 j_1} x_{j_1} + \left(\Delta M_2 \omega^2 - \Delta K_2\right)\beta_{j_1 j_2} x_{j_2} + \cdots \\ &\quad + \left(\Delta M_m \omega^2 - \Delta K_m\right)\beta_{j_1 j_m} x_{j_m} + x_{j_1}^* \\ x_{j_2} &= \left(\Delta M_1 \omega^2 - \Delta K_1\right)\beta_{j_2 j_1} x_{j_1} + \left(\Delta M_2 \omega^2 - \Delta K_2\right)\beta_{j_2 j_2} x_{j_2} + \cdots \\ &\quad + \left(\Delta M_m \omega^2 - \Delta K_m\right)\beta_{j_2 j_m} x_{j_m} + x_{j_2}^* \\ &\cdots\cdots\cdots\cdots \\ x_{j_m} &= \left(\Delta M_1 \omega^2 - \Delta K_1\right)\beta_{j_m j_1} x_{j_1} + \left(\Delta M_2 \omega^2 - \Delta K_2\right)\beta_{j_m j_2} x_{j_2} + \cdots \\ &\quad + \left(\Delta M_m \omega^2 - \Delta K_m\right)\beta_{j_m j_m} x_{j_m} + x_{j_m}^* \end{aligned} \tag{7.1.58}$$

in which x_{j_r} (r=1,2,...,m) denotes the amplitude of displacement for the disordered subsystem (*i.e.*,the j_r-th subsystem), ΔM_r, ΔK_r (r=1,2,...,m) denote the magnitudes of disorder in mass and stiffness for the j_r-th subsystem and x_j^* denotes the particular solution for the ordered system subjected to the same loads as those acting on the disordered system. Equation (7.1.58) is simultaneous linear equations with m unknown variables. In order to mention this method, consider a specific example. The system considered is the same as that described in the above example. The amplitudes of harmonic loads can be expressed as

$$F_j^* = P, \qquad F_k^* = 0, \quad k=1,2,...,j-1,j+1,...,n. \tag{7.1.59}$$

At the outset, the steady state response of the ordered system subjected to the same loading as that shown in equation (7.1.59) is studied. The solution of the forced vibration can be expressed by using the harmonic influence coefficients as

$$x_j^* = P\beta_{j,j}, \qquad x_{j+1}^* = P\beta_{j+1,j}, \qquad x_k^* = P\beta_{k,j}, \qquad k=1,2,... \tag{7.1.60}$$

For the present case, the governing equation (7.1.58) becomes

$$\begin{aligned}(\Delta K\beta_{j,j}+1)x_j + \Delta K\beta_{j,j+1}x_{j+1} &= P\beta_{j,j},\\ \Delta K\beta_{j+1,j}x_j + (\Delta K\beta_{j+1,j+1}+1)x_{j+1} &= P\beta_{j+1,j}\,,\end{aligned} \tag{7.1.61}$$

and then the solution can be obtained easily as

$$\begin{aligned}x_j &= \frac{P}{\Delta K}(\beta_{j,j}^2 - \beta_{j+1,j}^2 + \beta_{j,j}/\Delta K)\Big/[(\beta_{j,j}+1/\Delta K)^2 - \beta_{j+1,j}^2],\\ x_{j+1} &= \frac{P}{\Delta K^2}\beta_{j+1,j}\Big/[(\beta_{j,j}+1/\Delta K)^2 - \beta_{j+1,j}^2].\end{aligned} \tag{7.1.62}$$

The other displacements may be expressed as

$$x_k = -\Delta K\beta_{k,j}x_j - \Delta K\beta_{k,j+1}x_{j+1} + P\beta_{k,j} \tag{7.1.63}$$

where x_j, x_{j+1} have been defined by equation (7.1.62). If the frequency of harmonic excitation lies in the stop band, *i.e.*, $\omega < \omega_L$ or $\omega > \omega_U$ the harmonic influence coefficients can be expressed by the explicit function of frequency parameter η, as shown in equations (7.1.27)-(7.1.29).

7.2 Two Dimensional System

The system to be considered is made up of both a rectangular cable network with $m \times n$ uniform meshes and $m \times n$ numbers of mass-spring systems distributed uniformly as shown in Figure 7.2.1 where each lumped mass is attached to a node of network. The nodes at four boundary edges are fixed. The pretensioned cables will act as the coupling springs but their masses will be neglected.

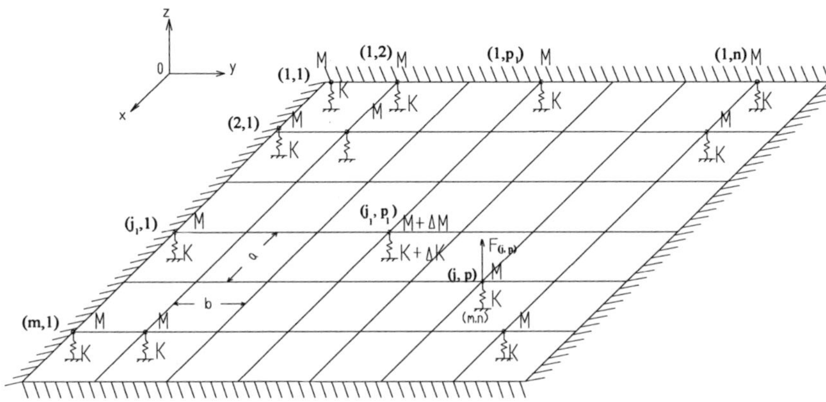

Figure 7.2.1 The $m \times n$ numbers of mass-spring systems

In a similar manner to that given in chapter 5, the equivalent system with cyclic periodicity in x- and y-directions may be obtained as shown in Figure 7.2.2. The equivalent system is an extended system whose length and width are twice the actual size. Moreover we regard the extended system as one having cyclic periodicity in x-and y-directions . The loading pattern is anti-symmetric with respect to two symmetric planes of the equivalent system so that the deformation is also anti-symmetric. The fixed end conditions for the actual system will be satisfied automatically.

Consider now the harmonic vibration of the equivalent system. The governing equation for all subsystems may be expressed as

$$(K + 2k_1 + 2k_2 - M\omega^2) W_{(j,p)} - k_1 (W_{(j+1,p)} + W_{(j-1,p)}) - k_2 \left(W_{(j,p+1)} + W_{(j,p-1)}\right) = F_{(j,p)}$$

$$j=1,2,..,2m \quad p=1,2,..,2n \tag{7.2.1}$$

where ω denotes the frequency of harmonic vibration, K,M denote the stiffness and mass of the mass-spring system; $W_{(j,p)}$ and $F_{(j,p)}$ denote the amplitudes of the transverse displacement and loading for the lumped mass of subsystem (j,p) and

$$k_1 = \frac{T_1}{a}, \qquad k_2 = \frac{T_2}{b} \tag{7.2.2}$$

in which T_1, T_2 denote the pretensions of the cables in x-and y-directions, and a, b denote the spacing of y-and x-cables.

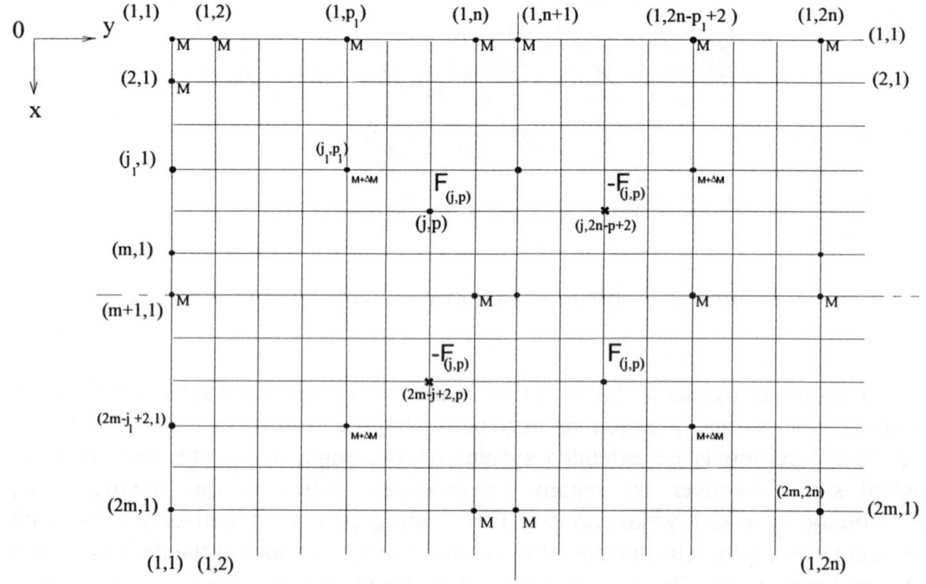

Figure 7.2.2 The equivalent systems with $2m \times 2n$ subsystems

In order to uncouple the simultaneous equation (7.2.1), the double U-transformation need be applied. Let

$$W_{(j,p)} = \frac{1}{\sqrt{2m}}\frac{1}{\sqrt{2n}}\sum_{r=1}^{2m}\sum_{s=1}^{2n} e^{i(j-1)r\psi_1} e^{i(p-1)s\psi_2} q_{(r,s)},$$

$$j=1,2,...,2m \quad p=1,2,...,2n \tag{7.2.3a}$$

and its inverse transformation

$$q_{(r,s)} = \frac{1}{\sqrt{2m}}\frac{1}{\sqrt{2n}}\sum_{j=1}^{2m}\sum_{p=1}^{2n} e^{-i(j-1)r\psi_1} e^{-i(p-1)s\psi_2} W_{(j,p)},$$

$$r=1,2,...,2m \quad s=1,2,...,2n \tag{7.2.3b}$$

with $\psi_1 = \pi/m$ and $\psi_2 = \pi/n$.

Applying the double U-transformation (7.2.3) to equation (7.2.1), yields

$$(K + 2k_1 + 2k_2 - M\omega^2 - 2k_1\cos r\psi_1 - 2k_2\cos s\psi_2)\,q_{(r,s)} = f_{(r,s)} \tag{7.2.4}$$

where

$$f_{(r,s)} = \frac{1}{\sqrt{2m}}\frac{1}{\sqrt{2n}}\sum_{j=1}^{2m}\sum_{p=1}^{2n} e^{-i(j-1)r\psi_1} e^{-i(p-1)s\psi_2} F_{(j,p)} \tag{7.2.5}$$

and then

$$q_{(r,s)} = f_{(r,s)} / (K + 2k_1 + 2k_2 - M\omega^2 - 2k_1\cos r\psi_1 - 2k_2\cos s\psi_2). \tag{7.2.6}$$

The frequency equation for ordered system may be obtained by inserting $f_{(r,s)} = 0$ in equation (7.2.4), as

$$(K + 2k_1 + 2k_2 - M\omega^2 - 2k_1\cos r\psi_1 - 2k_2\cos s\psi_2) = 0,$$

$$r=1,2,..,2m \quad s=1,2,...,2n. \tag{7.2.7}$$

The lower and upper limits of the pass band may be expressed as

$$\omega_L^2 = K/M, \qquad \omega_U^2 = (K + 4k_1 + 4k_2)/M. \tag{7.2.8}$$

Now let us consider the disordered system where only one subsystem is departing from the regularity. With no loss of generality, let (j_1, p_1) denotes the ordinal number of the disordered subsystem whose stiffness and mass are of $K+\Delta K$ and $M+\Delta M$. For this case, the "loads" satisfying the anti-symmetry condition may be given as follows:

$$F_{(j_1,p_1)} = F_{(2m-j_1+2,2n-p_1+2)} = (\Delta M\,\omega^2 - \Delta K)\,W_{(j_1,p_1)}$$

$$F_{(j_1,2n-p_1+2)} = F_{(2m-j_1+2,p_1)} = -F_{(j_1,p_1)} \tag{7.2.9}$$

with other "loads" being equal to zero. Substituting equation (7.2.9) into equation (7.2.5) results in

$$f_{(r,s)} = \frac{-4\,F_{(j_1,p_1)}}{\sqrt{2m}\sqrt{2n}}\sin(j_1-1)r\,\psi_1\sin(p_1-1)s\,\psi_2$$

$$r=1,2,\ldots,2m \quad s=1,2,\ldots,2n. \tag{7.2.10}$$

Substituting equations (7.2.6) and (7.2.10) into equation (7.2.3a), results in

$$W_{(j,p)} = \frac{4\,F_{(j_1,p_1)}}{mn}\sum_{r=1}^{m}\sum_{s=1}^{n}\frac{\sin(j_1-1)r\,\psi_1\sin(j-1)r\,\psi_1\sin(p_1-1)s\,\psi_2\sin(p-1)s\,\psi_2}{K+2k_1+2k_2-M\omega^2-2k_1\cos r\,\psi_1-2k_2\cos s\,\psi_2}$$

$$j=1,2,\ldots,2m \quad p=1,2,\ldots,2n. \tag{7.2.11}$$

It is obvious that the displacements shown in equation (7.2.11) satisfy the boundary condition for the original system ,*i.e.*,

$$W_{(j,1)}=0, \qquad W_{(j,n+1)}=0, \qquad j=1,2,\ldots,m+1;$$

$$W_{(1,p)}=0, \qquad W_{(m+1,p)}=0, \qquad p=1,2,\ldots,n+1. \tag{7.2.12}$$

By letting m and n approach infinity, the equation (7.2.11) becomes

$$W_{(j,p)} = \frac{4F_{(j_1,p_1)}}{\pi^2}\int_0^{\pi}\!\!\int_0^{\pi}\frac{\sin(j_1-1)\theta_1\sin(j-1)\theta_1\sin(p_1-1)\theta_2\sin(p-1)\theta_2}{K+2k_1+2k_2-M\omega^2-2k_1\cos\theta_1-2k_2\cos\theta_2}\,d\theta_1 d\theta_2$$

$$j=1,2,\ldots,2m \quad p=1,2,\ldots,2n. \tag{7.2.13}$$

If and only if ω in equation (7.2.13) lies in the stop band, *i.e.*, $\omega < \omega_L$ or $\omega > \omega_U$, the definite integral shown in the right hand side of equation (7.2.13) is in existence. Equation (7.2.13) may be rewritten as

$$W_{(j,p)} = \frac{F_{(j_1,p_1)}}{\pi^2}$$

$$\times \int_0^{\pi}\int_0^{\pi} \frac{[\cos(j-j_1)\theta_1 - \cos(j+j_1-2)\theta_1][\cos(p-p_1)\theta_2 - \cos(p+p_1-2)\theta_2]}{K+2k_1+2k_2-M\omega^2-2k_1\cos\theta_1-2k_2\cos\theta_2} d\theta_1 d\theta_2$$

$$j=1,2,\ldots,2m \quad p=1,2,\ldots,2n. \tag{7.2.14}$$

In general , the disordered subsystem is located at an infinite distance from the boundary (at infinity), but the interested subsystems are located within definite distance from the disordered subsystem. Letting (j_1+j) and (p_1+p) approach infinity and $(j-j_1),(p-p_1)$ keep invariable, and applying the well known Riemann Lemma to equation (7.2.14) yields

$$W_{(j,p)} = F_{(j_1,p_1)}\beta_{(j,p)(j_1,p_1)} \tag{7.2.15}$$

where

$$\beta_{(j,p)(j_1,p_1)} = \frac{1}{\pi^2}\int_0^{\pi}\int_0^{\pi} \frac{\cos(j-j_1)\theta_1\cos(p-p_1)\theta_2}{K+2k_1+2k_2-M\omega^2-2k_1\cos\theta_1-2k_2\cos\theta_2} d\theta_1 d\theta_2$$

(7.2.16)

which is referred to as the harmonic influence coefficient for the ordered system. By substituting equation (7.2.9) and $j=j_1, p=p_1$ into equation (7.2.15), the frequency equation for the disordered system may be found as

$$\Delta M\omega^2 - \Delta K = 1/\beta_{(j_1,p_1)(j_1,p_1)} \tag{7.2.17}$$

with

$$\beta_{(j_1,p_1)(j_1,p_1)} = \frac{1}{\pi^2}\int_0^{\pi}\int_0^{\pi}(K+2k_1+2k_2-M\omega^2-2k_1\cos\theta_1-2k_2\cos\theta_2)^{-1}d\theta_1 d\theta_2 \tag{7.2.18}$$

Because the frequency equation (7.2.17) is similar to that for one dimensional system. Therefore its roots may be obtained by graphical method. The relationship of the disorder, coupling parameters and the number of localized modes may also be expressed by Figure 7.1.3 in which the coupling parameter ε_1 must be defined as $(k_1+k_2)/K$ instead of k_1/K. The conclusion is the same as that for one dimensional system with single subsystem departing from the regularity, *i.e.*, arbitrarily small levels of disorder (ε_k or ε_m) can cause localization of one mode.

The attenuation rates in x-and y-directions are of different values which may be expressed as

$$\xi_1 = \frac{\beta_{(j_1+1,p_1)(j_1,p_1)}}{\beta_{(j_1,p_1)(j_1,p_1)}} \quad \text{and} \quad \xi_2 = \frac{\beta_{(j_1,p_1+1)(j_1,p_1)}}{\beta_{(j_1,p_1)(j_1,p_1)}}. \tag{7.2.19}$$

Consider now the nearly periodic system with multiple disordered subsystems whose ordinal numbers are denoted by $(j_1,p_1),(j_2,p_2),\ldots,(j_m,p_m)$. The governing equation of natural vibration can be expressed by means of the harmonic influence coefficients as

$$W_{(j_1,p_1)} = F_{(j_1,p_1)}\beta_{(j_1,p_1)(j_1,p_1)} + F_{(j_2,p_2)}\beta_{(j_1,p_1)(j_2,p_2)} + \cdots + F_{(j_m,p_m)}\beta_{(j_1,p_1)(j_m,p_m)},$$

$$W_{(j_2,p_2)} = F_{(j_1,p_1)}\beta_{(j_2,p_2)(j_1,p_1)} + F_{(j_2,p_2)}\beta_{(j_2,p_2)(j_2,p_2)} + \cdots + F_{(j_m,p_m)}\beta_{(j_2,p_2)(j_m,p_m)},$$

$$\cdots\cdots\cdots\cdots\cdots$$

$$W_{(j_m,p_m)} = F_{(j_1,p_1)}\beta_{(j_m,p_m)(j_1,p_1)} + F_{(j_2,p_2)}\beta_{(j_m,p_m)(j_2,p_2)} + \cdots + F_{(j_m,p_m)}\beta_{(j_m,p_m)(j_m,p_m)} \tag{7.2.20}$$

where

$$F_{(j_r p_r)} \equiv (\Delta M_r\omega^2 - \Delta K_r)W_{(j_r p_r)} \tag{7.2.21}$$

and $\Delta M_r, \Delta K_r$ denote the magnitudes of disorder in mass and stiffness for subsystem (j_r, p_r).

Let the coefficient determinant of the homogeneous linear equations be equal to zero, *i.e.*,

$$\begin{vmatrix} D_1\beta_{(j_1p_1)(j_1p_1)}-1, & D_2\beta_{(j_1p_1)(j_2p_2)}, & \cdots, & D_m\beta_{(j_1p_1)(j_mp_m)} \\ D_1\beta_{(j_2p_2)(j_1p_1)}, & D_2\beta_{(j_2p_2)(j_2p_2)}-1, & \cdots, & D_m\beta_{(j_2p_2)(j_mp_m)} \\ \cdots, & \cdots, & \cdots, & \cdots \\ D_1\beta_{(j_mp_m)(j_1p_1)}, & D_2\beta_{(j_mp_m)(j_2p_2)}, & \cdots, & D_m\beta_{(j_mp_m)(j_mp_m)}-1 \end{vmatrix} = 0 \quad (7.2.22)$$

with $D_r = \Delta M_r \omega^2 - \Delta K_r$ $(r = 1,2,\cdots,m)$ where the harmonic influence coefficients of the ordered system with infinite subsystems have been expressed in equation (7.2.16). Equation (7.2.22) is the frequency equation which is applicable to the natural vibration with localized modes. If ω lies in the pass band, there is no definition of harmonic influence coefficient. Therefore the ω in frequency equation (7.2.22) must lie in stop band.

In a similar way to discuss in section 7.1.2 the governing equation for forced vibration of two dimensional disordered system subjected to harmonic forces can be expressed as

$$W_{(j_1p_1)} = (\Delta M_1\omega^2 - \Delta K_1)\beta_{(j_1p_1)(j_1p_1)}W_{(j_1p_1)} + (\Delta M_2\omega^2 - \Delta K_2)\beta_{(j_1p_1)(j_2p_2)}W_{(j_2p_2)}$$
$$+\cdots+(\Delta M_m\omega^2 - \Delta K_m)\beta_{(j_1p_1)(j_mp_m)}W_{(j_mp_m)} + W^*_{(j_1p_1)}$$

$$W_{(j_2p_2)} = (\Delta M_1\omega^2 - \Delta K_1)\beta_{(j_2p_2)(j_1p_1)}W_{(j_1p_1)} + (\Delta M_2\omega^2 - \Delta K_2)\beta_{(j_2p_2)(j_2p_2)}W_{(j_2p_2)}$$
$$+\cdots+(\Delta M_m\omega^2 - \Delta K_m)\beta_{(j_2p_2)(j_mp_m)}W_{(j_mp_m)} + W^*_{(j_2p_2)}$$

$$\cdots\cdots\cdots\cdots$$

$$W_{(j_mp_m)} = (\Delta M_1\omega^2 - \Delta K_1)\beta_{(j_mp_m)(j_1p_1)}W_{(j_1p_1)} + (\Delta M_2\omega^2 - \Delta K_2)\beta_{(j_mp_m)(j_2p_2)}W_{(j_2p_2)}$$
$$+\cdots+(\Delta M_m\omega^2 - \Delta K_m)\beta_{(j_mp_m)(j_mp_m)}W_{(j_mp_m)} + W^*_{(j_mp_m)} \quad (7.2.23)$$

where $W^*_{(j,p)}$ denotes the steady state solution for ordered system subjected to the same harmonic excitation as that acting on the disordered system and ω denotes the frequency of harmonic loading.

Consider now the two dimensional disordered system with a single subsystem departing from the regularity and disordered subsystem, say $(j_1 p_1)$ subsystem, subjected to a harmonic force $Pe^{i\omega t}$. For this simple case , the governing equation (7.2.23) becomes

$$W_{(j_1p_1)} = (\Delta M_1\omega^2 - \Delta K_1)\beta_{(j_1p_1)(j_1p_1)} W_{(j_1p_1)} + P\beta_{(j_1p_1)(j_1p_1)} \tag{7.2.24}$$

and the solution for $W_{(j_1p_1)}$ of equation (7.2.24) is

$$W_{(j_1p_1)} = \frac{P\beta_{(j_1p_1)(j_1p_1)}}{1-(\Delta M_1\omega^2 - \Delta K_1)\beta_{(j_1p_1)(j_1p_1)}}\,. \tag{7.2.25}$$

The other displacement amplitude may be expressed as

$$W_{(j,p)} = (\Delta M_1\omega^2 - \Delta K_1)\beta_{(j,p)(j_1p_1)} W_{(j_1p_1)} + P\beta_{(j,p)(j_1p_1)}\,,$$
$$j - j_1 = \pm 1, \pm 2, \cdots \qquad p - p_1 = \pm 1, \pm 2, \cdots \tag{7.2.26}$$

with $W_{(j_1p_1)}$ shown in equation (7.2.25).

7.3 Three Dimensional System

The system considered consists of *l* number of two dimensional systems as described in the previous section in which the corresponding nodes of any two adjacent networks are coupled with spring k_3 and the nodes in the top and bottom layers are free in z-direction as shown in Figure 7.3.1. At the outset, let us consider the extended system whose length, width and height are twice those of the actual one as shown in Figure 7.3.2.

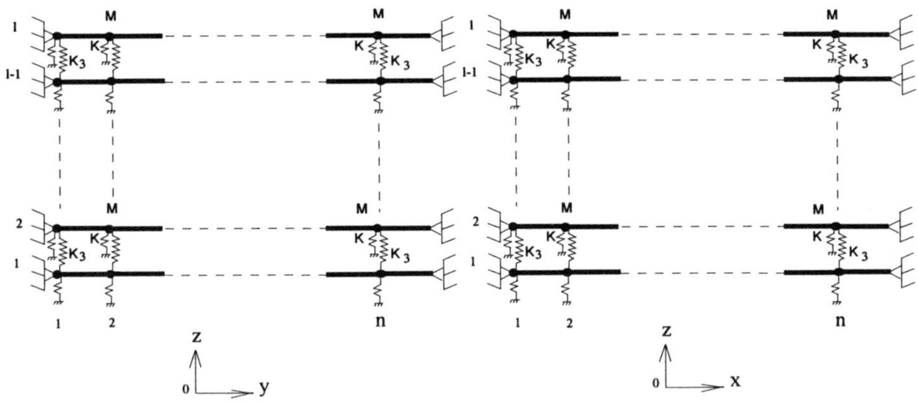

Figure 7.3.1 Three dimensional systems with $m \times n \times l$ subsystems

Moreover we regard the extended system as one having cyclic periodicity in x-, y- and z-directions. The corresponding nodes of both the top and bottom layers must be imagined to have been jointed by an additional spring with stiffness k_3. In order to satisfy automatically the boundary conditions of the actual system, the loading pattern must be anti-symmetric with respect to three symmetric planes of the equivalent system as shown in Figure 7.3.2.

In a similar manner as in analyzing the two dimensional system, the governing equation for harmonic vibration of the equivalent system may be given as

$$
\begin{gathered}
(K + 2k_1 + 2k_2 + 2k_3 - M\omega^2) W_{(j,p,h)} - k_1 (W_{(j+1,p,h)} + W_{(j-1,p,h)}) \\
- k_2 (W_{(j,p+1,h)} + W_{(j,p-1,h)}) - k_3 (W_{(j,p,h+1)} + W_{(j,p,h-1)}) = F_{(j,p,h)} \\
j=1,2,\ldots,2m, \quad p=1,2,\ldots,2n, \quad h=1,2,\ldots,2l \qquad (7.3.1)
\end{gathered}
$$

where the subscripts j, p, h in the round brackets denote the ordinal numbers of the subsystem along the x-, y-and z-directions respectively and the other notations have been defined previously.

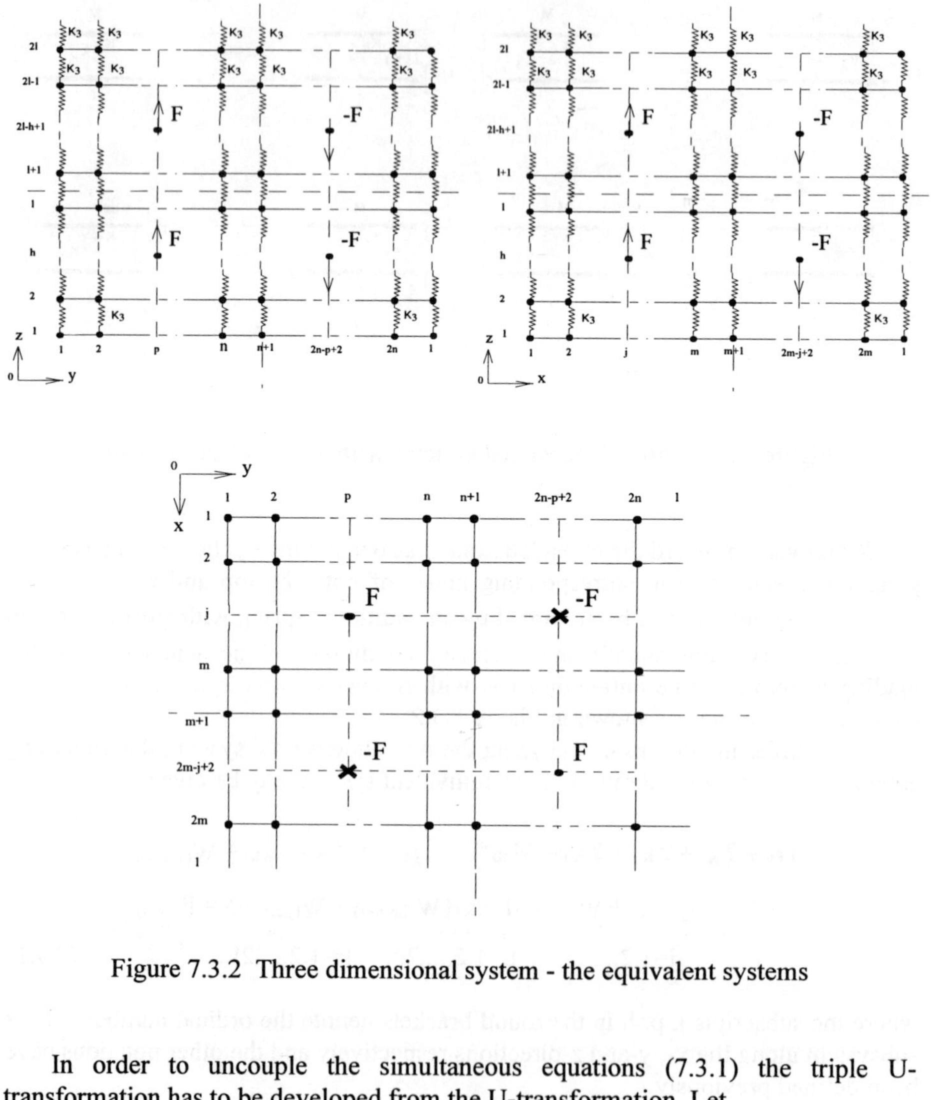

Figure 7.3.2 Three dimensional system - the equivalent systems

In order to uncouple the simultaneous equations (7.3.1) the triple U-transformation has to be developed from the U-transformation. Let

$$W_{(j,p,h)} = \frac{1}{\sqrt{2m}\sqrt{2n}\sqrt{2l}} \sum_{r=1}^{2m} \sum_{s=1}^{2n} \sum_{t=1}^{2l} e^{i(j-1)r\psi_1} e^{i(p-1)s\psi_2} e^{i(h-1)t\psi_3} q_{(r,s,t)} \quad (7.3.2a)$$

and its inverse transformation is

$$q_{(r,s,t)} = \frac{1}{\sqrt{2m}\sqrt{2n}\sqrt{2l}} \sum_{j=1}^{2m} \sum_{p=1}^{2n} \sum_{h=1}^{2l} e^{-i(j-1)r\psi_1} e^{-i(p-1)s\psi_2} e^{-i(h-1)t\psi_3} W_{(j,p,h)} \quad (7.3.2b)$$

with $\psi_1 = \pi/m, \psi_2 = \pi/n$ and $\psi_3 = \pi/l$.

Applying the triple U-transformation (7.3.2) to equation (7.3.1), results in

$$(K + 2k_1 + 2k_2 + 2k_3 - M\omega^2 - 2k_1 \cos r\psi_1 - 2k_2 \cos s\psi_2 - 2k_3 \cos t\psi_3) q_{(r,s,t)} = f_{(r,s,t)} \quad (7.3.3)$$

where

$$f_{(r,s,t)} = \frac{1}{\sqrt{2m}\sqrt{2n}\sqrt{2l}} \sum_{j=1}^{2m} \sum_{p=1}^{2n} \sum_{h=1}^{2l} e^{-i(j-1)r\psi_1} e^{-i(p-1)s\psi_2} e^{-i(h-1)t\psi_3} F_{(j,p,h)} \quad (7.3.4)$$

and then

$$q_{(r,s,t)} = \frac{f_{(r,s,t)}}{K + 2k_1 + 2k_2 + 2k_3 - M\omega^2 - 2k_1 \cos r\psi_1 - 2k_2 \cos s\psi_2 - 2k_3 \cos t\psi_3}. \quad (7.3.5)$$

The frequency equation for the ordered system may be obtained from equation (7.3.3) with $f_{(r,s,t)} = 0$, as

$$K + 2k_1 + 2k_2 + 2k_3 - M\omega^2 - 2k_1 \cos r\psi_1 - 2k_2 \cos s\psi_2 - 2k_3 \cos t\psi_3 = 0. \quad (7.3.6)$$

The lower and upper limits of the frequency band are

$$\omega_L^2 = K/M, \qquad \omega_U^2 = [K + 4(k_1 + k_2 + k_3)]/M. \quad (7.3.7)$$

Now, let us consider the disordered system with only one subsystem departing from the regularity. Let (j_1, p_1, h_1) denote the ordinal number of the disordered subsystem with stiffness $K+\Delta K$ and mass $M+\Delta M$. For such case, the " loads" with anti-symmetry may be given as

$$F_{(j_1,p_1,h_1)} = F_{(2m-j_1+2,2n-p_1+2,h_1)} = F_{(j_1,p_1,2l-h_1+1)} = F_{(2m-j_1+2,2n-p_1+2,2l-h_1+1)} ,$$

$$F_{(2m-j_1+2,p_1,h_1)} = F_{(j_1,2n-p_1+2,h_1)} = F_{(2m-j_1+2,p_1,2l-h_1+1)} = F_{(j_1,2n-p_1+2,2l-h_1+1)} = -F_{(j_1,p_1,h_1)} ,$$

$$F_{(j_1,p_1,h_1)} = (\Delta M\omega^2 - \Delta K)W_{(j_1,p_1,h_1)} \tag{7.3.8}$$

with the other loads being equal to zero.

Substituting equation (7.3.8) into equation (7.3.4) yields

$$f_{(r,s,t)} = -\frac{4\,F_{(j_1,p_1,h_1)}}{\sqrt{2m}\sqrt{2n}\sqrt{2l}}\sin(j_1-1)r\,\psi_1\sin(p_1-1)s\,\psi_2(e^{-i(h_1-1)t\psi_3}+e^{ih_1t\psi_3}) \tag{7.3.9}$$

Substituting equations (7.3.5) and (7.3.9) into equation (7.3.2a) results in

$$W_{(j,p,h)} = \frac{2F_{(j_1,p_1,h_1)}}{mnl}\sum_{r=1}^{m}\sum_{s=1}^{n}\sum_{t=1}^{2l}\{\sin(j_1-1)r\psi_1\sin(j-1)r\psi_1\sin(p_1-1)s\psi_2\sin(p-1)s\psi_2 \\ \times\frac{\cos(h-h_1)t\psi_3+\cos(h+h_1-1)t\psi_3}{K+2k_1+2k_2+2k_3-M\omega^2-2k_1\cos r\psi_1-2k_2\cos s\psi_2-2k_3\cos t\psi_3}\} \tag{7.3.10}$$

in which $\psi_1=\pi/m, \psi_2=\pi/n$ and $\psi_3=\pi/l$.

It can be verified that the displacements shown in equation (7.3.10) satisfy the boundary condition of the actual system, *i.e.*,

$$W_{(1,p,h)}=0, \qquad W_{(m+1,p,h)}=0, \qquad p=1,2,...,2n \quad h=1,2,...,2l$$

$$W_{(j,1,h)}=0, \qquad W_{(j,n+1,h)}=0, \qquad j=1,2,...,2m \quad h=1,2,...,2l$$

$$W_{(j,p,2l)}=W_{(j,p,1)}, \qquad W_{(j,p,l)}=W_{(j,p,l+1)}, \qquad j=1,2,...,2m \quad p=1,2,...,2n \tag{7.3.11}$$

By letting m, n and l approach infinity, the series on the right hand side of equation (7.3.10) becomes a triple definite integral, *i.e.*,

$$W_{(j,p,h)} = \frac{4\,F_{(j_1,p_1,h_1)}}{\pi^3}\int_0^{\pi}\int_0^{\pi}\int_0^{\pi} g_1(\theta_1,\,\theta_2,\,\theta_3)\,d\theta_1 d\theta_2 d\theta_3 \qquad (7.3.12a)$$

where

$$g_1(\theta_1,\theta_2,\theta_3) \equiv \sin(j_1-1)\theta_1\,\sin(j-1)\theta_1\,\sin(p_1-1)\theta_2\,\sin(p-1)\theta_2$$
$$\times\frac{\cos(h-h_1)\theta_3+\cos(h+h_1-1)\theta_3}{K+2k_1+2k_2+2k_3-M\omega^2-2k_1\cos\theta_1-2k_2\cos\theta_2-2k_3\cos\theta_3} \qquad (7.3.12b)$$

In a similar way used to derive the equations (7.2.15) and (7.2.16), letting $(j_1+j),(p+p_1)$ and (h_1+h) approach infinity and $(j_1-j),(p-p_1),(h-h_1)$ keep invariable, equations (7.3.12a.b) becomes

$$W_{(j,p,h)} = F_{(j_1,p_1,h_1)}\,\beta_{(j,p,h)(j_1,p_1,h_1)} \qquad (7.3.13)$$

where

$$\beta_{(j,p,h)(j_1,p_1,h_1)} = \frac{1}{\pi^3}\int_0^{\pi}\int_0^{\pi}\int_0^{\pi} g_2(\theta_1,\theta_2,\theta_3)\,d\theta_1 d\theta_2 d\theta_3 \qquad (7.3.14a)$$

in which

$$g_2(\theta_1,\theta_2,\theta_3) \equiv \frac{\cos(j-j_1)\theta_1\,\cos(p-p_1)\theta_2\,\cos(h-h_1)\theta_3}{K+2k_1+2k_2+2k_3-M\omega^2-2k_1\cos\theta_1-2k_2\cos\theta_2-2k_3\cos\theta_3} \qquad (7.3.14b)$$

which is the three-dimensional analog of equation (7.2.16).

By substituting equation (7.3.8) and $(j,p,h)=(j_1,p_1,h_1)$ into equation (7.3.13), the frequency equation for disordered system may be expressed as

$$\Delta M\omega^2 - \Delta K = \frac{1}{\beta_{(j_1,p_1,h_1)(j_1,p_1,h_1)}}, \qquad (\omega < \omega_L, \omega > \omega_U) \tag{7.3.15}$$

with

$$\beta_{(j_1,p_1,h_1)(j_1,p_1,h_1)} =$$

$$\frac{1}{\pi^3}\int_0^{\pi}\int_0^{\pi}\int_0^{\pi}(K+2k_1+2k_2+2k_3-M\omega^2-2k_1\cos\theta_1-2k_2\cos\theta_2-2k_3\cos\theta_3)^{-1}d\theta_1 d\theta_2 d\theta_3 \tag{7.3.16}$$

It is of interest to note that the frequency equations in one, two and three dimensions are of the same form, however they have an essential distinction, *i.e.*, when ω in the stop band approaches ω_L or ω_U, the dynamic flexibility $\beta_{(j_1,p_1,h_1)(j_1,p_1,h_1)}$ in three dimensional system approaches a finite value as a limit but $\beta_{(j_1,p_1)(j_1,p_1)}$ and β_{j_1,j_1} approach infinity.

$\beta_{(j_1,p_1,h_1)(j_1,p_1,h_1)}$ for both cases of $\omega = \omega_L - 0$ and $\omega = \omega_U + 0$ are of the same magnitude with opposite sign. Considering the specific case of $k_1 = k_2 = k_3$, by numerical integration following result can be obtained as

$$\beta_{(j_1,p_1,h_1)(j_1,p_1,h_1)}\Big|_{\omega=\omega_L-0} = -\beta_{(j_1,p_1,h_1)(j_1,p_1,h_1)}\Big|_{\omega=\omega_U+0}$$

$$= \frac{1}{2k_1}\frac{1}{\pi^3}\int_0^{\pi}\int_0^{\pi}\int_0^{\pi}(3-\cos\theta_1-\cos\theta_2-\cos\theta_3)^{-1}d\theta_1 d\theta_2 d\theta_3 \approx \frac{0.504}{2k_1} \tag{7.3.17}$$

in which ω_L and ω_U are defined as equation (7.3.7). It indicates that in three dimensions, when the ordered system is subjected to a harmonic force with frequency ω_L or ω_U, the amplitude of the forced mass approaches a finite value as a limit. Let

$$D(\omega^2) \equiv 1/\beta_{(j_1,p_1,h_1)(j_1,p_1,h_1)}, \qquad \text{when } \omega \geq \omega_U \text{ or } \omega \leq \omega_L \tag{7.3.18}$$

where D represents the dynamic stiffness of the ordered system. It may be proved that if

$$\Delta M\,\omega_L^2 - \Delta K < D(\omega_L^2) \qquad \text{and} \qquad \Delta M\,\omega_U^2 - \Delta K > D(\omega_U^2) \tag{7.3.19}$$

the frequency equation (7.3.15) does not have any solution for ω in the stop band, *i.e.*, it is impossible for any localized mode to occur. Substituting equation (7.3.7) into equation (7.3.19) yields

$$\varepsilon_m - \varepsilon_k < D(\omega_L^2)/K \quad \text{and} \quad \varepsilon_m\left[1 + 4(\varepsilon_1 + \varepsilon_2 + \varepsilon_3)\right] - \varepsilon_k > D(\omega_U^2)/K \tag{7.3.20}$$

where

$$\varepsilon_j = k_j/K \quad j = 1,2,3, \quad \varepsilon_m = \Delta M/M, \quad \varepsilon_k = \Delta K/K \tag{7.3.21}$$

and

$$D(\omega_L^2)/K = -D(\omega_U^2)/K$$

$$= 2\pi^3 \Big/ \int_0^\pi \int_0^\pi \int_0^\pi \left[\varepsilon_1(1-\cos\theta_1) + \varepsilon_2(1-\cos\theta_2) + \varepsilon_3(1-\cos\theta_3)\right]^{-1} d\theta_1\, d\theta_2\, d\theta_3\,. \tag{7.3.22}$$

For the special case of $k_1 = k_2 = k_3$, equation (7.3.20) becomes

$$\varepsilon_m - \varepsilon_k < \varepsilon_1/0.252 \quad \text{and} \quad (1 + 12\,\varepsilon_1)\,\varepsilon_m - \varepsilon_k > -\varepsilon_1/0.252 \tag{7.3.23}$$

If $\varepsilon_m = 0$, equation (7.3.23) is equivalent to

$$|\varepsilon_k| < \varepsilon_1/0.252 = 3.97\,\varepsilon_1\,. \tag{7.3.24}$$

It is shown that a finite threshold of disorder is needed for a given magnitude of coupling in order to cause localization of one mode. For the present case , the ratio of the disorder (ε_k) to coupling (ε_1) must be greater than 3.97 in order to localize one mode.

If $\varepsilon_k = 0$, equation (7.3.23) becomes

$$-\frac{3.97\varepsilon_1}{1 + 12\varepsilon_1} < \varepsilon_m < 3.97\varepsilon_1\,. \tag{7.3.25}$$

Then the same conclusion may also apply.

Consider now the three dimensional periodicity system with m numbers of disordered subsystems , say (j_1, p_1, h_1), $(j_2, p_2, h_2), \cdots, (j_m, p_m, h_m)$ subsystems. The governing equation of natural vibration takes the similar form to equation (7.2.20), *i.e.*,

$$W_{(j_1,p_1,h_1)} = F_{(j_1,p_1,h_1)} \beta_{(j_1,p_1,h_1)(j_1,p_1,h_1)} + F_{(j_2,p_2,h_2)} \beta_{(j_1,p_1,h_1)(j_2,p_2,h_2)} + \cdots + F_{(j_m,p_m,h_m)} \beta_{(j_1,p_1,h_1)(j_m,p_m,h_m)},$$

$$W_{(j_2,p_2,h_2)} = F_{(j_1,p_1,h_1)} \beta_{(j_2,p_2,h_2)(j_1,p_1,h_1)} + F_{(j_2,p_2,h_2)} \beta_{(j_2,p_2,h_2)(j_2,p_2,h_2)} + \cdots + F_{(j_m,p_m,h_m)} \beta_{(j_2,p_2,h_2)(j_m,p_m,h_m)}$$

$$\cdots\cdots\cdots\cdots$$

$$W_{(j_m,p_m,h_m)} = F_{(j_1,p_1,h_1)} \beta_{(j_m,p_m,h_m)(j_1,p_1,h_1)} + F_{(j_2,p_2,h_2)} \beta_{(j_m,p_m,h_m)(j_2,p_2,h_2)} + \cdots + F_{(j_m,p_m,h_m)} \beta_{(j_m,p_m,h_m)(j_m,p_m,h_m)} \tag{7.3.26}$$

with

$$F_{(j_r p_r h_r)} \equiv (\Delta M_r \omega^2 - \Delta K_r) W_{(j_r p_r h_r)} \qquad r=1,2,...,m \tag{7.3.27}$$

in which $\Delta M_r, \Delta K_r$ denote the magnitudes of disorder in mass and stiffness for $(j_r p_r h_r)$-subsystem. The frequency equation can be obtained by letting the coefficient determinant of simultaneous equations (7.3.26) be equal to zero , as

$$\begin{vmatrix} (\Delta M_1 \omega^2 - \Delta K_1)\beta_{x_1x_1} - 1, & (\Delta M_2 \omega^2 - \Delta K_2)\beta_{x_1x_2}, & \cdots, & (\Delta M_m \omega^2 - \Delta K_m)\beta_{x_1x_m} \\ (\Delta M_1 \omega^2 - \Delta K_1)\beta_{x_2x_1}, & (\Delta M_2 \omega^2 - \Delta K_2)\beta_{x_2x_2} - 1, & \cdots, & (\Delta M_m \omega^2 - \Delta K_m)\beta_{x_2x_m} \\ \cdots, & \cdots, & \cdots, & \cdots \\ (\Delta M_1 \omega^2 - \Delta K_1)\beta_{x_mx_1}, & (\Delta M_2 \omega^2 - \Delta K_2)\beta_{x_mx_2}, & \cdots, & (\Delta M_m \omega^2 - \Delta K_m)\beta_{x_mx_m} - 1 \end{vmatrix} = 0 \tag{7.3.28}$$

where ω lies in the stop band and the subscripts $x_r \equiv (j_r p_r h_r)$ r=1,2,...,m. The solutions for ω of equation (7.3.28) are corresponding to the localized modes.

The forced vibration equation for three dimensional disordered system subjected to harmonic loading can also be expressed, by means of the harmonic influence coefficients for the ordered system, as

$$[1 - (\Delta M_1 \omega^2 - \Delta K_1)\beta_{x_1x_1}] W_{x_1} - (\Delta M_2 \omega^2 - \Delta K_2)\beta_{x_1x_2} W_{x_2} - \cdots$$

$$-(\Delta M_m \omega^2 - \Delta K_m)\beta_{x_1x_m} W_{x_m} = W^*_{x_1}$$

$$-(\Delta M_1 \omega^2 - \Delta K_1)\beta_{x_2x_1} W_{x_1} + [1 - (\Delta M_2 \omega^2 - \Delta K_2)\beta_{x_2x_2}] W_{x_2} - \cdots$$

$$-(\Delta M_m \omega^2 - \Delta K_m)\beta_{x_2x_m} W_{x_m} = W^*_{x_2}$$

$$\cdots\cdots\cdots\cdots$$

$$-(\Delta M_1 \omega^2 - \Delta K_1)\beta_{x_mx_1} W_{x_1} - (\Delta M_2 \omega^2 - \Delta K_2)\beta_{x_mx_2} W_{x_2} - \cdots$$

$$+[1 - (\Delta M_m \omega^2 - \Delta K_m)\beta_{x_mx_m}] W_{x_m} = W^*_{x_m}$$

(7.3.29)

where $W^*_{x_r}$ denotes the steady state solution for the ordered system subjected to the same harmonic loading as that acting on the disordered system and ω denotes the frequency of the harmonic excitation.

APPENDIX

U-TRANSFORMATION

A.1 Definitions

U Matrix

If a square matrix **A** satisfies

$$\overline{\mathbf{A}}^{\mathrm{T}}\mathbf{A} = \mathbf{I} \tag{A.1}$$

where **I** denotes the unit matrix, then **A** is called an unitary matrix or the U matrix for short .

U-Transformation

In general, the complex linear transformation may be expressed as

$$\begin{Bmatrix} x_1 \\ x_2 \\ \vdots \\ x_N \end{Bmatrix} = \begin{bmatrix} a_{11}, a_{12}, \cdots, a_{1N} \\ a_{21}, a_{22}, \cdots, a_{2N} \\ \vdots \quad \vdots \qquad \vdots \\ a_{N1}, a_{N2} \cdots a_{NN} \end{bmatrix} \begin{Bmatrix} q_1 \\ q_2 \\ \vdots \\ q_N \end{Bmatrix} \qquad \text{or} \qquad \mathbf{x} = \mathbf{Aq} \tag{A.2}$$

If **A** is a **U** matrix, then the linear transformation (A.2) is referred to as the U-transformation.

U Matrix for the Special Purpose

Let the square matrix of order N

$$\mathbf{U} \equiv [\mathbf{U}_1, \mathbf{U}_2, \cdots, \mathbf{U}_N] \tag{A.3}$$

with the vector components

$$\mathbf{U}_m = \frac{1}{\sqrt{N}}\left[1, e^{im\psi}, e^{i2m\psi}, \cdots, e^{i(N-1)m\psi}\right]^T \qquad m=1,2,...,N \tag{A.4}$$

in which $\psi = 2\pi/N$ and $i = \sqrt{-1}$.

$\mathbf{U}_m$ (m=1,2,...,N) is an orthonormal basis, *i.e.*,

$$\overline{\mathbf{U}}_j^T \mathbf{U}_k = \delta_{jk} \tag{A.5}$$

which can be verified, by substituting equation (A.4) into equation (A.5), as

$$\overline{\mathbf{U}}_j^T \mathbf{U}_k = \frac{1}{N}\left[1 + e^{i(k-j)\psi} + e^{i2(k-j)\psi} + \cdots + e^{i(N-1)(k-j)\psi}\right]$$

$$\text{when } j = k, \quad \overline{\mathbf{U}}_j^T \mathbf{U}_k = 1;$$

$$\text{when } j \neq k, \quad \overline{\mathbf{U}}_j^T \mathbf{U}_k = \frac{1}{N}\frac{1 - e^{iN(k-j)\psi}}{1 - e^{i(k-j)\psi}} = 0. \tag{A.6}$$

Then it can be proved that

$$\overline{\mathbf{U}}^T \mathbf{U} = \mathbf{I} \tag{A.7}$$

by using equation (A.5). Therefore the matrix **U** defined by equations (A.3) and (A.4) is a U matrix and denoted by **U** which is in common use in this book.

A.2 Property of U Matrix

The elementary cyclic matrices can be defined as

$$\varepsilon_j = \begin{array}{c} \begin{matrix} 1 & 2 & \ldots & j & j+1 & & \end{matrix} \\ \begin{bmatrix} 0 & 0 & \cdots & 0 & 1 & & \\ & & & & & \ddots & \\ & & & & & & 1 \\ 1 & & & & & & \\ & \ddots & & & & & \\ & & & 1 & 0 & \cdots & 0 \end{bmatrix} \\ \begin{matrix} 1 & 2 & \ldots & j & & \ldots & N \end{matrix} \end{array}, \qquad j=0,1,2,...,N-1 \qquad (A.8)$$

with the empty elements being equal to zeros. ε_0 is an unit matrix. Noting the cyclic periodicity of vector $\mathbf{U}_m$ and

$$\varepsilon_j \mathbf{U}_m = e^{ijm\psi} \mathbf{U}_m \qquad (A.9)$$

it can be verified that

$$\overline{\mathbf{U}}^T \varepsilon_j \mathbf{U} = \Phi_j \qquad (A.10)$$

where

$$\Phi_j = \begin{bmatrix} e^{ij\psi} & & & 0 \\ & e^{i2j\psi} & & \\ & & \ddots & \\ 0 & & & e^{iNj\psi} \end{bmatrix}. \qquad (A.11)$$

In general, the cyclic matrix can be written as

$$\mathbf{K} = \begin{bmatrix} K_{11} & K_{12} & \cdots K_{1N} \\ K_{21} & K_{22} & \cdots K_{2N} \\ \vdots & \vdots & \vdots \\ K_{N1} & K_{N2} & \cdots K_{NN} \end{bmatrix} \qquad (A.12)$$

with

$$K_{11} = K_{22} = \cdots = K_{NN} \ ,$$

$$K_{1j} = K_{2,j+1} = \cdots = K_{N-j+1,N} = K_{N-j+2,1} = \cdots = K_{N,j-1} \ , \qquad j=2,3,\ldots,N. \tag{A.13}$$

The cyclic matrix can be expressed as the linear combination of the elementary cyclic matrices, *i.e.*,

$$\mathbf{K} = \sum_{j=1}^{N} K_{1j} \varepsilon_{j-1} \ . \tag{A.14}$$

Therefore

$$\overline{\mathbf{U}}^T \mathbf{K} \mathbf{U} = \sum_{j=1}^{N} K_{1j} \Phi_{j-1} = \begin{bmatrix} k_1 & & & 0 \\ & k_2 & & \\ & & \ddots & \\ 0 & & & k_N \end{bmatrix} \tag{A.15}$$

where

$$k_r = \sum_{j=1}^{N} K_{1j} e^{i(j-1)r\psi} \ , \qquad r=1,2,\ldots,N. \tag{A.16}$$

A.3 The Component Forms of U-Transformation and Inverse U-Transformation

The linear transformation (A.2) can be written as

$$x_j = \sum_{m=1}^{N} a_{jm} q_m \qquad j=1,2,\ldots,N. \tag{A.17}$$

Inserting the corresponding elements of matrix **U**, *i.e.*,

$$a_{jm} = \frac{1}{\sqrt{N}} e^{i(j-1)m\psi} \tag{A.18}$$

in equation (A.17) yields the U-transformation

$$x_j = \frac{1}{\sqrt{N}} \sum_{m=1}^{N} e^{i(j-1)m\psi} q_m \tag{A.19}$$

with $\psi = 2\pi/N$.

The inverse U-transformation may be expressed as

$$\mathbf{q} = \overline{\mathbf{U}}^{\mathbf{T}} \mathbf{x} \tag{A.20}$$

whose component form is

$$q_m = \frac{1}{\sqrt{N}} \sum_{j=1}^{N} e^{-i(j-1)m\psi} x_j \qquad m=1,2,...,N. \tag{A.21}$$

REFERENCES

1. D.L.Thomas, *Dynamics of rotationally periodic structures*, International Journal for Numerical Methods in Engineering Vol. 14, 81-102 (1979).
2. C.W.Cai and F.G.Wu, *On the vibration of rotationally periodic structures*, Acta Scientiarum Naturalium Universitatis Sunyatseni, Vol.22, No.3, 1-9, (1983).
3. C.W.Cai, Y.K.Cheung and H.C.Chan, *Uncoupling of dynamic equations for periodic structures*, Journal of Sound and Vibration, Vol. 139, No.2, 253-263 (1990).
4. C.W.Cai, *On analytical solution of rotationally periodic structures,* Acta Scientiarum Naturalium Universitatis Sunyatseni, Vol. 25, No.2, 64-67, (1986).
5. F.B.He, *The application of U-transform method in the analysis of linear periodic structures*, Journal of The Graduates Sun Yat-Sen University (Natural Sciences) Vol.13, N0.4, 51-61 (1992).
6. C.W.Cai, Y.K.Cheung and H.C. Chan, *Transverse vibration analysis of plane trusses by analytical method*, Journal of Sound and Vibration, Vol. 133, No.1, 139-150 (1989).
7. Y.K.Cheung, H.C.Chan and C.W.Cai, *Exact method for static analysis of periodic structures*, ASCE, Journal of Engineering Mechanics, Vol. 115, No.2, 415-434 (1989).
8. C.W.Cai, Y.K.Cheung and H.C.Chan, *Dynamic response of infinite continuous beams subjected to a moving force-an exact method*, Journal of Sound and Vibration, Vol.123, No.3, 461-472 (1988).
9. C.W.Cai, J.H.Zhang and Y.Li, *On exact solutions for an infinite uniform beam with periodic elastic supports under arbitrary loads,* Acta Scientiarum Naturalium Universitatis Sunyatseni, Vol.30, No.1, 1-6 (1991).
10. Y.Li, W.S.Cai and C.W.Cai, *Dynamic response of infinite continuous beams with periodic elasticity supports - an exact method,* Proceedings of ASIA Vibration Conference'89, Shenzhen, China, 251-255(1989).
11. S.Y.Chen and C.W.Cai, *Exact solution for static analysis of continuous deep beams*, Acta Scientiarum Natrualium Universitatis Sunyatseni, Vol.33, No.3, 7-14 (1994).
12. H.C.Chan, C.W.Cai and Y.K.Cheung, *A static solution of stiffened plates*, Thin-Walled Structures, 11, 291-303 (1991).

13. C.W.Cai and W.L.Huang, *Stability analysis of stiffened plate using the exact method*, Advances in Engineering Mechanics, Peking University Press, 182-186 (1992).
14. R.Narayanan, *Plated structures, stability and strength, Applied Science Publishers*, London and New York (1983).
15. M.S Troitsky, *Stiffened plates, bending, stability and vibrations*, ELSEVIER Scientific Publishing Company (1976).
16. Y.K.Cheung, H.C.Chan and C.W.Cai, *Natural vibration analysis of rectangular networks* , International Journal of Space Structures, Vol.3, No.3, 139-152 (1988).
17. H.C.Chan, C.W.Cai and Y.K.Cheung, *Moments and deflections of simply supported rectangular grids-an exact method*, International Journal of Space Structures, Vol.4, No.3, 163-173 (1989).
18. H.C.Chan, C.W.Cai and Y.K.Cheung, *An analytical method for static analysis of double layer grids*, International Journal of Space Structures, Vol.4, No.2,107-116 (1989).
19. B.L.Dhooper, P.C.Gupta and B.P.Singh, *Vibration analysis of orthogonal cable networks by transfer matrix method*, Journal of Sound and Vibration, Vol.101, No.4, 575-584 (1985).
20. Y.K.Cheung, H.C.Chan and C.W.Cai, *Dynamic response of orthogonal cable networks subjected to a moving force*, Journal of Sound and Vibration, Vol. 156, No.2, 337-347 (1992).
21. W.S.Cai, Y.Li and C.W.Cai, *Natural vibration analysis of diagonal networks*, Proceedings of International Conference On Vibration Problems in Engineering Vol.1, 111-115 Wuhan- Chongqing China (1991).
22. C.W.Cai and H.C.Chan, *On the convergence of static and buckling analysis of beams by using a cubic beam element*, Proceedings of The International Conference on Computational Methods in Structural and Geotechnical Engineering, Hong Kong, Vol. 4, 1384-1389 (1994).
23. H.C.Chan, C.W.Cai and Y.K.Cheung, *Convergence studies of dynamic analysis by using the finite element method with lumped mass matrix*, Journal of Sound and Vibration, Vol. 165, No. 2, 193-207 (1993).
24. H.C.Chan and C.W.Cai, *On the convergence of static and buckling analysis of square plates by using a non-conforming plate element*, Proceedings of The Third Asian-Pacific Conference on Computational Mechanics, Seoul Korea, Vol. 1, 191-196 (1996).

25. C.W.Cai, Y.K.Cheung and H.C.Chan, *Mode localization phenomena in nearly periodic systems*, ASME Journal of Applied Mechanics, Vol. 62, No. 1, 141-149 (1995).
26. C.W. Cai, H.C. Chan and Y.K. Cheung, *Localized modes in periodic systems with nonlinear disorders*, ASME Journal of Applied Mechanics, Vol.64, 940-945 (1997)
27. H.C.Chan and C.W.Cai, *Dynamics of nearly periodic structures,* Journal of Sound and Vibration (1998, in press).
28. C.W. Cai, H.C. Chan and Y.K. Cheung, *Exact method for static and natural vibration analysis of bi-periodic structures*, ASCE Journal of Engineering Mechanics (1998, in press).

INDEX

E

F

G

H

I

L

M

N

O

P

R

S

T

U